Physics of amorphous materials

To my parents

Physics of amorphous materials

S R ELLIOTT

Department of Physical Chemistry,
University of Cambridge

Copublished in the United States with
John Wiley & Sons, Inc., New York

Longman Scientific & Technical
Longman Group UK Limited
Longman House, Burnt Mill, Harlow,
Essex CM20 2JE, England
and Associated Companies throughout the world.

Copublished in the United States with
John Wiley & Sons Inc., 605 Third Avenue, New York, NY 10158

© Longman Group Limited 1983

All rights reserved; no part of this publication may be
reproduced, stored in a retrieval system, or transmitted in any
form or by any means, electronic, mechanical, photocopying,
recording, or otherwise, without either the prior written
permission of the Publishers, or a licence permitting restricted
copying in the United Kingdom issued by the Copyright
Licensing Agency Ltd, 33–34 Alfred Place, London WC1E
7DP

First published in 1984
Reprinted 1989

British Library Cataloguing in Publication Data
Elliott, S. R.
   Physics of amorphous materials.
   1. Amorphous substances
   I. Title
   530.4′1   QC176.8.A44

   ISBN 0-582-44636-8

Library of Congress Cataloging in Publication Data Available

Printed in Great Britain at The Bath Press, Avon

Contents

Preface and acknowledgements vii
Glossary of symbols x
Glossary of abbreviations xiii

1 Preparation 1

1.1 Introduction 1
1.2 Definitions 3
1.3 Preparation of amorphous materials 6
1.4 Summary 22

2 Glasses 23

2.1 Introduction 23
2.2 The glass transition 24
2.3 Theories for the glass transition 26
2.4 Factors that determine the glass-transition temperature 37
2.5 Glass-forming systems and ease of glass formation 38
  Problems 50
  Bibliography 51

3 Structure 53

3.1 Introduction 54

I MICROSCOPIC STRUCTURE 55
3.2 Experimental techniques and short-range order 55
3.3 Structural modelling 94
3.4 Medium-range structure 110

II MACROSCOPIC STRUCTURE 118
3.5 Experimental techniques 118
3.6 Examples of macroscopic structure 122
  Problems 131
  Bibliography 133

4 Vibrations — 135

4.1	Introduction	135
4.2	Vibrational excitations	136
4.3	Computational methods	142
4.4	Experimental probes	150
4.5	Low-temperature properties	158
	Problems	169
	Bibliography	172

5 Electrons — 173

5.1	Introduction	174
5.2	Electronic density of states	178
5.3	Localization	192
5.4	Transport properties	208
5.5	Small polarons	228
5.6	Optical properties	234
	Problems	245
	Bibliography	245

6 Defects — 247

6.1	Introduction	247
6.2	Types of defect	248
6.3	Defect-controlled properties	274
	Problems	310
	Bibliography	310

7 Amorphous metals — 311

7.1	Introduction	311
7.2	Structure	316
7.3	Electronic properties	331
7.4	Magnetic properties	350
7.5	Mechanical properties	357
	Problems	359
	Bibliography	360

Appendix: Suppliers of amorphous materials	361
References	362
Index	375

Preface and acknowledgements

This book has been written to be an introduction to the science of amorphous materials, a subject which receives scant mention in conventional solid-state texts. It is aimed at final year undergraduates, beginning graduate students and researchers in solid-state physics or chemistry, materials science or engineering. Some background of (crystalline) solid-state physics is assumed, say between the levels of Kittel's *Introduction to Solid-State Physics* and Ziman's *Principles of the Theory of Solids*, because it was felt that in a book of this length such background materials could not easily be incorporated; however, no knowledge of amorphous materials is assumed.

When I first considered the idea of writing such a book as this, it seemed to me that a general understanding of amorphous materials had been established, and that much of what remained to be done in the field lay in the realms of 'engineering' new materials and filling in the details of the theories proposed to account for their properties. It seemed to me then to be an appropriate time to take stock and write a general introduction to the field of amorphous solids, covering a wide range of their properties, and bridging the gap for the newcomer to the field to such specialized monographs on the subject as the classic work by Mott and Davis, *Electronic Processes in Non-Crystalline Materials*. However, as is usual in these matters, I have found myself overtaken by events. Recent work has thrown into the melting pot many of the 'established' ideas – the example of scaling theories of electron localization and their relation to mobility edges and the so-called minimum metallic conductivity, springs immediately to mind. I felt that such important developments could not be left out, but this immediately led to problems since it is difficult in an introductory textbook to deal with recent controversial ideas. I have tried to resolve this dilemma by mentioning the new concepts, certainly, but putting them into historical perspective, as it were, and pointing out their relation and relevance to previous ideas. In this way, I hope I will have given the reader a flavour of the vitality of this rapidly expanding branch of solid-state science.

I have also tried to make this book into something of a reference work by the inclusion of several tables of useful parameters, such as glass-transition, melting, superconducting and Curie temperatures, electrical and optical gaps for amorphous semiconductors, etc. These are all quantities which practising researchers use on a day-to-day basis, and in this small way, I hope that this book might be of some value to them, since I believe that such collections of data are not readily available together elsewhere. I have also given perhaps more references to the original literature than is usual in a textbook, but have kept the number down by referring, where possible, to reviews although references to original works have been given where these have been seminal to the development of their aspect of the field. I hope that in this way, the beginning researcher will find his way into the literature. A collection of problems and exercises are given at the end of each chapter which illustrate or extend points raised in the text.

The layout of this book differs somewhat from that conventionally adopted in texts dealing with the crystalline solid state. The first chapter begins with a series of definitions of the terms 'amorphous', 'glassy' etc., followed by a review of various preparative techniques which can be used to produce non-crystalline materials. Such a discussion is necessary because there is no unique structure of an amorphous solid (unlike a single crystal); non-crystalline materials produced in different ways can have very different structures and hence physical properties, even though the composition might be the same. One of the most common and important methods of producing amorphous materials is to freeze a liquid sufficiently rapidly that crystallization is precluded; in this way a 'glass' is formed. Chapter 2 is devoted to a study of the glassy state, considering the kinetic, thermodynamic and other factors which control the glass transition, and discussing those material parameters which can facilitate glass formation.

The structure of amorphous materials is one of the recurrent themes of this book. The determination of the structure of a solid is a difficult task when there is no periodicity; special experimental techniques need to be employed and these form the subject of the first part of Chapter 3. Even when such techniques are used, only a limited amount of local structural information is generally obtainable, and the construction of structural models can be a most useful route to a further understanding of the structure, particularly medium-range order; modelling is also discussed in Chapter 3. Amorphous materials are seldom structurally homogeneous, and different preparative techniques can result in various growth morphologies, e.g. phase separation in melt-quenched glasses or the inclusion of voids in vapour-deposited films; this 'macroscopic' structure, and the experimental techniques which probe it, form the last part of the chapter.

The absence of a periodic lattice in amorphous materials has several effects; an immediate consequence is that there is no reciprocal lattice and so k is no longer a good quantum number for excitations in the solid, such as phonons or electrons. Thus the Bloch formalism cannot be used to simplify the mathematical description of the excitations, and the phonon (or electron) states cannot be described in terms of dispersion curves, $\omega(q)$ (or band structures, $E(k)$); instead the only quantity which is a valid description of excitations in a non-crystalline solid is the 'density of states', and the various methods which have been used to compute it are described in Chapter 4 for the case of phonons, and again in Chapter 5 for electrons.

One seemingly universal feature of amorphous solids is the existence of very low-energy excitations, describable as 'two-level systems', which affect thermal properties in a variety of ways: they can act as internal degrees of freedom, thereby giving rise to an anomalous specific heat, and they can act as scattering centres for phonons, thereby controlling the thermal conductivity and ultrasonic absorption. Two-level systems form the subject of the last part of Chapter 4.

Electronic excitations, manifested as transport or optical properties, are the other dominant theme of this book. It is convenient to divide the discussion, on structural grounds, into two: Chapter 5 deals with the electronic behaviour of 'ideal' (i.e. defect-free) amorphous semiconductors, and Chapter 6 considers the electronic (and other) properties associated with a variety of structural defects. A consequence of the presence of structural disorder is the possibility that excitations, whether vibrational or electronic, may become spatially localized. Different models for electron localization, such as scaling theories or that giving rise to a minimum metallic conductivity and a mobility edge, are introduced in Chapter 5. Electrical transport in amorphous semiconductors, discussed in terms of these models, and in terms of a rival theory involving localized, self-trapped charge carriers (small polarons), also form part of Chapter 5.

The structural, electronic, magnetic and mechanical properties of amorphous metals form the subject of Chapter 7. It was felt that amorphous metals merited a separate chapter in view of the fact that, while they exhibit certain features in common with other amorphous solids (electron localization effects for example), many of their properties are very different to those shown by amorphous semiconductors and insulators discussed earlier in the book.

Finally, I owe an enormous debt of gratitude to those in the Cavendish Laboratory in Dr A. D. Yoffe's group, principally Professors Davis and Mott, who guided my first forays in this field and whose influence has been incalculable, and more recently to those in Professor J. M. Thomas's group in the Department of Physical Chemistry, from whom I have gained many new insights. I am also extremely grateful to Professor E. A. Davis, Drs P. H. Gaskell, M. R. J. Gibbs, J. Klinowski, E. Marseglia, W. A. Phillips and A. C. Wright and Messrs T. G. Fowler and A. J. Lowe for reading parts of this book, for their valuable comments and for pointing out many errors. I am also especially grateful to Dr P. Extance for reading the galley proofs. Any mistakes that remain, however, are my sole responsibility. I am also grateful to Mrs M. Pomery, Mrs M. Crosbie and Miss J. Rowe for typing various parts of the manuscript, to Mr E. Smith and his colleagues for preparing the photographs, to Mrs K. I. Johnson for preparing the index and Mrs W. Roberts for running the literature search to generate Fig. 1.1. Last, but not least, I am grateful to Penelope for putting up with a rather unsociable author during the time that it has taken to write this book.

It is well known that in the writing of any text, the person who benefits most from the process is the author himself. I think this to be true in this case, but I hope a little rubs off on others.

The following have kindly consented to the reproduction of material originally published by them:

Academic Press, Akademie-Verlag, American Institute of Physics, American Physical Society, Cambridge University Press, Centre for Industrial Consultancy and Liaison (University of Edinburgh), Elsevier Sequoia, Institutue of Physics, International Union of Crystallography, Japanese Institute of Metals, John Wiley and Sons, Les Editions de Physique, Macmillan, Marcel Dekker, Metals Society, North-Holland Publishing, Oxford University Press, Pergamon Press, Plenum Publishing, Physical Society of Japan, Royal Microscopy Society, Royal Society, Scientific American, Society of Glass Technology, Springer-Verlag, Stanford University Synchrotron Radiation Laboratory, Taylor and Francis.

Glossary of symbols

α	optical absorption coefficient
α^{-1}	localization length
α_d	disorder parameter of CTRW theory
α_T	coefficient of thermal expansion
α_ρ	temperature coefficient of resistivity
b	neutron scattering length
C_p	heat capacity (constant pressure)
C_v	heat capacity (constant volume)
γ	temperature coefficient of the electrical gap
$\gamma(r)$	shape factor (SAXS)
γ_g	gyromagnetic ratio
Γ	Urbach edge parameter
D	diffusion constant
$\mathbf{D}$	force constant matrix
Δ	energy level splitting
ΔE_σ	activation energy of electrical conductivity
E_A	valence band edge
E_B	conduction band edge
E_c	conduction band mobility edge
E_F	Fermi energy (level)
E_g	band gap
E_o	optical band gap (mobility gap)
E_v	valence band mobility edge
ε_1	real part of dielectric constant
ε_2	imaginary part of dielectric constant
$f(E)$	Fermi–Dirac distribution function
$f(k)$	atomic scattering (form) factor
F	Helmholtz free energy
$F(k)$	reduced scattering intensity
g	Lande g-factor
g_s	Lande g-factor (free spin)

Glossary of symbols

$g(E)$	density of states ($eV^{-1}\ cm^{-3}$)
$\mathbf{G}$	Green function
$G(r)$	reduced radial distribution function
η	viscosity
η_p	atomic packing fraction
η_q	quantum efficiency
$\mathbf{H}$	magnetic field
H_c	superconducting critical magnetic field
H_{co}	coercive magnetic field
θ	bond angle
θ_C	Curie temperature
θ_D	Debye temperature
I	nuclear spin
$\mathbf{I}$	identity tensor
$I(\mathbf{k})$	scattering intensity
$J(r)$	radial distribution function
$\mathbf{k}$	scattering (or momentum) vector
k_B	Boltzmann constant
K_u	uniaxial magnetic anisotropy constant
κ	thermal conductivity
κ_d	Ginzberg–Landau parameter (dirty limit)
κ_T	isothermal compressibility
ℓ	mean free path
λ	electron–phonon coupling constant
λ_s	magnetostrictive constant
$\mathbf{\Lambda}$	spin–orbit interaction tensor
m_e	electron mass
$\mathbf{M}$	dipole moment vector
M_s	saturation magnetization
$M(\mathbf{k})$	modification function
$\boldsymbol{\mu}$	magnetic moment
μ_B	Bohr magneton
μ_d	drift mobility
μ_e	average band conduction mobility
μ_H	Hall mobility
$\mu(E)$	mobility at energy E
$\mu_X(E)$	X-ray absorption coefficient
$n(\omega)$	Bose occupation number
$N(E)$	integrated density of states ($eV^{-1}\ cm^{-3}$)

Glossary of symbols

v_{el}	characteristic electronic frequency
v_0	characteristic phonon frequency
$p(\omega)$	Raman depolarization ratio
$P(\phi)$	dihedral-angle distribution
$P(\omega)$	participation ratio
Π	Peltier coefficient
q	configuration coordinate
$\boldsymbol{Q}$	scattering vector
Q_{cc}	quadrupole coupling constant
R_H	coefficient
ρ^0	average density
$\rho(E)$	density of states (eV^{-1})
$\rho(r)$	atomic density function
$\rho(T)$	electrical resistivity
S	thermopower
$S(\boldsymbol{Q},\omega)$	dynamic structure factor
σ_E	electrical conductivity at energy E
σ_f	fracture strength
σ_{min}	minimum metallic conductivity
$\sigma(\omega)$	a.c. conductivity
T_c	superconducting transition temperature
T_f	fictive temperature
T_g	glass-transition temperature
T_m	melting temperature
U	Hubbard (correlation) energy
U_{eff}	effective Hubbard energy
W_D	disorder energy
W_H	polaron hopping energy
W_p	polaron energy
$W(T)$	Debye–Waller factor
ϕ	dihedral angle
$\chi(k)$	EXAFS intensity
$\chi'(\omega)$	real part of dielectric susceptibility
$\chi''(\omega)$	imaginary part of dielectric susceptibility
Z	atomic number
ω	radial frequency

Glossary of abbreviations

AC	actinide metal
AE	alkaline earth metal
BCS	Bardeen, Cooper and Schrieffer (theory)
CBH	correlated barrier hopping
CFO	Cohen, Fritzsche and Ovshinsky (model)
CON	chemically ordered network
CRN	continuous random network
CTRW	continuous time random walk
CVD	chemical vapour deposition
DRP	dense random packing
DRPSS	dense random packing of soft spheres
DSC	differential scanning calorimetry
DTA	differential thermal analysis
EDC	energy distribution curve
ESCA	electron spectroscopy for chemical analysis
ESR	electron spin resonance
EXAFS	extended X-ray absorption fine structure
GD	glow discharge (decomposition)
HREM	high-resolution electron microscopy
IR	infra-red
IVAP	intimate valence alternation pair
LA	longitudinal acoustic
LCAO	linear combination of atomic orbitals
LEED	low-energy electron diffraction
LO	longitudinal optic
M	metalloid
MD	molecular dynamics
NMR	nuclear magnetic resonance
NQR	nuclear quadrupole resonance
ODMR	optically detected magnetic resonance
OHC	oxygen hole centre
PL	photoluminescence
PLE	photoluminescence excitation (spectrum)
QMT	quantum-mechanical tunnelling
RCN	random covalent network
RDF	radial distribution function
RE	rare earth metal

Glossary of abbreviations

RPA	random-phase approximation
S	simple metal
SANS	small-angle neutron scattering
SAXS	small-angle X-ray scattering
T	transition metal
TA	transverse acoustic
TCR	temperature coefficient of resistivity
TE	early transition metal
TL	late transition metal
TLS	two-level system
TO	transverse optic
UPS	ultra-violet photoemission spectroscopy
VAP	valence alternation pair
XANES	X-ray absorption near-edge structure
XPS	X-ray photoemission spectroscopy

1 Preparation

1.1	**Introduction**
1.2	**Definitions**
1.3	**Preparation of amorphous materials**
1.3.1	Thermal evaporation
1.3.2	Sputtering
1.3.3	Glow-discharge decomposition
1.3.4	Chemical vapour deposition
1.3.5	Melt quenching
1.3.6	Gel desiccation
1.3.7	Electrolytic deposition
1.3.8	Chemical reaction
1.3.9	Reaction amorphization
1.3.10	Irradiation
1.3.11	Shock-wave transformation
1.3.12	Shear amorphization
1.4	**Summary**

1.1 Introduction

Amorphous materials *per se*, are not new; the iron-rich siliceous glassy materials recovered from the moon by the Apollo missions are some billions of years old, and man has been manufacturing glassy materials (principally from silica) for thousands of years. What is new, however, is the *scientific study* of amorphous materials, and there has been an explosion of interest recently as more new materials are produced in an amorphous form, some of which have considerable technological promise. This can be seen perhaps most strikingly in the number of scientific papers published on the subject each year, and this is shown in Fig. 1.1 for the period 1967 to 1981. This is obviously a fast-moving field, and any textbook on the subject runs the risk of becoming out-dated rather rapidly. Nevertheless, it is felt that sufficient is now known about the amorphous state to warrant a general, introductory text on the subject, particularly in view of the fact that amorphous materials rarely receive more than a passing mention in the conventional solid-state texts which deal almost wholly with *crystalline* materials.

Preparation

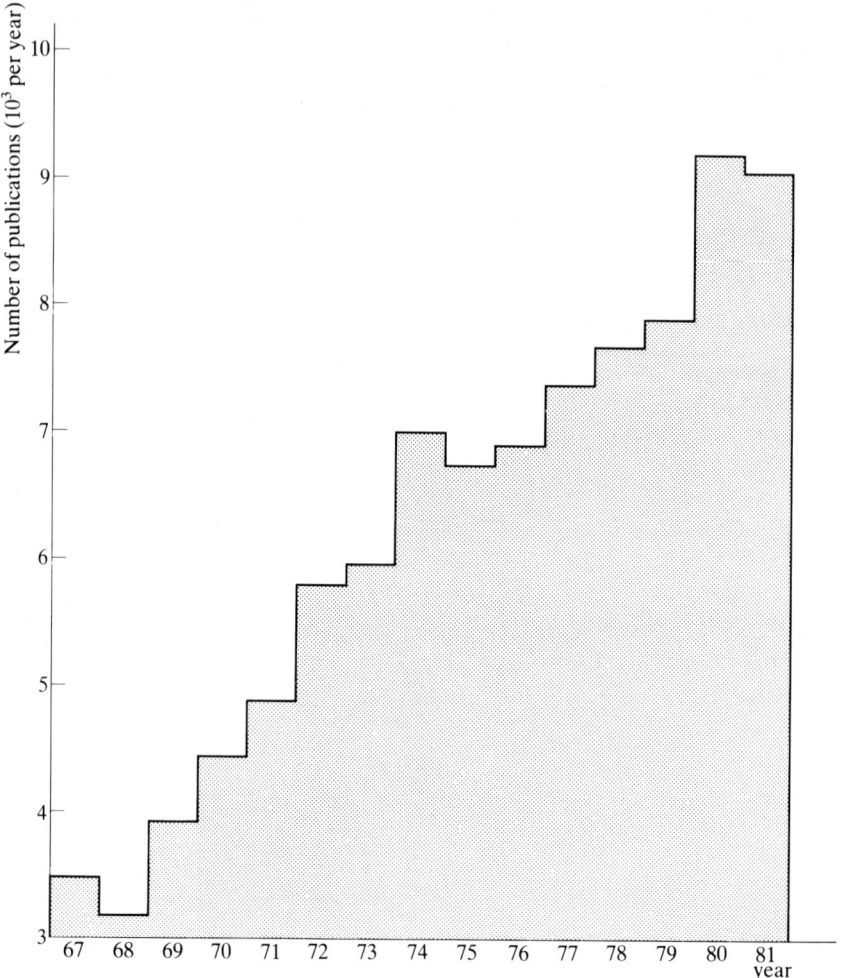

Fig. 1.1 Number of publications on amorphous materials published each year for the period 1967–1981. (Source: *Chemical Abstracts*).

The interest in amorphous materials is perhaps twofold. Firstly, there is the materials science aspect; a wide diversity of materials can be rendered amorphous – indeed almost all materials can. This is in sharp contrast to the knowledge of the average layman, for whom the word 'glass' signifies only that transparent material (made from silica with the addition of a few alkali oxides) which is placed in windows.

The second interest in amorphous materials is in the fundamental physics of such systems: Why is window 'glass' transparent when the conventional solid-state explanation of band gaps depends crucially on the assumption of periodicity in the underlying lattice and hence on the presence of Bloch electron wavefunctions? Furthermore, amorphous materials exhibit many properties which are unique to them and are not shared by crystalline solids at all.

It is the purpose of this book, then, to guide the reader through the murky waters of the subject, beginning with a description of various ways of producing amorphous materials, giving a discussion of their structure and properties, and incorporating some technological applications. In this way, it is hoped that interest in, and appreciation of, the fascinating field of the amorphous state will be quickened.

1.2 Definitions

This book is concerned with non-crystalline (or amorphous) materials which possess randomness to some degree. But what kind of randomness, and how much? We will try to answer the first question here and leave the second for discussion in Chapter 3.

Randomness can occur in several forms, of which topological, spin, substitutional, or vibrational disorder are the most important. These types of disorder are illustrated schematically in Fig. 1.2. Disorder is not a unique property, it must be compared to some standard, and that standard is the perfect crystal. This can be defined in the following way:

A perfect crystal is that in which the atoms (or groups of atoms or 'motifs') are arranged in a pattern that repeats periodically in three dimensions to an infinite extent.

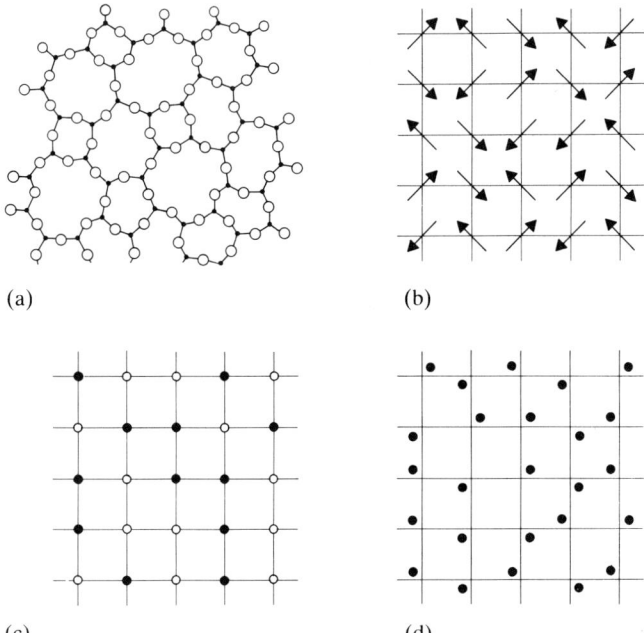

Fig. 1.2 Types of disorder:
(a) topological disorder (no long-range order);
(b) spin disorder (on regular lattice);
(c) substitutional disorder (on regular lattice);
(d) vibrational disorder (about equilibrium positions of a regular lattice).

With this definition, an imperfect crystal can simply be one which is finite and ends at surfaces (non-infinite extent), or one which possesses a defect, e.g. a vacancy or interstitial foreign atom or a dislocation (non-periodic). However, the forms of disorder with which we are concerned are more drastic than these small perturbations to perfect crystallinity.

Topological (or geometric) disorder is that form of randomness in which there is no translational periodicity whatsoever (Fig. 1.2(a)), and this type of positional disorder forms the theme of this book. Nevertheless, there are 'degrees' of topological disorder: certain amorphous materials have considerable short-range (or local) order while others have little; both have no long-range order however. All amorphous or glassy solids are therefore distinguished by their lack of periodicity.

Another variety of randomness is spin (or magnetic) disorder, in which the underlying perfect crystalline lattice is preserved, but each atomic site possesses a spin or magnetic moment, and this is oriented randomly (Fig. 1.2(b)). This situation occurs in some dilute magnetic alloys such as Cu–Mn or Au–Fe, with from 0.1 to 10 at. % magnetic component. The local moments are frozen into particular, but random, orientations because the 'exchange' interaction between moments in a metal has an oscillatory spatial dependence, and so randomly distributed moments in a *metallic* host suffer a corresponding distribution of exchange interactions, and hence become randomly oriented. These, and those materials which are topologically disordered *and* possess a randomly oriented spin, are termed 'spin glasses', and must not be confused with true glasses which are defined below.

A further kind of randomness is substitutional disorder (Fig. 1.2(c)) in which, although the underlying crystalline lattice is preserved, the material is in fact an alloy (say Cu–Au) with one type of atom randomly substituting for the other in the lattice. These systems are of great importance in metallurgy and other branches of materials science, and for the case of binary alloys are described by the Bragg–Williams theory.

The final category of randomness considered here is vibrational disorder of a crystalline lattice (Fig. 1.2(d)). Of course the concept of a perfect crystal is only valid at the absolute zero of temperature (if zero-point motion is ignored), and at any finite temperature the random motion of atoms about their equilibrium positions destroys the perfect periodicity. It is important to note, however, that vibrational disorder is *not* another form of topological disorder, since although the atoms are vibrating, they do so about their equilibrium *crystalline* positions which of course are not topologically disordered.

There is a considerable amount of confusion in the literature concerning the terms 'amorphous', 'non-crystalline' and 'glassy' and there are as yet no universally accepted definitions. At the outset, therefore, we will define what in this book we mean by these terms. We begin with the word 'amorphous':

Amorphous materials do not possess the long-range order (periodicity) characteristic of a crystal.

The terms amorphous and non-crystalline are synonymous under this definition, and can be used interchangeably. The precise degree of randomness which an amorphous solid possesses is the subject of Chapter 3. The term 'glass' is more restricted and as a consequence varies most in its definition. We will adopt the following definition:

A glass is an amorphous solid which exhibits a glass transition

The glass transition is the phenomenon in which a solid amorphous phase exhibits a more or less abrupt change in *derivative* thermodynamic properties (e.g. heat capacity or thermal expansivity) from crystal-like to liquid-like values with change of temperature. This is discussed in more detail in the next chapter. It is to be noted that this definition differs from a commonly accepted version in which glass is defined simply as the amorphous product produced by the rapid quenching of a melt. The definition offered here has the advantage that the term glassy is confined to those materials which can be obtained in a reproducible state (even after temperature cycling) since the material can be in a state of internal equilibrium above the glass transition. 'Glassy' materials, therefore, need not be prepared solely by quenching from the melt. Note that glassy (or equivalently, vitreous) solids are a special sub-set of amorphous materials or, to express it in a different way, all glasses are amorphous but not all amorphous solids are necessarily glasses.

The definition of the term amorphous given above does not explicitly exclude liquids, although in this book we will almost always be concerned with the solid state. Therefore, we must also define what we mean by the term 'solid', particularly in view of the fact that in the next chapter we will discuss the changes that take place when a glassy solid forms on cooling a liquid:

A solid is a material whose shear viscosity exceeds $10^{14.6}$ poise (or $10^{13.6}$ N s m^{-2}).

This rather arbitrary division of viscosity corresponds to a relaxation time of one day. This may be seen as follows. The expression for the viscosity can be written as

$$\eta = G_x/(\mathrm{d}v_x/\mathrm{d}z)$$

for the application of a shear stress G_x in the x direction causing a velocity gradient $\mathrm{d}v_x/\mathrm{d}z$, where $\mathrm{d}z$ is the thickness of an element perpendicular to the direction of the applied stress. A force of 100 N applied for one day to 1 cm^3 of material having a viscosity of $10^{13.6}$ N s m^{-2} yields a deformation of 0.02 mm, a value which is just measurable. Hence, a material is a solid if the application of a small force for one day produces no permanent deformation. For comparison, the viscosities of most common liquids at room temperature are of the order 10^{-2} poise.

The absence of long-range order, or periodicity, characteristic of an amorphous solid is most clearly evinced in a diffraction experiment, for example using X-rays, where instead of the sharp Bragg spots or rings produced by single crystal or polycrystalline samples respectively, broad diffuse haloes are observed (see Fig. 3.2). The origin and interpretation of such diffraction patterns form the subject of Chapter 3; suffice it to say here that the observation of diffuse haloes in a diffraction experiment is a prerequisite to characterization of an amorphous material.

1.3 Preparation of amorphous materials

There are at least a dozen different techniques that can be used to prepare materials in an amorphous state. Of these, five are commonly used in one form or another to produce most non-crystalline materials of commercial or academic interest. This section details these various methods, in approximate order of importance, together with their advantages and disadvantages and the types of material that are commonly produced using them. The different techniques use variously all three phases of matter, vapour, liquid and solid, as the starting materials in producing amorphous solids, but only deposition from vapour or liquid phases is really important.

Since the amorphous phase is less thermodynamically stable than the corresponding crystalline form (i.e. it possesses a greater free energy), the preparation of amorphous materials can be regarded as the addition of excess free energy in some manner to the crystalline polymorph. How this is done can vary widely, but it is a rule of thumb that the faster the rate of deposition (or cooling), the further the amorphous solid lies from 'equilibrium'.

1.3.1 Thermal evaporation

This technique is perhaps conceptually the most easy to understand, and is possibly the most widely used method for producing amorphous thin films. It is one of several ways of producing amorphous solids by deposition from a vapour. In essence, it is very simple; the starting compound is vaporized and the material is collected on a substrate. In practice, the evaporation is performed *in vacuo* to reduce contamination; a typical arrangement is shown in Fig. 1.3. An oil diffusion pump, together with a liquid nitrogen cooled trap (to eliminate water vapour and to a lesser extent hydrocarbon contamination) can produce a base pressure typically of the order of 10^{-6} Torr in the chamber. The starting compound can be heated to vaporization in one of two ways. A 'boat' containing the powdered material acts as a

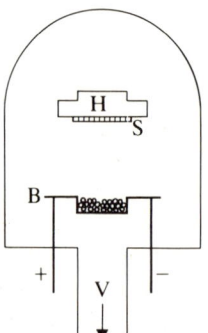

Fig. 1.3 Schematic illustration of a thermal evaporation chamber. The material to be evaporated lies in a high-melting point metallic boat (B) through which a heating current is passed, and the evaporated film is collected on a substrate (S) attached to a heater (H). The entire apparatus is placed in a vacuum chamber pumped to a high vacuum (V).

resistive element between two electrodes for the passage of a large d.c. current; high melting-point metals, e.g. molybdenum or tungsten, are used as boat materials. This method is suitable for relatively low melting-point compounds; those materials with higher melting points are best vaporized by bombardment with high-energy electrons from an electron gun placed in the chamber. The purpose of the electron-beam bombardment is simply to heat the sample to its melting point; compare this with the mechanism that takes place in 'sputtering' which is the topic of the next section. One major problem with the evaporation technique is differential evaporation, which is discussed more fully later; a remedy is to use 'flash evaporation', in which powdered material is dropped steadily on to a heated ribbon, thereby almost instantaneously vaporizing it.

Many materials can be rendered amorphous using thermal evaporation. The essential feature is that atomic surface mobility is greatly diminished because of the cold substrate, causing the adatoms to be 'frozen' in the random positions at which they arrive. Most amorphous semiconductors of interest to be discussed later can be produced in this way on substrates held at, or near, room temperature. Examples include Si and Ge and the tetrahedrally bonded III–V alloys (e.g. GaAs), 'chalcogenide' materials (those containing wholly or mainly chalcogen group VI elements), and SiO_x (where x can be varied between 0 and 2). These are all examples of solids containing principally covalent bonds. Certain metals can also be produced in an amorphous form by thermal evaporation without the need of stabilizing 'metalloid' atoms (see Chs 2 and 7), but this must be done using substrates held at very low temperatures (4 K), otherwise polycrystalline films are produced. Amorphous Ga and Bi can be made in this way, although other elements invariably crystallize even at these low temperatures, and other methods must be employed.

The principal disadvantage of thermal evaporation as a preparative technique lies in the variability in purity and composition of the resulting films (in the case of alloys). Many factors combine to produce this variability, and some are listed below:

(a) Substrate temperature.
(b) Substrate–source separation and orientation.
(c) Base gas pressure in the chamber.
(d) Impurities from the evaporation boat, or desorbed gases from the chamber surfaces.
(e) Boat or filament temperature.

Some of these parameters the experimenter can control (and should then specify in a description of the material preparation), others, including several not listed, cannot be varied at will, and so can change in a random fashion from one preparation to another of ostensibly the *same* material. This was the principal reason why so many experimental data were apparently contradictory in the early days of the subject. Some of the variables listed above are self-evident, others are more subtle. The substrate temperature is important since the higher the temperature, the higher is the surface mobility which can lead to a material with considerably fewer structural defects. However, if the substrate temperature is too high, the material will crystallize. The relative orientation of substrate with respect to source is of importance since evaporation at oblique angles can produce materials which are macroscopically inhomogeneous (Leamy *et al.* 1980). A 'shadowing' effect causes a

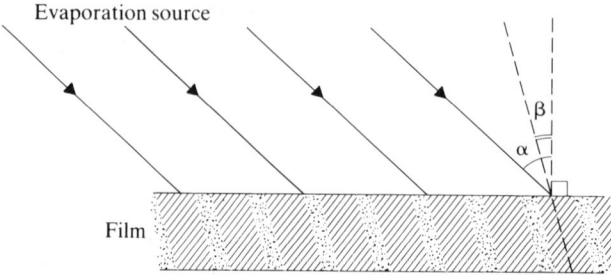

Fig. 1.4 Schematic illustration of the formation of columnar growth morphology of thin films evaporated at oblique angles of incidence. Note that in general the columns do not lie parallel to the evaporant beam direction.

columnar growth morphology to be formed with dense regions parallel to the source–substrate direction (see Fig. 1.4); this has been observed by scanning electron microscopy in e.g. thin films of a-Ge and a-Ge–Se alloys. In general, the column direction and the evaporant beam direction do not lie parallel to each other; rather, the columns lie more nearly perpendicular to the film surface. The respective angles of incidence, β and α, are found to obey the empirical equation (Leamy *et al.* 1980):

$$\tan \alpha = 2 \tan \beta. \qquad [1.1]$$

A major problem with conventional thermal evaporation is differential evaporation and the consequent lack of control over the resulting composition of the deposited film. In conventional vacuum evaporation, the vapour is removed immediately from the vicinity of the molten source, thereby preventing the establishment of equilibrium. In this case the element in the source alloy with the highest vapour pressure evaporates preferentially, depleting the source and leading to compositional inhomogeneities in the deposited film. Flash evaporation can be used to circumvent this problem to a certain extent, as mentioned earlier. A further complication which can arise in both flash and conventional evaporation is that the dominant (equilibrium) vapour species may not have the same composition as the source material. A case in point is that of arsenic chalcogenide alloys (e.g. As–S), in which the dominant vapour species are molecules of As_4S_4 even though the starting material may be stoichiometric As_2S_3. If the substrate is at a low enough temperature (and room temperature might be sufficient for this purpose) the deposited film will tend to consist of discrete As_4S_4 molecules bonded together by Van der Waals' forces. Annealing the film at a higher temperature will polymerize and cross-link these units yielding a fully coordinated random structure similar to that obtained by quenching from the melt (see sect. 1.3.5). Alternatively, the substrate can be held at an elevated temperature and the annealing then occurs simultaneously with the deposition process. Certain vapour species (e.g. As_4) have very low 'sticking coefficients' because of their shape, and consequently produce poor quality films.

It must be stressed that thermal evaporation, like other vacuum techniques is a *thin-film* preparative technique; typically film thicknesses of the order of tens of microns are produced for deposition rates of the order of 0.1–1.0 μm per second.

Nevertheless, this is sufficient for certain applications, and indeed the amorphous selenium at the heart of the Xerox process is produced by evaporation (see sect. 6.3.5).

1.3.2 Sputtering

The sputtering process is rather more complicated than thermal evaporation but enjoys the advantage of being far more flexible. In essence, it consists of the bombardment of a target by energetic ions from a low-pressure plasma, causing erosion of material, either atom by atom or as clusters of atoms, and subsequent deposition of a film on a substrate. The simplest way to induce sputtering is to apply a high negative voltage to the target surface, thereby attracting positive ions from the plasma. However, this d.c. sputtering process is only feasible for targets composed of metals, or at least consisting of material sufficiently electrically conducting that the target can act as an electrode; certain of the *crystalline* arsenic chalcogenides are sufficiently conducting to act in this way to produce amorphous films.

A more common approach is to apply an r.f. field (at typically $\simeq 13$ MHz), to the target (Fig. 1.5). In this case, both metallic and insulating materials may be used

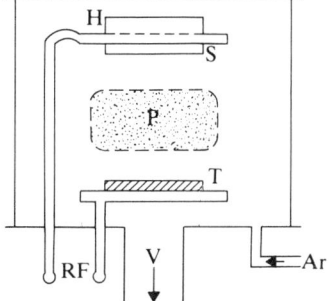

Fig. 1.5 Schematic illustration of an r.f. sputtering apparatus. The r.f. field is applied between the target (T) electrode and the substrate (S) electrode. A sputtering gas (e.g. Ar) is introduced into the chamber and a plasma (P) is struck. The gas is pumped away by a vacuum system (V); the substrate may also be heated by an electrically insulated heater (H).

as targets; all that is required is for the r.f. voltage to be capacitatively coupled to the target surface. For metallic targets, this is achieved by connecting a capacitor in series with the target; insulating targets are normally bonded to a metal backing electrode which itself acts as a capacitive component. As in evaporation a residual gas pressure in the chamber of less than 10^{-6} Torr is desirable and achieved by similar means. The sputtering gas, commonly an inert gas such as argon, is introduced at a higher pressure, say 1–20 mTorr. Application of the r.f. voltage causes the striking of a plasma and positive ions are attracted to the target during each negative half-cycle. However, since the mobility of electrons is higher than that of ions in the plasma, more electrons than ions are attracted to the front surface of the target in their respective half-cycles, resulting in the build-up of a

9

negative bias. After some time, therefore, the ions are attracted from the plasma essentially by the bias potential, rather than the r.f. potential. The ejected material from the target is then carried to the substrate by the following half-cycle of the r.f. field.

The material deposited on the substrate can form an amorphous film for the same reason as in evaporation; namely, adatom mobility is sufficiently low, because of the relatively low temperature of the substrate, that crystallization is prevented. However, the substrate must be cooled and the r.f. power limited so that heating due to electron bombardment is minimized. This method generally produces reasonably homogeneous and uniformly thick samples; for $\pm 5\%$ uniformity of film thickness the diameter of the target needs to be about 5 cm larger than the maximum linear dimension of the substrate. The process can be operated in either sputter-up or sputter-down modes, the former having the advantage of being able to use targets in powder form.

Sputtering, particularly using r.f. fields, can produce amorphous samples of most materials of interest, Si, Ge, SiO_2, chalcogenide glasses and amorphous metals being examples. Sputtering is significantly superior to evaporation for the production of multicomponent systems. For targets already in compound form this is because the sputtering rates for different elements do not vary widely, unlike melting points, vapour pressures and, therefore, evaporation rates. As a consequence, sputtered films tend to preserve the stoichiometry of the starting material. In addition, amorphous alloys can be sputtered simply by using a target composed of the individual components of the alloy. For targets comprising, say, small circles of one component placed uniformly over the surface of the circular target of the predominant component, films possessing a good degree of compositional homogeneity can be achieved for sufficiently large substrate-target distances; for 'split' targets consisting of two semicircles of different components, on the other hand, films having a composition gradient (perpendicular to the division of the target) can be obtained. Usual sputtering rates for insulating target materials in an inert gas atmosphere (e.g. Ar) are in the range 1–10 Å s^{-1}, and perhaps a little higher for metallic targets. These rates, it will be noticed, are significantly lower than those encountered in evaporation.

A further option offered by sputtering is the use of gases other than Ar which chemically *react* with the target, resulting in 'reactive' sputtering. This can significantly increase the sputtering rate, as well as incorporate chosen additives into the films. This has been used in producing hydrogenated amorphous silicon (by sputtering in an $Ar:H_2$ mixture), doped amorphous silicon films ($Ar:H_2:N_2/PH_3$), and amorphous transition-metal oxide materials ($Ar:O_2$).

Most of the factors which cause variations in the films produced by thermal evaporation are also significant in the sputtering process. Additional factors which need to be considered are the following:

(a) Sputtering gas pressure.
(b) Ratio of partial pressures of reactive gas to inert gas (in reactive sputtering).
(c) R.F. power applied to target.
(d) Bias voltage of target or substrate.

Close control over all these parameters is necessary to ensure that films with reproducible characteristics are produced.

1.3.3 Glow-discharge decomposition

Another technique which can produce thin films of amorphous materials is glow-discharge (GD) decomposition in the vapour phase. This has sprung into prominence recently with the discovery by Spear and Le Comber (1975) that prepared in this manner, certain amorphous semiconductors can be substitutionally doped to control their electronic properties.

This technique, like sputtering, relies on the production of a plasma in a low-pressure gas, but instead of ions from the plasma ejecting (sputtering) material from a target, chemical decomposition of the gas itself takes place, leading to deposition of a solid film on a substrate placed in the plasma (Fig. 1.6(a)). The plasma is produced by the application of an r.f. field (as in sputtering) in one of two ways, either using inductive coupling or capacitive coupling; these configurations are shown schematically in Fig. 1.6(b). The capacitive configuration is the more important since it can readily be scaled up to produce large surface area material for device applications. The reactant gas may be used either by itself, or mixed with a carrier gas (e.g. Ar), or another reactant gas to form alloys, and is admitted to the reaction chamber through a mixing tank and flowmeter. Typically, reactant gas pressures between 0.05 and 1.0 Torr are used with about 1–10 W of, say, 14 MHz r.f. power being supplied to the plasma, resulting in deposition rates of $1-10$ Å s^{-1}.

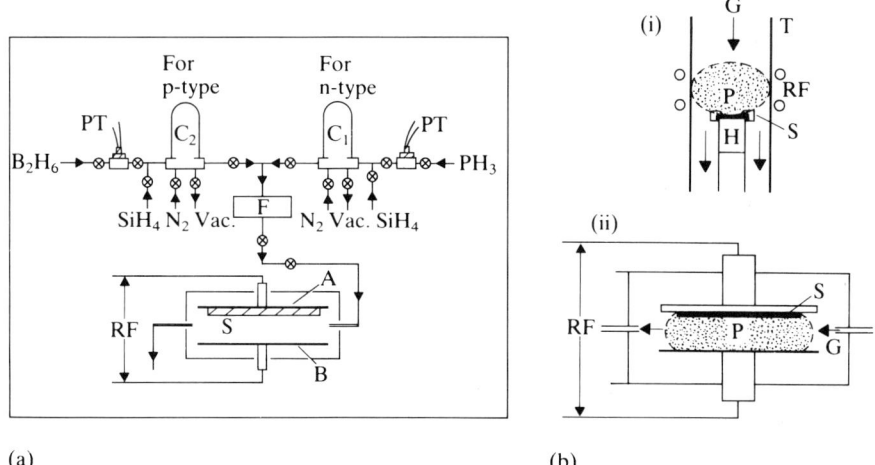

Fig. 1.6 Illustration of the glow-discharge decomposition preparative technique.
(a) Schematic illustration of the preparative unit for the deposition of n- and p-type a-Si:H (C_1, C_2, glass cylinders; PT, pressure transducer; F, flow meter; S, substrate).
(b) Methods of coupling the r.f. power to the plasma: (i) inductive, (ii) capacitive. (G, gas flow; P, plasma; S, substrate; H, heater; T, glass tube) (Spear 1977).

The reactant gases that are used are generally hydrides and produce polymers under certain conditions; this is the 'glow-discharge polymerization' of monomers utilized in chemistry. Solid-state physicists refer to the process as 'glow-discharge decomposition', particularly when non-polymeric material is produced. The material most often produced by the GD technique is amorphous silicon (a-Si); this is important because of its device prospects in view of the fact that it can be readily doped (see sect. 5.4.1). Silane (SiH_4) has long been the reactant gas used to produce a-Si, with or without Ar as diluent. This produces *hydrogenated* amorphous silicon (a-Si:H) films under optimum conditions (with of the order of 5–10 at. % H contained within the material), but polymeric polysilane or amorphous silicon hydride (a-SiH_x) films with much higher hydrogen contents (and significantly worse electronic characteristics) are formed under other conditions. The a-Si:H materials are doped either n-type or p-type by admitting phosphine (PH_3) or diborane (B_2H_6) respectively, into the silane gas stream; the dopant atoms are incorporated into the amorphous structure substitutionally, thereby liberating excess electrons (for P) or holes (for B). Recently, fluorinated material produced by the GD decomposition of SiF_4 and H_2 (a-Si:F:H) has been shown to possess superior properties to simply hydrogenated silicon (Ovshinsky and Madan 1978). In the same manner, hydrogenated germanium (a-Ge:H) can be obtained using germane (GeH_4) and can be doped in a similar fashion. Amorphous SiO_2 can be obtained from the GD decomposition of a mixture of SiH_4 and N_2O, and amorphous Si_3N_4 is produced from a mixture of SiH_4 and NH_3.

The GD process, depending in essence on plasma chemistry, is more complicated than either thermal evaporation or sputtering. As a result, the significant parameters that can affect the characteristics of a particular film are more varied. Some of these are given below:

(a) Gas pressure and temperature.
(b) Ratio of reactant gas (e.g. SiH_4) to carrier gas (e.g. Ar).
(c) Gas flow rate and consequent dwell time in the chamber.
(d) Chamber geometry and substrate position relative to the plasma.
(e) R.F. power applied to plasma.
(f) Method of coupling of r.f. power to plasma (inductive or capacitive).
(g) Substrate bias.
(h) Substrate temperature.

The substrate temperature is a critical parameter, and experience has shown that, at least in the preparation of a-Si:H, higher temperatures are advantageous, and substrate temperatures of 520 K are often used. Such high temperatures help to ensure that volatile side-products are not incorporated into the films. Another crucial parameter is the pressure of the reactant gas. In an informative review of the GD process, Brodsky (1978) points out that there are two distinct pressure regimes that govern the behaviour of the plasma discharge; these are shown schematically in Fig. 1.7. This behaviour is analogous to Paschen's law, which applies strictly only for the threshold characteristics of the GD, but can be applied to the operating region if the r.f. power levels are sufficiently low. At high pressures, the electron energies within the plasma are collision-limited within the plasma itself, and this is found to promote polymerization within the vapour phase leading to (amorphous) silicon

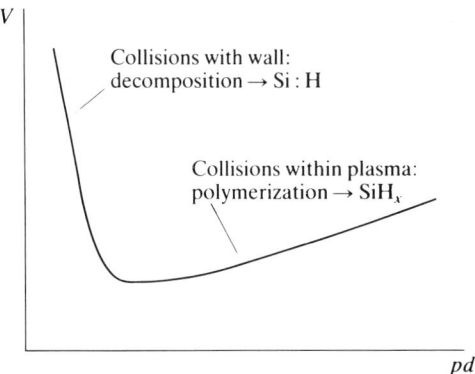

Fig. 1.7 Schematic illustration of the characteristics of a glow-discharge plasma (Paschen's law): V is the applied r.f. voltage, p is the pressure and d is a characteristic dimension of the chamber (Brodsky 1978).

hydrides (a-SiH$_x$). In contrast, at low pressures the electron energies are limited by collisions near or at surfaces, and this is found to enhance decomposition resulting in the desired hydrogenated material (a-Si:H).

1.3.4 Chemical vapour deposition

The technique of chemical vapour deposition (CVD) is analogous to the GD method in that both depend on the decomposition of a vapour species. The difference is that the CVD process relies on thermal energy for the decomposition (i.e. it is pyrolytic), and the applied r.f. field (if used) simply serves to heat up the substrate upon which the vapour decomposes. Temperatures of the order of 1000 K are commonly used. The CVD process is often used to produce polycrystalline material, particularly silicon from SiH$_4$. However, under the appropriate conditions true amorphous hydrogenated silicon (a-Si:H) can also be deposited by the CVD technique from SiH$_4$, and doped samples using mixtures with PH$_3$ or B$_2$H$_6$.

Another important application which can be regarded as a variant of CVD is the production of amorphous silica (SiO$_2$) by the reaction:

$$SiCl_4 + 2H_2O \rightarrow SiO_2 + 4HCl$$

This results in so-called 'vapour phase hydrolysis', which is carried out in an oxyhydrogen flame producing 'synthetic grade' high purity amorphous silica.

1.3.5 Melt quenching

The oldest established method of producing an amorphous solid is to cool the molten form of the material sufficiently quickly; what is meant by 'quick cooling' will emerge later. Amorphous materials produced in this manner have often in the past been termed 'glasses'; we prefer here to define such materials as those which exhibit glass-transition phenomena (discontinuities in specific heat, etc.) and hence are not necessarily formed only by melt quenching, although the great majority of melt-quenched amorphous solids do indeed show glass-transition behaviour. The distinguishing feature of the melt-quenching process of producing amorphous materials is that the amorphous solid is formed by the *continuous* hardening (i.e. increase in viscosity) of the melt. In contrast, crystallization of the melt occurs as a *discontinuous* solidification, solid growth taking place only at the liquid–solid interface, with the result that crystallites grow in the body of the melt.

An essential prerequisite for 'glass' formation from the melt, therefore, is that the cooling be sufficiently fast to preclude crystal nucleation and growth; the crystalline phase is thermodynamically more stable and crystal growth will always dominate over the formation of the amorphous phase if allowed to take place. A good discussion of crystal nucleation and growth kinetics can be found in Turnbull (1969).

The crystallization rate of an undercooled liquid depends on the rate of crystal nucleation and on the speed, u, with which the crystal–liquid interface moves. Both in turn are strongly dependent on the reduced temperature $T_r = T/T_m$, and the undercooling $\Delta T_r = (T_m - T)/T_m$. It is expected that u should be proportional to some function of ΔT_r (linear for small ΔT_r), and inversely proportional to an average jump time, τ, of atoms in the interfacial region; τ in turn might be expected to scale as the viscosity. Thus, high values of viscosity reduce the crystallization-front velocity and hence the crystallization rate itself; this is in accordance with common experience in which it is often found that the liquids of 'easy' glass-formers have high viscosities for temperatures at and below T_m (e.g. the viscosity of silica $= 10^7$ poise at the melting point of crystobalite). Nucleation of crystallites in a liquid may be either 'heterogeneous' or 'homogeneous'; the former arises when 'seeds', such as crystals of the material suspended in the liquid or the container walls are introduced into the system, and the latter takes place when all heterogeneous centres have been removed.

What then is a condition for glass formation? It is that the nucleation rate I_n should be less than a certain value, say, 10^{-6} cm^{-3} s^{-1}, not to be observable in practicable time-scales and liquid volumes. Simple nucleation theory dictates that crystal nucleation is *promoted* by the change in free energy on crystallization (dependent on the *volume* of the crystallite), and at the same time *retarded* by the work needed to form the crystal–liquid interface (proportional to the *area* times the surface tension γ); the interplay between these two competing factors ensures that only those crystals which contain more than some critical number of atoms can exist as crystal nuclei; all others remelt. Turnbull (1969) has shown that the homogeneous nucleation rate can be expressed in the form

$$I_n = \frac{k}{\eta} \exp\left[-b\alpha_n^3 \beta_n / T_r (\Delta T_r)^2\right] \qquad [1.2]$$

where k is a constant, b is a constant dependent on the shape of the nucleus ($16\pi/3$ for a sphere) and α_n and β_n are dimensionless parameters defined as:

$$\alpha_n = \frac{(N\bar{V}^2)^{1/3} \gamma}{\Delta H_m} \qquad [1.3]$$

(where N is Avogadro's number, $\bar{V}$ is the molar volume of the crystal and ΔH_m is the molar heat of fusion), and as:

Preparation of amorphous materials

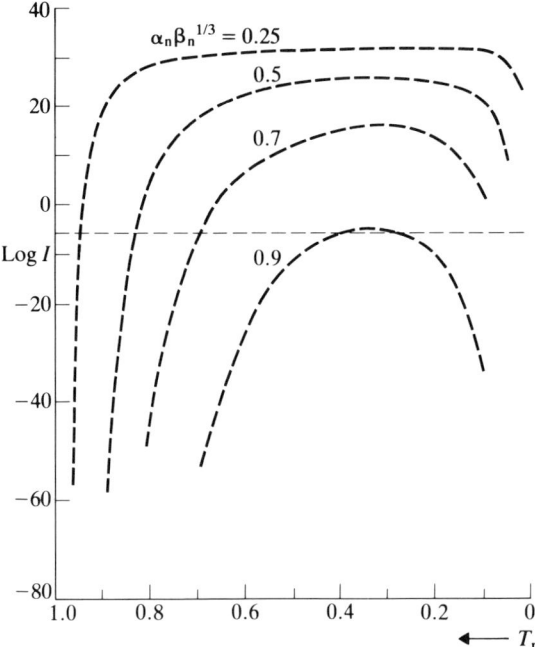

Fig. 1.8 Homogeneous nucleation rate as a function of reduced temperature $T_r = T/T_m$ calculated for various values of $\alpha_n \beta_n^{1/3}$ (Turnbull 1969).

$$\beta_n = \frac{\Delta H_m}{RT_m} = \frac{\Delta S_m}{R} \qquad [1.4]$$

The quantity α_n represents the number of monolayers per unit area of crystal which would be melted at T_m by an enthalpy equal in magnitude to γ. A plot of [1.2] is shown in Fig. 1.8, showing the nucleation rate I_n as a function of reduced temperature T_r for different values of the parameter $\alpha_n \beta_n^{1/3}$. The value chosen for the pre-exponential factor in [1.2] was 10^{32} with a temperature independent value of $\eta = 10^{-2}$ poise; this factor can be written equivalently as the product of the atomic density (10^{21} cm^{-3}) and a typical atomic jump frequency (10^{11} s^{-1}). It can be seen from Fig. 1.8 that for values of $\alpha_n \beta_n^{1/3} \gtrsim 0.9$ the crystal nucleation rate becomes less than that which is readily observable, and these liquids would not crystallize (unless seeded), forming glasses if sufficiently undercooled. Using a more realistic model for the temperature dependence of the viscosity (see [2.6]), it is found that the peak in the I–T_r relation moves to higher values of T_r and becomes lower and narrower for increasing values of the reduced glass-transition temperature T_g/T_m. The consequence of this is that glass-forming tendency should increase with T_g/T_m.

In this respect, then, we may regard 'glass' formation as a kinetic phenomenon; a glass is simply a supercooled liquid in which no time for crystal growth has been allowed. In fact, 'glass' is considerably more complicated than this simple picture implies, and many factors contribute to whether a particular glass will form; this topic forms the subject of Chapter 2.

We have seen that the cooling rate is often a critical factor in determining glass formation. Certain 'easy' glass-formers such as B_2O_3, will form an amorphous solid even under conditions of very slow cooling (say 1 K s^{-1}), whereas other materials, notably metallic glasses, require very high rates of cooling indeed before they form an amorphous phase; otherwise polycrystalline material is produced. The most usual way of producing samples of 'easy' glass-formers is to seal a charge (1–10 g) in a fused silica ampoule under a good vacuum (10^{-6} Torr), and keep the ampoule in a rocking or rotating oven at a sufficiently elevated temperature that the constituents become molten and can react; the rocking motion ensures that a thorough mixing of the mixture takes place. The melt can then be quenched, either slowly by simply switching off the oven, or more rapidly by bringing the ampoule into the air, or yet more rapidly by plunging the ampoule immediately from the oven into a liquid (preferably one with a high thermal conductivity and high latent heat of vaporization so that heat is conducted away from the sample as fast as possible without the formation of a thermally insulating vapour layer around the ampoule). In this manner cooling rates of the order 10^{-2} to 10^3 K s^{-1} are achievable (see Table 1.1), and are sufficient for most purposes. Some parameters which are important in the ampoule method of melt-quenching are the following:

(a) Temperature of the oven.
(b) Rate (or equivalently, method) of cooling.
(c) Volume of charge in ampoule.
(d) Thickness of the wall of the ampoule.

Material produced in this manner is often in the form of a 'plug' or rod, but can be made into thinner sections by either sawing and polishing or by heating the glass up to its softening point (glass-transition temperature) and then compressing it between two parallel plates or by blowing a thin film.

For certain materials however, notably metals, even the fastest cooling rates achieved by the above methods are insufficient to prevent crystallization, and still faster cooling methods are required. Such methods have been pioneered by Duwez who was the first to demonstrate, in 1959, that cooling rates in excess of 10^6 K s^{-1} can be achieved. The technique relies on the rapid cooling afforded by a 'chill-block' in intimate contact with a thin film of the liquid, in which the heat is swiftly and continually transported away from the interface; chill-blocks of copper are commonly used. The liquid melt may be brought into contact with the chill-block in one of several ways. The first that was used projected small droplets of liquid at a Cu sheet ('splat cooling'); in this manner the first amorphous metal $Au_{75}Si_{25}$ was produced by melt quenching (Klement et al. 1960). This and a related technique, the aptly named hammer-and-anvil 'drop smasher', can give cooling rates of the order of 10^5 to 10^6 K s^{-1}, but produces thin foils of material of irregular area. These amorphous metallic materials have many interesting properties (e.g. see Cahn 1980) which lend themselves to several technological applications (see Ch. 7), but material produced by the hammer-and-anvil or projection techniques is not generally in a usable form. However, there are certain chill-block cooling techniques more recently developed which do produce amorphous metals in a useful geometry, and with a faster cooling rate as an added bonus; these are the 'melt-spinning' and 'melt-extraction' techniques (Figs 1.9(a) and (b)). Both use a rapidly spinning copper disc as the chill-block. The former uses a jet of liquid impinging on the disc producing

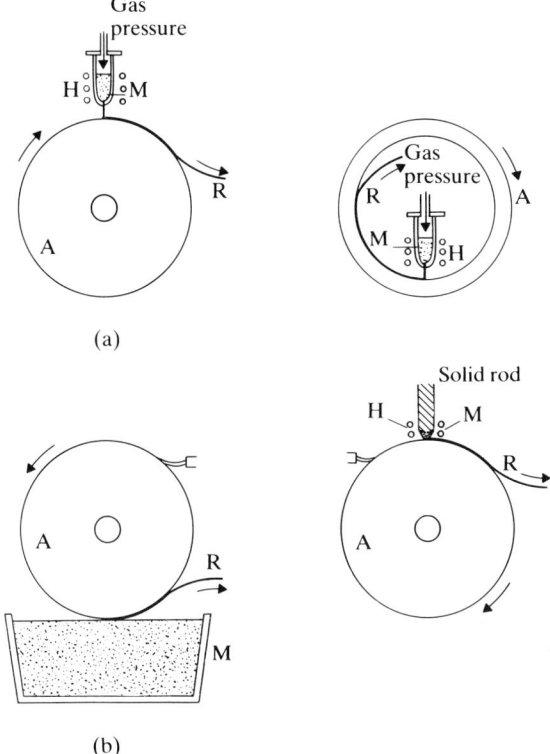

Fig. 1.9 Schematic illustration of the different techniques to form continuous ribbons of glassy metals (Cahn 1980).
(a) Melt spinning, both external and internal.
(b) Melt extraction, both crucible and pendant drop.
(A, rotating Cu disc or cylinder; M, melt; H, induction heater; R, ribbon of metallic glass.)

thin ribbons, whereas the latter removes liquid from a reservoir producing fine wires; in both cases, cooling rates of the order 10^6–10^8 K s^{-1} can be achieved. These deceptively simple techniques yield cooling rates which depend in a complicated way on a variety of parameters, some of which for melt spinning as an example, are given below:

(a) Rotation speed of Cu disc.
(b) Coating and surface finish of disc.
(c) Melt jet diameter, velocity and angle of attack.
(d) Ribbon geometry and uniformity, particularly surface smoothness.
(e) Nature and pressure of ambient gas.

Despite this apparent complexity, ribbons up to 15 cm wide and 30 μm thick can be produced by careful attention to conditions of preparation, although 1 cm by 15 μm is more usual. Notice that in contrast to other melt-quenching methods, chill-block techniques cannot produce material in bulk form, but only in the form of thin ribbons (or wires). If attempts are made to produce ribbons of thickness greater than a few tens of microns, only polycrystalline material is produced, since beyond a

Table 1.1 Quenching techniques and their characteristic rates of cooling

Technique	Cooling rate ($K\ s^{-1}$)
Annealing	
large telescope mirror	10^{-5}
optical 'glass'	3×10^{-4}
ordinary 'glass'	10^{-3}–10^{-2}
Air quenching	1–10
Liquid quenching	10^2–10^3
Chill-block	
splat-cooling	10^5
melt-spinning, extraction	10^6–10^8
Evaporation, sputtering	10^9 ?

critical thickness conduction of heat *through* the ribbon becomes the critical factor, and this rate cannot be made large enough to produce amorphous metals (see sec. 7.1).

A summary of various melt-quenching techniques and their cooling rates are given in Table 1.1 together for comparison with those processes normally regarded as annealing. It will be seen that thermal evaporation and sputtering (i.e. *vapour quenching*) are also included; the precise rates of 'cooling' are impossible to quantify in these cases, but must certainly be well in excess of those achievable using chill-block methods. This makes understandable the fact that if a material cannot be rendered amorphous any other way, then some form of vapour deposition will almost certainly be successful.

1.3.6 Gel desiccation

A recently developed technique for producing amorphous materials via the sol–gel process has considerable technological promise (e.g. see Mukherjee 1980). The method has its greatest usefulness for those systems which give rise to very viscous melts near the melting point (and consequently have considerable difficulty in achieving homogeneous mixing), or alternatively which have extremely high melting points and hence pose considerable technical problems in actually being able to make a glass by melt quenching.

There are two essentially different routes to produce an amorphous material via a gel, using either aqueous or organic materials as the starting components:

(a) Destabilization of a sol (commonly of silica), with other components being added in the form of appropriate aqueous solutions.
(b) Hydrolysis and polymerization of mixtures of organometallic compounds.

In both methods, however, a multicomponent 'gel' (the elastic solid product produced abruptly from a viscous liquid by a process of continuing polymerization), being non-crystalline and homogeneous, is heated to remove volatile components and cause an initial densification, which is completed by a final process of sintering

Preparation of amorphous materials

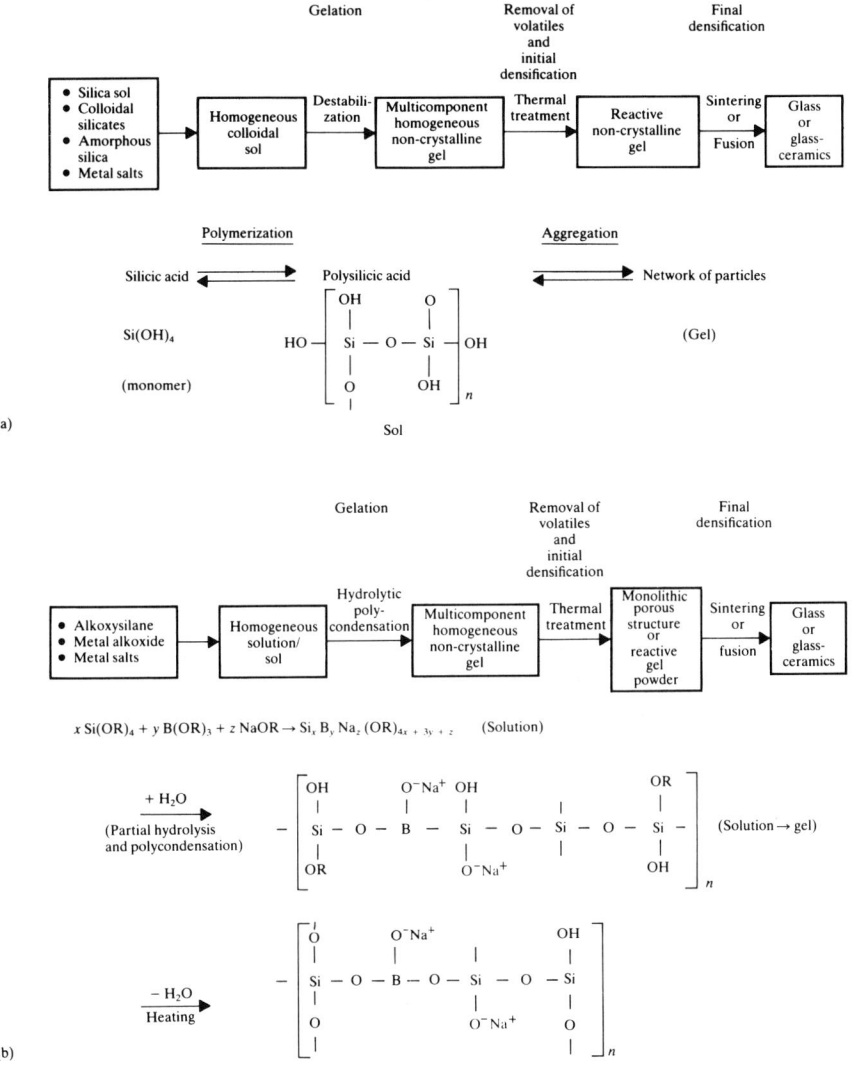

Fig. 1.10 The sol–gel process for glass manufacture (Mukherjee 1980).
(a) Aqueous process. Silica (in the form of silicic acid), together with any desired network modifiers, are formed into an aqueous sol. This is then converted into a rigid gel by progressive polymerization and formation of a branched network, which is then transformed into a dense glass by sintering or fusion.
(b) Organo-metallic process. Alkoxysilane, together with any desired network modifiers in the form of metal alkoxides or salts, are formed into a sol which is then polymerized and gelled by partial hydrolysis. Heat treatment removes any volatiles and produces a dense glass.

or fusion to produce the amorphous solid. The two methods are represented schematically in Figs 1.10(a) and (b).

The method employing aqueous components starts with a silica 'sol' of silicic acid, $Si(OH)_4$ (a dispersion at the molecular level in aqueous solution) to which are added other components (e.g. metal salts in aqueous solution). The formation of the gel then involves polymerization of the monomeric $Si(OH)_4$ to form particles, growth and interconnection of particles to form branched chains, finally forming a 3-D network. The gelling process is influenced by several factors, e.g. pH of the medium, particle size and temperature. The gel can then be converted to an amorphous solid in one of three ways: heating at temperatures *below* the glass-transition temperature, T_g, resulting in chemical polymerization; sintering (with or without the application of pressure) at temperatures above T_g but much below the melting point; fusing the gel particulates to form a glass. In this manner, amorphous materials in, for example, the SiO_2–P_2O_5 or SiO_2–B_2O_3 series may readily be produced by using phosphoric acid or ammonium tetraborate additions respectively.

The second approach uses alkoxysilanes (e.g. tetraethoxysilane $Si(OC_2H_5)_4$) mixed with other metal alkoxides (or salts) as the starting materials which, in the presence of a small amount of water, partially hydrolyse replacing alkoxide groups by hydroxyl progressively, while polymerizing and incorporating the various components homogeneously into the network. As an example, consider the reaction between silicon, boron and sodium alkoxides, $Si(OR)_4$, $B(OR)_3$ and $NaOR$ (where R is an alkyl group). On addition of water these hydrolyse partially and polymerize to give, say,

$$\begin{bmatrix} \text{OH} & \text{O}^- & \text{OH} & \text{OH} & \text{OR} \\ | & | \; \text{Na}^+ & | & | & | \\ \text{Si}-\text{O}-\text{B} & -\text{Si} & -\text{Si}-\text{O} & -\text{Si} \\ | & & | & | & | \\ \text{OH} & & \text{OR} & \text{O}^-\text{Na}^+ & \text{OH} \end{bmatrix}$$

which then gels. Heating then causes dehydration and loss of alkyl groups leaving a mixed oxide gel, which can be transformed to an amorphous material by one of the processes mentioned earlier. Compounds such as $Si(OC_2H_5)_4$, $B(OCH_3)_3$, $Al(OC_4H_9)_3$ and $Na(OCH_3)$ can be used in this way.

The principal advantage of the gel desiccation technique is that very refractory materials, such as those in the TiO_2–SiO_2 system, can be prepared homogeneously at relative *low* temperatures, in some cases much below the melting point. Very pure products can also be made, since the starting components can readily be purified. The final amorphous products, although not prepared by melt quenching, behave in a very similar manner to glasses, exhibiting the same kinds of glass-transition phenomena.

1.3.7 Electrolytic deposition

Another technique which uses solutions of the desired material is electrolytic deposition. In this manner, for example, amorphous Ge films can be deposited on a Cu cathode by the electrolysis of $GeCl_4$ in glycol. Although comparatively thick films ($\sim 30 \, \mu m$) can be prepared in this way, contamination by the solution and reaction products is a problem.

1.3.8 Chemical reaction

In certain cases the precipitate thrown out of solution as a result of a chemical reaction proves to be amorphous. For example, amorphous As_2S_3 can be formed by passing H_2S gas through a solution of As_2O_3 in dilute hydrochloric acid. The product shows very similar glass-transition behaviour to that observed in melt-quenched glasses.

1.3.9 Reaction amorphization

Thin films of amorphous SiO_2 can result from the anodic oxidation of Si; the use of non-aqueous electrolytes prevents the formation of porous films that result when aqueous electrolytes are used. Mixed silicate systems, such as SiO_2–P_2O_5 and SiO_2–B_2O_3, can be formed using appropriate electrolyte solutions. Thermal oxidation of Si with dry oxygen or steam at high temperatures also leads to the formation of a-SiO_2 thin films.

1.3.10 Irradiation

The interaction between high-energy ionizing particles and crystalline solids can often produce enough structural damage to amorphize the material. As an example, crystalline quartz is transformed to a-SiO_2 by irradiation with $\sim 10^{20}$ neutrons cm^{-2}. This form, interestingly, has a density which is *higher* (2.26 g cm^{-3}) than that produced by conventional melt quenching (2.205 g cm^{-3}); the density decreases to the lower value on annealing.

Ion bombardment of crystalline solids produces amorphous surface layers, some hundreds of ångströms in thickness. This is of great significance in the current semiconductor industry (based on *crystalline* silicon), where doping to control the electronic properties is often accomplished by implanting the dopant atoms (e.g. P, B) by ion bombardment. The resulting amorphous layers are very deleterious to the performance of the device, but can be removed by 'laser annealing'. Illumination of the amorphous surface by pulsed, high-power laser light causes the surface to *recrystallize* and heal the bombardment damage; the laser-annealing process is complicated, and believed to be not simply a surface melting phenomenon, but possibly due to bonds reforming as a result of the creation of an intense electron-hole plasma caused by the absorption of the photons.

Electron-beam irradiation can also result in amorphization in certain cases. Some organic crystals became amorphous on exposure to electrons in an electron microscope. The same effect is also seen in silicate-based materials; quartz itself is rendered amorphous as are various complicated alumino-silicate minerals (zeolites) under the action of electron irradiation.

The crystalline-to-amorphous transition as a result of irradiation can also occur naturally. Certain minerals, although possessing the external form of crystals (facets, etc.) are, in fact, *amorphous* as determined by X-ray diffraction, etc. These materials, termed 'metamict' (Pabst 1952), are always found to possess radioactive elements such as uranium or thorium, sometimes only in trace amounts, and over geological time have been rendered amorphous by the radioactive decay products of the U or Th.

1.3.11 Shock-wave transformation

Shock waves produced by explosion can cause amorphization. For example, quartz is rendered amorphous by the impact of a shock-wave front producing pressures of the order of 600 kbar and temperatures of the order of 1200 °C. Local melting is likely to be responsible.

1.3.12 Shear amorphization

Many crystalline materials are rendered amorphous by the simple act of grinding. Quartz, crystobalite and the arsenic chalcogenides (S, Se, Te) are among those susceptible to this form of transformation.

1.4 Summary

Despite the welter of techniques that can produce the amorphous state, only the first five, or possibly six, of those described here are at all important commercially or otherwise. It is very important at the outset of a study of the amorphous state to notice that although many materials can be prepared in an amorphous form by more than one technique, *the different forms need not, and indeed often do not, have the same properties.* As we shall see in Chapter 3, there is no *unique* structure for an amorphous solid (an infinite number of possible structures being in principle possible), and hence there is no reason *a priori* to suppose that different preparative techniques produce materials with the same or similar structure. In addition, the various techniques differ in the degree to which their products depart from equilibrium, mainly as a result of the rate of (effective) cooling; vapour-deposited material is therefore expected to be less near an equilibrium state than melt-quenched material. This difference can be evinced at two levels: *macroscopic* structural inhomogeneities (e.g. voids, density fluctuations) are more prevalent in vapour-deposited material as are *microscopic* structural defects, such as broken bonds and vacancies. These can have dramatic effects on various properties, particularly electronic, as we shall see later.

A further important point to note about the preparation of amorphous solids is that, *even for the same preparative technique*, there may be differences in the nature of the material produced unless great care and attention is paid to control of the various experimental parameters, some of which we have seen. This non-reproducibility of samples bedevilled the field in its early days, especially as far as vapour-deposited materials were concerned. It is therefore of paramount importance to give as many details as possible of the controllable parameters in the preparation process in order that future studies may attempt to reproduce the material in question.

Finally in this section on the preparation of amorphous materials, the reader is referred to the Appendix, in which a list (by no means exhaustive) of some suppliers of a variety of amorphous materials is given.

2 Glasses

2.1	Introduction
2.2	The glass transition
2.3	Theories for the glass transition
2.3.1	Thermodynamic phase transition
2.3.2	Entropy
2.3.3	Relaxation processes
2.3.4	Free volume
2.3.5	Summary
2.4	Factors that determine the glass-transition temperature
2.5	Glass-forming systems and ease of glass formation
2.5.1	Structure and topology
2.5.2	Eutectic compositions
2.5.3	Crystalline polymorphs
2.5.4	Coordination number
2.5.5	Summary
	Problems
	Bibliography

2.1 Introduction

We have seen in the last chapter that many different techniques can be used to prepare amorphous materials. Perhaps the most widely used, and certainly the most important historically, is melt quenching. Materials prepared in this way are often referred to as 'glasses' because they exhibit glass-transition phenomena (of which more later), although it should be noted that the widely used term 'metallic glass' is somewhat of a misnomer, since these materials often crystallize *before* exhibiting a glass transition on heating.

Materials which can be quenched from a melt to form an amorphous solid are represented by all the major types of bonding found in solids, covalent, ionic, metallic, Van der Waals' and hydrogen; some representative examples of glass-formers from each category are listed in Table 2.1. Inevitably, these are some examples which do not fit into any one category. The most widely known glass of all, silica, has some ionic contribution to the predominant covalent bonding, and Pd–Si,

Table 2.1 Examples of glass-formers for the major types of bonding

Type of bond	Material
Covalent	As_2Se_3, Se
Ionic	KNO_3–$Ca(NO_3)_2$
Metallic	Zr–Cu, Ni–Nb
Van der Waals'	*ortho* terphenyl
Hydrogen bond	$KHSO_4$, ice, aqueous solution of ionic salts (e.g. LiCl)

alloys, for example, have some covalent character (precisely how much is disputed) in addition to the prevalent metallic bonding. Since many materials can be rendered glassy (some authors would maintain that practically all liquids, with the exception of the quantum liquid of helium, can be made glassy given high enough cooling rates), it is obviously of interest to discuss these particular materials in a little more detail.

The first part of this chapter, then, attempts to answer the question '*how* do glasses form?' In so doing, the fascinating and complex field of glass-transition phenomena is introduced. The second part of the chapter tackles the knotty problem of '*why* do glasses form?' In particular, why are some materials so much better glass-formers than others?

The beginning reader may prefer at first to peruse the section describing the glass transition (section 2.2), followed by the sections dealing with the various theories (entropic, kinetic) for the transition (sections 2.3.2, 2.3.3 and 2.3.5). Finally, sections 2.5.1 and 2.5.5 describing some aspects which influence ease of glass formation may be read on a first reading.

2.2 The glass transition

When a liquid is cooled, one of two events may occur. Either *crystallization* may take place at the melting point T_m, or else the liquid will become 'supercooled' for temperatures below T_m, becoming more viscous with decreasing temperature, and may ultimately form a *glass*. These changes can be observed readily by monitoring the volume as a function of temperature by means of a dilatometer, and a typical result is shown schematically in Fig. 2.1. The crystallization process is manifested by an *abrupt* change in volume at T_m, whereas glass formation is characterized by a *gradual* break in slope. The *region* over which the change of slope occurs is termed the 'glass-transition temperature' T_g. (Similar behaviour would also be exhibited by other extensive thermodynamic variables, entropy S, and enthalpy H.) Since the transition to the glassy state is continuous, and the glass-transition temperature is not well defined (albeit often used), it is often convenient also to use the 'fictive' temperature T_f (Jones 1956), which is defined as that specific temperature obtained as the intersection of the extrapolated liquid and glass curves (Fig. 2.2(a)). It is the temperature at which the glass would be in metastable equilibrium if it could be brought to T_f instantaneously. Although it appears that T_f defined in this way is a precise temperature, in fact this is not so; it depends on the *rate* of cooling of the

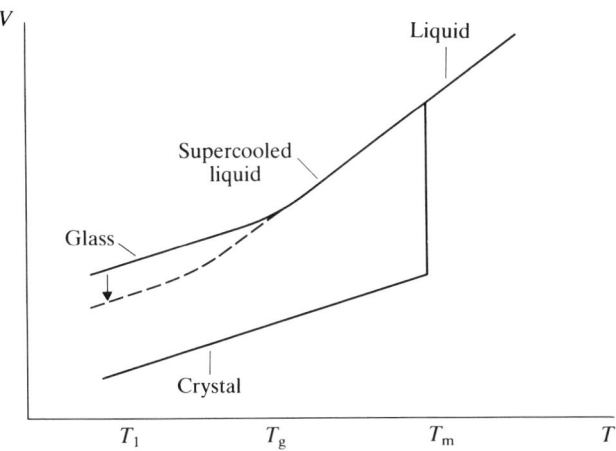

Fig. 2.1 Schematic illustration of the change in volume with temperature as a supercooled liquid is cooled through the glass-transition temperature T_g. The first-order phase transition accompanying crystallization from the melt is also shown. The vertical arrow illustrates the volume change accompanying the structural relaxation or stabilization of the glass if held at temperature T_1.

supercooled liquid. It is found that the *slower* the rate of cooling, the larger is the region for which the liquid may be supercooled, and hence the *lower* is the fictive or glass-transition temperature; this is shown schematically in Fig. 2.2(b). Thus, the glass-transition temperature of a particular material depends on its thermal history, and is not an intrinsic property in this respect; if the T_g of a glass is determined by heating the material through the transition and observing a given parameter, spurious results may be obtained unless the heating rate is equal to the original cooling rate. The actual value of the glass-transition temperature may vary by as much as 10 to 20% for widely differing cooling rates. As an example, for silica-based glasses, the change in T_g for different cooling rates may be as much as 100–200 K for values of T_g in the range 600–900 K. An expression relating T_g and the cooling rate q can be derived using free-volume theory (see problem 2.2) and is:

$$q = q_0 \exp[-1/C(1/T_g - 1/T_m)] \qquad [2.1]$$

where C is a constant ($\sim 3 \times 10^{-5}$), and for chalcogenide glasses (Se, As_2Se_3 and As_2S_3), q_0 varies between 10^{23} K s^{-1} (Se) and 10^4 K s^{-1} (As_2S_3) (Owen 1973). The outcome of this discussion, therefore, is that one should not expect glasses prepared in different ways, with different cooling rates, to exhibit a glass transition at the same temperature. This is particularly true when comparing the same material prepared either by melt quenching or by vacuum deposition.

We have seen that certain thermodynamic variables (volume, entropy, enthalpy) are continuous through the glass transition, but exhibit a change of slope there. This implies that at T_g there should be a discontinuity in *derivative* (or intensive) variables, such as coefficient of thermal expansion $\alpha_T = (\partial \ln V/\partial T)$, compressibility $\kappa_T = -(\partial \ln V/\partial p)_T$ and heat capacity $C_p = (\partial H/\partial T)_p$. This is indeed the case, and the behaviour of the heat capacity is shown schematically in Fig. 2.3(a); the value of the heat capacity for the glass is generally comparable to that of the crystal, but considerably smaller than that of the liquid. It can be seen that such

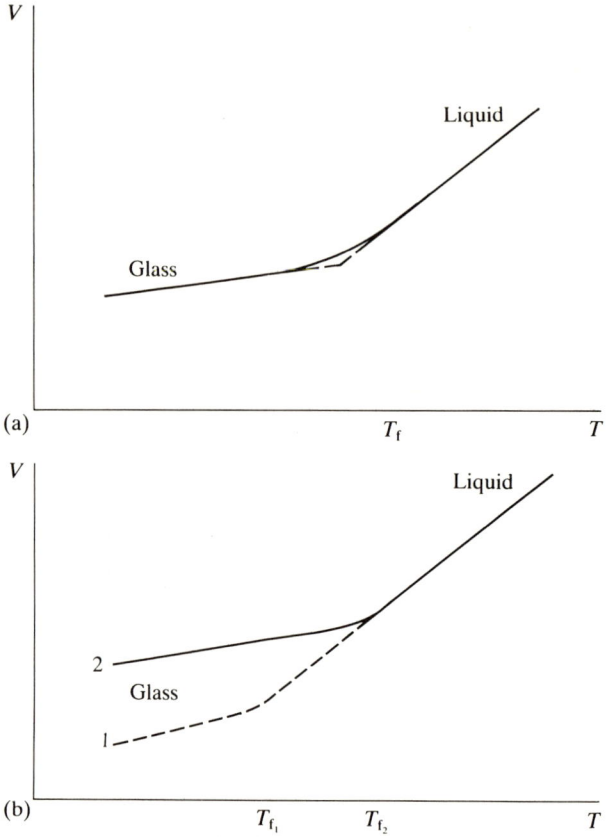

Fig. 2.2 (a) Schematic illustration of the experimental determination, by extrapolation, of the fictive temperature T_f.
(b) Illustration of the change in fictive temperature (or glass-transition temperature) with cooling rate (cooling rate of curve 1 is less than that of curve 2).

calorimetric experiments offer a much better marker of the glass-transition temperature than, for example, simple dilatometric measurements. A convenient way of monitoring glass-transition phenomena is by means of differential scanning calorimetry (DSC) or differential thermal analysis (DTA), in which the sample is heated at a constant rate and the changes in heat (DSC) or temperature (DTA) with respect to an empty reference pan are measured; a schematic diagram of a typical DTA trace is shown in Fig. 2.3(b), exhibiting both a glass transition and subsequent crystallization and melting. Controlled cooling is also possible with such instruments, and hence detailed studies of the dependence of T_g on the heating rate and on the previous thermal history are possible.

2.3 Theories for the glass transition

The nature of the glass transition is very complex and even now is poorly understood. Nevertheless, we shall review the various aspects of this phenomenon,

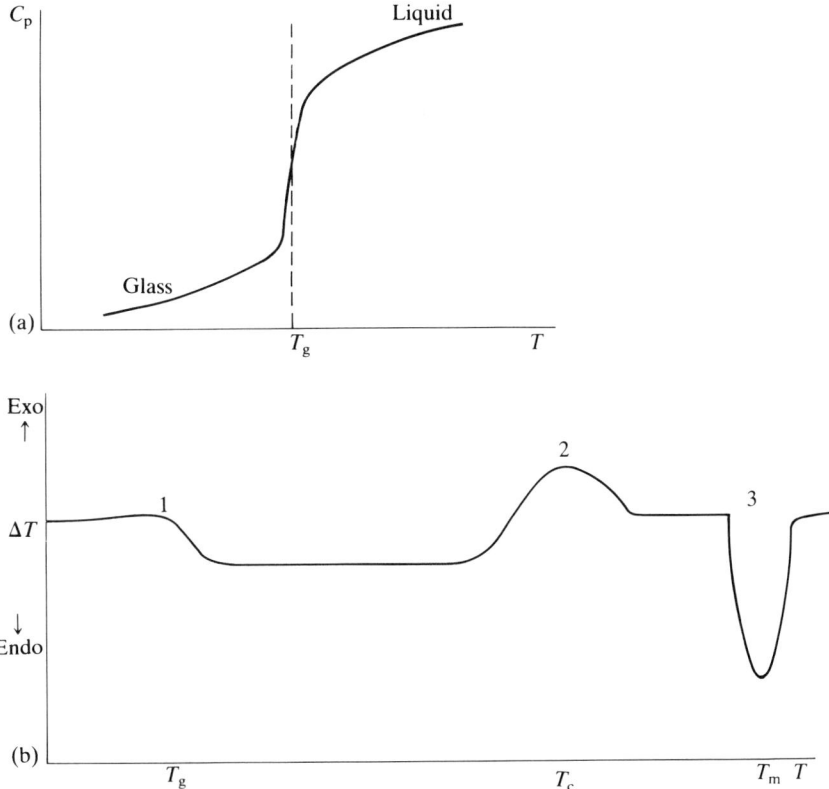

Fig. 2.3 Manifestations of the glass transition in thermodynamic measurements.
(a) Schematic illustration of the change in specific heat at constant pressure (C_p) on cooling through the glass-transition temperature T_g.
(b) Schematic DTA trace showing the glass transition (1), crystallization (2) and melting (3).

and the approaches that have been made towards an understanding of it. Many theories assume at the outset that there is a *single* parameter or property which characterizes the glass; we shall see that this is almost certainly not the case, but such one-parameter models do possess a certain virtue in being able to explain at least some of the data in a simple fashion.

2.3.1 Thermodynamic phase transition

The fact that certain extensive thermodynamic variables (e.g. V, S, H) are continuous (Fig. 2.1), yet the differential quantities α_T, C_p, κ_T are discontinuous at the transition (Fig. 2.3(a)) immediately suggests that the glass transition is a manifestation of a *second-order* phase transition. The definition of the order of a transition in the Ehrenfest scheme is the order of the lowest derivative of the Gibbs free energy which shows a discontinuity at the transition point. Thus the liquid–*crystal* transition is an example of a first-order transition, since $V = (\partial G/\partial p)_T$ and the volume changes discontinuously at T_m (see Fig. 2.1), whereas for a glass $C_p (= T(\partial S/\partial T)_p =$

$-T(\partial^2 G/\partial T^2)_p)$ is discontinuous at T_g. Unfortunately, this rather appealingly simple model of the glass transition fails on several counts. We have already seen that the glass-transition temperature (or T_f) is a function of the thermal history or the cooling rate to which the melt is subjected; variations by several tens of degrees are not uncommon. However, one would not expect changes of this magnitude in the transition temperature due to simply kinetic factors if this were a genuine thermodynamic phase transition.

A further difficulty arises when one examines in more detail the predictions made for second-order phase transitions, in particular the relationship between the various parameters in the theory. A second-order phase transition is distinguished by virtue of the fact that first-order extensive thermodynamic variables are continuous at the transition. Thus, for example the entropies of the high- and low-temperature forms (liquid and glass, respectively) must be equal at the transition, i.e. $S_1 = S_2$. Changes in temperature or pressure must be such that the two forms remain in equilibrium so that $dS_1 = dS_2$, or in terms of partial derivatives:

$$\left(\frac{\partial S_1}{\partial T}\right)_p dT + \left(\frac{\partial S_1}{\partial p}\right)_T dp = \left(\frac{\partial S_2}{\partial T}\right)_p dT + \left(\frac{\partial S_2}{\partial p}\right)_T dp \qquad [2.2]$$

Since $C_p = T(\partial S/\partial T)_p$, and using one of Maxwell's thermodynamic relations $(\partial S/\partial p)_T = -(\partial V/\partial T)_p$, where $\alpha_T = 1/V(\partial V/\partial T)_p$, one obtains finally an expression for the shift in transition temperature with pressure:

$$\frac{dT_g}{dp} = \frac{TV(\alpha_{T_2} - \alpha_{T_1})}{(C_{p_2} - C_{p_1})} = TV\frac{\Delta\alpha_T}{\Delta C_p} \qquad [2.3]$$

Similarly, by considering continuity of volume at the transition, a relation involving the bulk compressibility $\kappa_T = -(1/V)(\partial V/\partial p)_T$ is obtained:

$$\frac{dT_g}{dp} = \frac{\Delta\kappa_T}{\Delta\alpha_T} \qquad [2.4]$$

It is found by measuring the discontinuities $\Delta\alpha_T, \Delta C_p, \Delta\kappa_T$ at the glass transition that [2.3] is almost always obeyed within experimental error, but that values for $\Delta\kappa_T/\Delta\alpha_T$ are generally appreciably higher than those of dT_g/dp ([2.4]). It therefore appears on this evidence that the glass transition is not a simple second-order phase transition. However, without involving a second-order thermodynamic transition, Prigogine and Defay (1954) showed generally that the ratio

$$R = \frac{\Delta\kappa_T \Delta C_p}{TV(\Delta\alpha_T)^2} \qquad [2.5]$$

is equal to unity if a single ordering parameter determines the position of equilibrium in a relaxing system, but if more than one ordering parameter is responsible, then $R > 1$; the latter case seems to describe most glasses. It should be noted in passing that an additional consequence of the experimental verification of [2.3] but not [2.4], is that one expects glasses prepared under higher pressures to have higher than normal densities but normal entropies or enthalpies, since continuity of entropy leads to [2.3] whereas continuity of volume leads to [2.4].

2.3.2 Entropy

One would expect intuitively that entropy might play a role in determining the glass transition since we are considering disordered systems. We have already seen one instance of this in the previous section in which glass formation may be assumed to occur at a given value of *excess* entropy, and by further supposing that the entropy is continuous at the transition, [2.3] results.

A further influence of entropy was first pointed out by Kauzmann (1948), and concerns the precipitous decrease in heat capacity at the glass-transition temperature (Fig. 2.3(a)). The heat capacities for both the crystalline and glassy ($T < T_g$) states of most materials are essentially the same, and arise from vibrational contributions. The excess heat capacity measured for the glass above T_g is due to configurational degrees of freedom which the material possesses in the supercooled liquid state. We have already seen that the rate of cooling influences the glass-transition temperature; the slower the rate, the lower T_g. The point at issue is whether there is a lower limit to this decrease in T_g – i.e. is there an 'ideal' glass-transition temperature? The argument advanced by Kauzmann suggests that such a limit does exist. The area under the curve of C_p plotted versus the *logarithm* of temperature yields the entropy, and such a plot for the case of an ionic system, lithium acetate, is shown in Fig. 2.4 (Wong and Angell 1976). The change in entropy upon fusion of the crystal which occurs at T_m is ΔS_f (5.48 entropy units (e.u.) in this case). The *lowest* temperature to which the liquid could possibly be supercooled by progressively increasing the experimental time-scale, yet maintaining the large liquid-like value of C_p, is governed by the requirement that the area under the $C_p - \ln T$ curve for the supercooled liquid be equal to or greater than ΔS_f. The temperature, defined by the equality of the areas, is called the calorimetric ideal glass-transition temperature, T_{oc}. The experimental glass-transition temperature cannot decrease below T_{oc}, since if it did, the total entropy of the liquid would fall *below* that of the crystal, thereby violating the Third Law of Thermodynamics; this situation is termed 'Kauzmann's paradox'. It can be seen that for the example shown in Fig. 2.4, the experimentally determined T_g occurs just 20 K above T_{oc}; in many cases, the difference is considerably larger, particularly for silicate or similarly based glasses. We conclude from this discussion, therefore, that even if no other factor contributed to the glass transition, the heat capacity should decrease sharply at a certain temperature as a result solely of *equilibrium* thermodynamic considerations. The existence of an ideal glassy state possessing zero residual entropy, having a glass-transition temperature T_{oc}, is probably hypothetical. This is because near T_g the probability of crystal growth and nucleation increases very rapidly, and thus for very long experimental time-scales crystallization is more probable thatn relaxation to a lower energy, fully amorphous state. Nevertheless, there are certain materials, 'atactic' polymers, which have a random arrangement of side-groups, which *cannot* crystallize and which, therefore, offer a practical prospect of investigating whether an 'ideal' amorphous solid may exist; unfortunately, the relevant data needed to determine the excess entropy in the liquid are not available.

The ideas above have been employed in a sophisticated discussion limited to linear polymer chains (Gibbs and Di Marzio 1958, Gibbs 1960), and the transition temperature was defined there as that point at which the configurational entropy

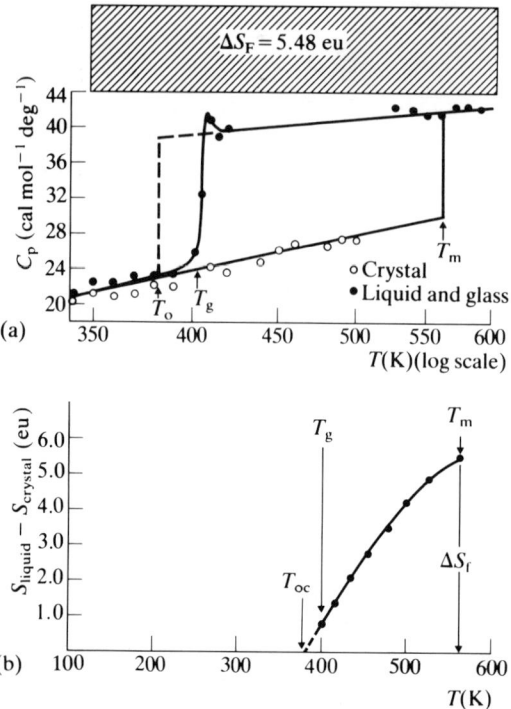

Fig. 2.4 (a) Heat capacities of glassy, liquid and crystalline phases of lithium acetate. The data are plotted against log T so that integrated areas under the curves yield entropies directly, and the entropy of fusion is shown shaded in the upper part of the figure.
(b) The difference in entropy between liquid and crystalline phases as a function of temperature. The vanishing excess entropy temperature is termed the 'ideal' glass-transition temperature T_{oc} (Wong and Angell 1976).

should vanish. The number of configurations available for a polymer chain decreases with decreasing temperature because of the hindrance to intramolecular rotation caused by the presence of side-groups which produce rotational potential barriers; a chain becomes effectively stiffer as the temperature decreases. The configurational entropy was calculated using the partition function for a quasi-lattice model and a relationship between the transition temperature and the potential barriers was established, leading to the qualitative result that T_g should increase with increasing barrier height. This entropy theory concentrates solely on the stiffness factor of a polymer chain, and does not take into account intermolecular interactions which are almost certainly important. A more complete discussion of glass-transition phenomena in polymers can be found in Gee (1970).

2.3.3 Relaxation processes

Another important facet of the glass transition – some would argue the most important aspect – concerns the relaxation processes that occur as a supercooled liquid cools. We have already seen that the experimentally measured value of T_g is not unique; it depends on the time-scale of the experiment used to observe it. This is

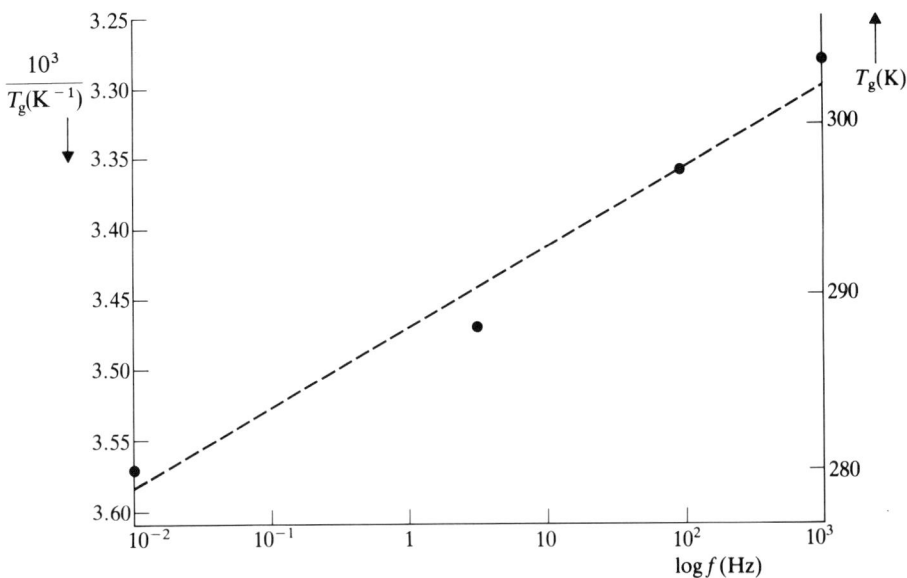

Fig. 2.5 Plot of dependence of the measured glass-transition temperature on the experimental time-scale for poly-3: 3-bischloromethyloxacyclobutane (Brydson 1972). The experimental probes used were electrical (dielectric loss, $f = 10^3$ Hz), mechanical vibrational (mechanical loss, $f = 89$ Hz), slow tensile ($f = 3$ Hz) and dilatometry ($f = 10^{-2}$ Hz). The straight line is drawn assuming that the structural relaxation time τ_R obeys the equation $\tau_R \propto \exp(E/k_B T_g)$, where E is a constant activation energy (3.7 eV).

strikingly shown in Fig. 2.5 for a glassy polymer where T_g varies appreciably when measured by experimental techniques having widely different characteristic time-scales. The configurational changes that cause the relaxation of the supercooled liquid become increasingly slow with decreasing temperature until, at a given temperature (the glass-transition temperature), the material behaves as a solid; for times of observation, t_o, long compared with the 'structural' relaxation time τ_r ($t_o > \tau_r$) the material appears 'liquid-like' whereas for $t_o < \tau_r$ the material behaves as if it were 'solid-like'. A 'transition' will appear to have taken place if values of liquid-like parameters differ significantly from solid-like ones, as is the case for the heat capacity. T_g occurs therefore when $t_o \simeq \tau_r$, viewed in this way. A typical example of the kind of relaxation readily observed in glasses is depicted schematically in Fig. 2.1 in which 'stabilization' of the glass occurs, the structure tending to approach the equilibrium state of the supercooled liquid for temperatures below T_g. This process can occur in times of the order of minutes for $T \lesssim T_g$, but may take years for $T \ll T_g$.

Another (but rather unsatisfactory) definition of T_g, again in terms of the experimental time-scale, is that temperature at which the liquid attains a certain viscosity (say, 10^{13} or $10^{14.6}$ poise, a value characteristic of solids – sect. 1.2). This brings us to the subject of viscosity of fluids, which is important since many glass-forming liquids have high viscosities at elevated temperatures. The viscosity of these liquids changes very rapidly in the region of the glass-transition temperature, as can be seen from Fig. 2.6. The temperature dependence of the viscosity for the two types of behaviour commonly encountered differs quite markedly. 'Strong' network-formers, such as SiO_2, GeO_2 or BeF_2 obey accurately an Arrhenius behaviour over a

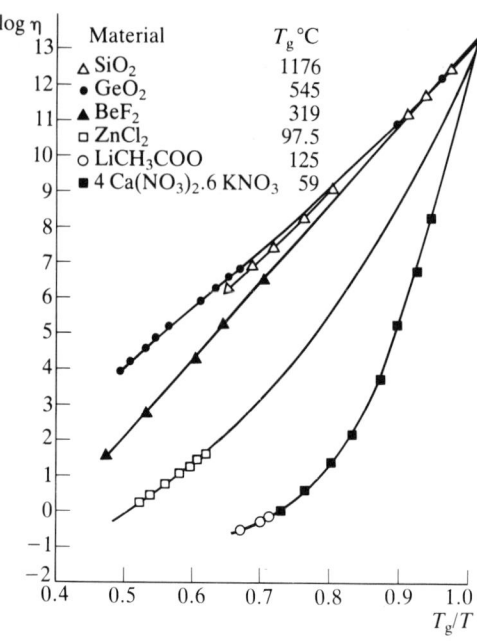

Fig. 2.6 Viscosities of a variety of glass-forming systems as a function of reduced reciprocal temperature. The curves have been normalized such that the glass-transition temperature is defined to be when $\eta = 10^{13}$ poise (Wong and Angell 1976).

large temperature range, whereas many ionic glasses, such as $Ca(NO_3)_2$ and organic glasses, such as o-terphenyl or i-butyl bromide, show strong deviations from a simple Arrhenius behaviour. Over much of the range, the empirical Fulcher (also known as the Vogel–Tammann–Fulcher) equation for the 'fluidity' (inverse viscosity)

$$\eta^{-1} = A \exp[-a/(T-T_{o\eta})] \qquad [2.6]$$

is obeyed, although for very high viscosities near T_g, the dependence is often observed to become Arrhenius-like again. It is a matter of considerable interest to note that in those glasses for which Fulcher's law is obeyed, $T_{o\eta} \simeq T_{oc}$, seeming to indicate that the ideal glass-transition temperatures observed kinetically and thermodynamically might have a common origin. It is important to note that the viscosity is continuous through T_g, exhibiting none of the discontinuities observed in heat capacity, etc., although there is a singularity at the (inaccessible) temperature $T_{o\eta}$. The particular temperature dependence embodied in Fulcher's law is rather unusual, and as Anderson (1979) has pointed out, it cannot be obtained by an appropriate averaging of the energy in an Arrhenius expression nor by a suitable temperature dependence of the energy, and its physical origin remains relatively obscure. The form of [2.6] is in fact common to many transport phenomena of the liquid phase, e.g. diffusion rate, nuclear spin relaxation rate and ionic conduction, and it is found that the values for the different T_o (and a) are very similar in all cases. An important source of information on this area is 'relaxation' spectroscopy for temperatures near T_g, in which the decay of the response to an applied field is monitored as a function of time or equivalently (through a Kramers–Kronig or Hilbert transform) the continuous dissipation caused by an externally applied sinusoidal field. Examples of this technique are stress relaxation (or mechanical loss) and dielectric loss; a fuller discussion with many examples is given by Wong and Angell (1976).

A theory which attempts to link relaxation aspects with entropy considerations has been proposed by Adam and Gibbs (1965). They accounted for mass transport (diffusion, viscosity, etc.) by considering cooperative atomic rearrangements, and by finding the smallest size of a molecular group capable of rearrangement at a given temperature, showed that the group size and therefore the probability of rearrangement and hence, say, the viscosity could be expressed as a function of the total configurational entropy S_c:

$$\eta = \eta_o \exp(B/TS_c) \qquad [2.7]$$

where η_o and B are constants. S_c is given by the equation:

$$S_c = \int_{T_o}^{T} \Delta C_p \, d \ln T \qquad [2.8]$$

In this manner, a natural relationship between the temperature dependence of the viscosity ($\eta \simeq 10^{14}$ poise at T_g) and the discontinuities in C_p observed also at T_g is obtained. Furthermore, the ideal glass-transition temperature enters into expressions for mass transport in a natural way. For the case when ΔC_p is very small, as it is for SiO_2, GeO_2, etc., then S_c is almost independent of temperature, and the viscosity should obey an Arrhenius equation. On the other hand, when ΔC_p is large, S_c contributes a temperature-dependent term to the equation for viscosity [2.7] at all temperatures above T_o, leading to strongly non-Arrhenius behaviour for the viscosity, as observed in many organic and ionic glasses.

An alternative approach to the relaxation behaviour of viscous liquids has been given by Goldstein (1969). In this, a fluid system composed of N particles is characterized by a $(3N+1)$ dimensional potential energy surface containing many minima of different depths. This space is then explored as the system evolves in time, but at low temperatures the system is restricted to the deeper minima which are fewer in number and hence contribute to a lower value of configurational entropy. The activation energy needed to move between minima is therefore temperature dependent. The glass transition, in this picture, then corresponds to the system becoming trapped in one of the deep minima. An important feature of the theory is the prediction of 'islands of mobility' in which local rearrangement is possible (between neighbouring similarly deep minima) but long-range flow is impossible. This circumstance can explain the weak secondary (or sub-glassy) relaxations which are observed experimentally in many glasses at temperatures well below the mamor glass-transition temperature T_g. These are most often observed in polymeric systems, where they are termed β-processes to differentiate secondary relaxations from the primary (α-) relaxations occurring at T_g. The β-relaxations are perhaps most clearly seen in plots of dielectric loss, ε_2 or loss tangent $\tan \delta = \varepsilon_2/\varepsilon_1$ (see sect. 6.3.4) versus temperature in which T increases through T_g; an example is shown schematically in Fig. 2.7. The β-processes can be observed also by other relaxation spectroscopic

Glasses

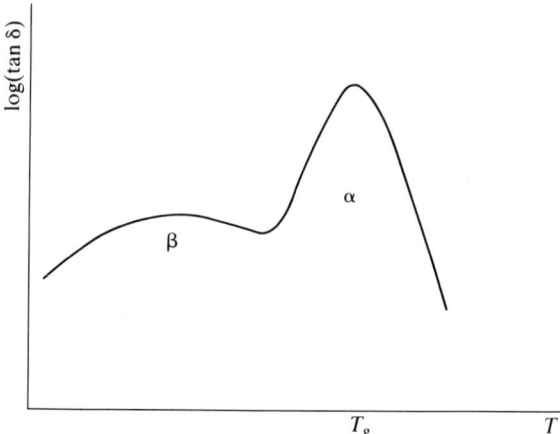

Fig. 2.7 Schematic illustration of the primary (α-) and secondary (β-) relaxations which can occur in glasses at and below T_g, manifested by measurement of the dielectric loss, $\tan \delta = \varepsilon_2/\varepsilon_1$, as a function of temperature at constant frequency.

techniques (McCrum *et al.* 1967). For the case of glassy polymers, β-relaxations can be understood physically as the (rotational or librational) rearrangement of small side-groups which can occur for $T < T_g$, in contrast to the conformational changes of the polymer backbone chain involving many tens of atoms, which occur at temperatures above T_g.

An intermediate type of motion has been advanced to account for the so-called 'γ-transition' in polyethylene (and other polymers containing more than four consecutive —CH$_2$ groups in the chain) in which four polymer groups rotate simultaneously around an axis parallel to the chain direction, termed crank-shaft motion. This is shown schematically in Fig. 2.8, together with the

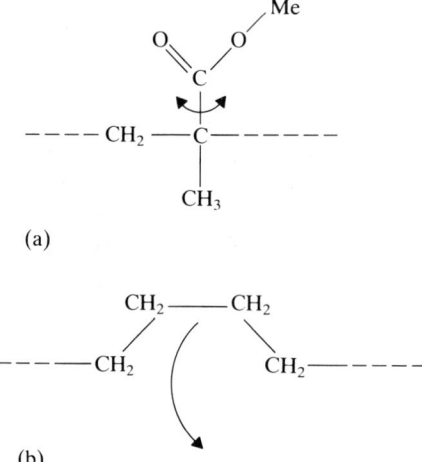

Fig. 2.8 Molecular motions giving rise to sub-glassy relaxations.
(a) β-Relaxation in polymethyl methacrylate caused by librational motion of side-groups.
(b) γ-Relaxation in polyethylene caused by 'crank-shaft' motion of the polymer backbone.

rotational or librational modes of side-groups believed to be responsible for β-relaxations. However, the fact that sub-glassy relaxations are not restricted to polymeric materials alone, but have also been observed experimentally in other systems, such as ionic or simple molecular liquids in which there are no freely rotating side-groups or sub-units (Johari and Goldstein 1970, 1971), lends support to the general notions of Goldstein's potential minima picture.

2.3.4 Free volume

A popular theory describing liquids and certain aspects of the glass transition is that concerned with 'free volume', which has many similarities with aspects of the relaxation theories discussed in the previous section. Nevertheless, we accord it the honour of a separate section in view of its historical importance in the development of the subject. The model was originally developed for fluids assumed to be composed of hard spheres (Cohen and Turnbull 1959, Turnbull and Cohen 1961, 1970). The spheres oscillate thermally in their own cages corresponding to their Voronoi polyhedra.† The total volume of a liquid is supposed to be divided into the part 'occupied' by the spheres or molecules V_{occ}, and that part in which the molecules are free to move. The latter volume permitting the diffusive motion is termed the 'free volume' V_f. It is further assumed that molecular transport can occur only when the voids have a volume greater than a critical volume. The free volume is shared communally; in other words it is divided into holes or voids of varying size and position by the random motion of the molecules, and no local free energy is required for redistribution of the free volume. As the temperature of the liquid is lowered, both the occupied volume and the free volume are expected to contract. The essential difference between a glass and a liquid on this model is, therefore, that (a) the free volume is independent of temperature, and (b) redistribution of free volume no longer occurs, i.e. the free volume is 'frozen-in' in the locations occupied when the glass was formed. The idea that the glass transition occurred when the free volume decreased below some critical value was first proposed by Fox and Flory (1950, 1951, 1954).

Some idea of the magnitude of the free volume can be gauged by consideration of the coefficients of cubic expansion α_{Tl} and α_{Tg} of the liquid and glass, respectively (Gee 1970). The expansion of the free volume is $\Delta \alpha_T = \alpha_{Tl} - \alpha_{Tg}$, since α_{Tg} represents the expansion of the occupied volume. The occupied volume (at $T=0$) can be obtained by extrapolating the liquid volume, giving $V_{occ}(T=0) = V_g(1 - \alpha_{Tl} T_g)$, where V_g is the total volume of the glass at T_g. Similarly, the total volume of the glass at $T=0$ is $V = V_g(1 - \alpha_{Tg} T_g)$ and hence the free volume of the glass is $V_{fg} = V - V_{occ}$, giving an expression for the fractional free volume $\mathcal{R}$:

$$\mathcal{R} = \frac{V_{fg}}{V_g} = T_g \Delta \alpha_T \qquad [2.9]$$

Application of this formula to a variety of glasses, including organic high polymers,

---

†A Voronoi polyhedron (or equivalently the Wigner–Seitz cell) is defined as that cell surrounding a given site which contains all the points closer to that site than to any other. The concept is particularly suited to simple metals, rare-gas solids or liquids, rather than those materials in which covalent bonding predominates.

and chalcogenide glasses, indicates that at T_g some 10% of the total volume is 'free'; the value for B_2O_3 (a very good glass-former which has never been observed to devitrify), interestingly, is much higher at 34% (Suzuki and Abe 1981).

Free-volume theory has also been applied to the problem of viscosity of liquids. This stems from the discovery of an empirical relation between viscosity and free volume, the Doolittle (1951) equation:

$$\eta = A \exp (BV_{occ}/V_f) \qquad [2.10]$$

where A and B are empirical constants. Assuming a linear form for the thermal expansion, such that the free volume of the *liquid* is $V_f = V_{fg} + V_g \Delta\alpha_T(T - T_g)$ at $T > T_g$ and inserting this into [2.10], one recovers the Vogel–Tammann–Fulcher expression [2.6]. Alternatively, by additional algebraic manipulation, substitution of the equation for the temperature dependence of the free volume of the liquid into the Doolittle expression yields an equation of the same form as that of the empirical Williams–Landel–Ferry (1955) equation (see Problem 2.1) which has been very successful in accounting for relaxational behaviour of many glasses:

$$\ln\left(\frac{\eta_g}{\eta}\right) = \frac{A(T - T_g)}{B + (T - T_g)} \qquad [2.11]$$

where A and B are empirical constants.

More recently, Cohen and Grest (1979) have enlarged on earlier work on the free-volume model, and used 'percolation' theory (see sect. 5.3.6) to account for the exchange of free volume between nearest-neighbour 'liquid-like' cells which have a sufficient number of neighbouring liquid-like cells that the volumes of any neighbouring *solid*-like cells do not change volume simultaneously. In this manner, they were able to calculate the communal entropy, associated with the accessibility of all the configurational volume inside the finite liquid clusters, and hence calculate the heat capacity. They conclude, rather surprisingly, that the glass transition is predicted to be more likely a *first*-order phase transition rather than second order.

Free-volume theory is a conceptually appealing and relatively simple theory for the liquid state and the glass transition. Nevertheless, it does have its drawbacks and inconsistencies, as indeed do all the theories mentioned so far. A serious objection is the assumption that the thermal expansion remains linear even when V_f becomes very small, and that although diffusive motion ceases when $V_f \to 0$ the thermal contraction is continuous up to that point and then ceases (Anderson 1979). A further problem lies in the fact that free-volume theory is derived on the basis of the hard-sphere model of liquids in which non-directional bonds are predominant, but is then applied to certain glass-formers which are strongly covalent and are almost certain to have directed bonds for $T > T_g$.

2.3.5 Summary

From the discussion above, it is apparent that the phenomenon of the glass transition is very complex and that no single theory that has been advanced so far is capable of accounting for all aspects of it. The glass transition, as observed experimentally, is certainly at least partly a relaxation phenomenon, in that the experimental time-scale profoundly affects the transition observed as the supercooled liquid falls out of equilibrium; very different results are obtained if the glass transition is probed by calorimetric (DSC or DTA) experiments in which the time-scale is of the order of seconds, than if, say, ultrasonic relaxation spectroscopy is employed. Nevertheless, there is also very definitely a thermodynamic aspect in addition to the relaxation character of the transition; this is clearly evinced by Kauzmann's paradox. Although the glass transition appears superficially to be a second-order thermodynamic phase transition, it is almost certainly not ideally second order. Any complete theory describing the formation of a glass from the liquid state must therefore combine both relaxation and thermodynamic aspects in a natural way; the recent developments of free-volume theory using percolation arguments perhaps point the way to this goal.

2.4 Factors that determine the glass-transition temperature

We have seen in the last section that the value of the glass-transition temperature, T_g, observed experimentally is, to a considerable degree, influenced by the prevailing experimental conditions (e.g. cooling rate of the melt). Nevertheless, if measured under standard conditions (e.g. calorimetrically at a fixed heating rate), is T_g dependent on any material parameter or the structure? This is a very important question technologically (for example, the glass–rubber transition for polymers marks the *lower* limit for the use of an amorphous rubber but the *upper* limit for a thermoplastic, such as polystyrene or PVC), and although the question has long been posed there is as yet no consensus as to which factors, if any, are responsible for determining T_g. Various proposals made in the past have suggested that T_g scales with the melting temperature T_m, the cohesive energy (scaling approximately as the normal boiling temperature T_b), the Debye temperature of the phonon spectrum, or the band gap (for amorphous semiconductors). For example, many glass-formers (particularly organic materials) have been observed to have $T_g \simeq 2/3 T_m$ (Kauzmann 1948). Many of the suggestions stem from an attempt to correlate the large-scale molecular or atomic motions taking place above T_g with characteristic energies needed to give rise to such motions, either through bond breaking or the creation of free volume. Other characteristic energies that have been suggested are less directly related. It is assumed that the optical band gap, for instance, scales with the cohesive energy; the stronger the chemical covalent bond, the larger the band gap (see Ch. 5). Such a correlation between T_g and optical gap has been found by de Neufville and Rockstad (1974) for a variety of chalcogenide-based amorphous semiconductors, but a simple comparison has to be treated with caution since, for example, amorphous Se, As_2Se_3 and GeAsSe all have approximately the *same* value for the optical gap, yet the glass-transition temperatures vary between 318 K and 677 K! However, within chemically similar families, e.g. the arsenic chalcogenide series As_2X_3, where X = S, Se and Te, a reasonably good correlation is found. De Neufville and Rockstad in fact grouped together materials with the same average number of outer valence electrons per atom, $\bar{N}$, or equivalently average coordination number (equal to $8 - \bar{N}$). Such attempts at correlation depend crucially therefore on subtle differences in the composition of the valence or conduction bands. It obviously makes a great difference whether the top of the valence band, say, is composed of bonding or non-bonding electron states, since the optical band gap is determined by the energy difference between the highest occupied (valence) and lowest unoccupied (conduction) states (see sect. 5.6), whereas the cohesive energy is related to the energy difference between bonding and antibonding states.

However this simple picture becomes more complicated if the variation in T_g with composition in a given alloy system is examined. For example in the simple binary chalcogenide alloy As_xSe_{1-x} ($0 < x < 1$), Myers and Felty (1967) found that T_g did not vary monotonically with composition; instead, a maximum in T_g was found for $x \simeq 0.4$, corresponding to the stoichiometric composition As_2Se_3. Obviously in this case T_g is sensitive to the chemistry involved, and a maximum in T_g at the stoichiometric composition may result from the formation of a 'chemically ordered network' (see sect. 3.4.2) in which only the stronger As–Se bonds (rather than the weaker As–As and Se–Se bonds) are present. A further note of caution must be sounded in correlating T_g with the cohesive energy (or a monitor of it) since it is likely that the weaker secondary 'back bonds' binding structural units together, particularly important in chalcogenide glasses, may be ruptured preferentially at T_g, rather than the primary covalent bonds, thereby allowing large-scale motion of clusters or groups of atoms to take place.

Alternatively, adopting the free-volume picture of the glass transition, it might be supposed that the important factor in determining T_g is the energy required to produce atomic mobility in a volume v (Gee 1970). This can be expressed as v times the cohesive energy density, where the cohesive energy density, U, of the liquid can be related to the molar latent heat of evaporation L_m and volume V_m by the equation $U = (L_m - RT)/V_m$. However, the flexibility of a polymer chain, for example, will be important; the stiffer the chain, the larger will be v and hence the higher the glass-transition temperature. Such a dependence is observed in glassy polymers, and many empirical rules relating chain parameters to T_g have been devised. (Note that the entropy theory for the glass transition of Gibbs *et al.* concentrates solely on the flexibility factor, and ignores rather unphysically the intermolecular interactions.)

In conclusion, there appear as yet to be no firm rules for predicting the glass transition, no ready prescriptions with which to calculate T_g for a given glassy material. With this in mind, we give in Table 2.2 for the benefit of the reader a table of values for T_g for a variety of materials in common use or which frequently form the subject of research. The values given in Table 2.2 must be treated with a certain degree of caution, since in practice appreciable departures from them may be observed for substantially differing heating rates, etc., but the values given are chosen to be as representative as possible.

2.5 Glass-forming systems and ease of glass formation

The question '*why* do certain materials readily form glasses on cooling a melt?' is one of considerable practical and technological importance. In many cases, this question may be reformulated as 'why do certain chemical compositions of materials have a

Table 2.2 Table of glass-transition and melting temperatures for some glassy materials
1. *Inorganic materials*

Material	T_g (K)	Reference	T_m (K)	Reference
S	246	1	392	1
Se	318	2	490	1
As_2S_3	478	2	573	5
As_2Se_3	468	2	633	5
As_2Te_3	379	3	633	6
$GeSe_2$	695	2	980	5
GeS_2	765	4	1073	6
SiO_2	1453	1	2003	1
GeO_2	853	4	1388	1
BeF_2	598	4	821	1
$ZnCl_2$	380	1	590	1
B_2O_3	530	1	793	1

References: (1) L. G. Van Uitert (1979), *J. Appl. Phys.* **50**, 8052; (2) J. P. de Neufville and H. K. Rockstad (1974), *Proc. 5th Int. Conf. on Amorphous and Liquid Semiconductors* J. Stuke and W. Brenig (eds) (Taylor and Francis) p. 419; (3) C. H. Seeger and R. K. Quinn (1975), *J. Non-Cryst. Sol.* **17**, 386; (4) A. C. Wright, G. Etherington, J. A. Erwin Desa, R. N. Sinclair, G. A. N. Connell and J. C. Mikkelson (1982), *J. Non-Cryst. Sol.* **49**, 63; (5) *Handbook of Chemistry and Physics* (61st edn), R. C. Weast and M. J. Astle (eds) (CRC:1980); (6) *Comprehensive Inorganic Chemistry* Vol. 2, A. F. Trotman-Dickensen (ed.) (Pergamon: 1973).

2. *Organic polymers*

Material	T_g (K)	T_m (K)
Polystyrene (isotactic)	373	503
Polymethyl methacrylate (isotactic)	318	433
Poly-*N*-vinyl carbazole	423	593
Nylon-6	323	488
Poly(ethylene oxide)	218	339
Poly(propylene oxide)	211	338
Cellulose triacetate	380	573
Polyethylene	253	393
Polypropylene	278	423
Polytetrafluoroethylene (PTFE)	388	600
Poly(vinyl chloride) (PVC)	353	—

Reference: J. A. Brydson in *Polymer Science*, A. D. Jenkins (ed.) (North Holland: 1972).

greater glass-forming tendency than others?' This remains one of the great unsolved mysteries of glass science, and although empirical prescriptions have been developed which are reasonably successful in accounting for the glass-forming tendencies in certain specific cases, there is no general rule which may be used universally to predict glass-forming ability in any given system. Of course, the question of the ease of glass formation on cooling a melt (i.e. the cooling rate required) is intimately related to, and inseparable from, the problem of *how* do glasses form. As we have seen in the previous sections, the subject of the glass transition is poorly understood, and then often only in a phenomenological rather than a quantitative sense. A comprehensive theory of the glass transition is needed before categorical answers to the questions surrounding glass-forming ability can be given. Nevertheless, there is a virtue in rather arbitrarily divorcing a discussion on the ease of glass formation from that of the glass transition itself in view of the fact that this is a very real practical problem, encountered at the laboratory bench. Profound consideration of the nature of the glass transition rather fades into insignificance when faced with the question of, say, which chemical composition to use in order that glass formation is facilitated and crystal growth is inhibited! This aspect of the subject, as the reader may well anticipate, is still very much at the 'alchemy' stage. Frequently, and particularly in technological applications, small amounts of additives are found to be necessary so that glass formation is improved in some way, but it is often not known precisely why they should act in the way they do. For example, trace amounts of As and Cl stabilize the Se films used in xerography against crystallization, as does the presence of As and Ge or Si in chalcogenides used as electronic 'switching' devices. Alternatively, the presence of a third (alkali) oxide is found to be necessary to prevent phase separation and immiscibility in binary oxide melts and hence in the resulting glasses (e.g. alkaline earth oxide–B_2O_3 or –SiO_2 systems – sec. 3.6.2).

Despite the apparent arbitrariness and uncertainty alluded to in the foregoing, none the less there are several factors which experience has shown do play a significant role in determining the ease of glass formation. These generally relate to chemical or structural properties of the system; thermodynamic or free-volume aspects have been discussed previously. They include the occurrence of eutectic compositions, the existence of several crystalline polymorphs of the material, the (average) atomic coordination and the topology of the amorphous structure. These are discussed in turn in what follows, but it must be stressed that the 'rules' enunciated often refer only to particular restricted systems of glassy materials, and are not generally universally applicable.

2.5.1 Structure and topology

The first really successful attempt to categorize materials into glass-formers and non-glass-formers was by Zachariasen (1932). This is one of the truly seminal papers in the field of glass science, and has had an enormous impact on the subject in areas other than just glass formation. Some 50 years ago, when Zachariasen wrote his paper, practically the only known glass-forming materials were oxides, of which only five formed glasses by themselves: SiO_2, GeO_2, B_2O_3, As_2O_3 and P_2O_5. In addition, these five oxides can also form glasses when mixed (up to a limiting percentage) with another oxide or oxides, which are not by themselves glass-formers. The rules that Zachariasen formulated were then capable of explaining why, for example, SiO_2 is a glass-former and Na_2O is not, but why $xNa_2O:SiO_2$ should also be a glass-former.

Zachariasen argued that those materials most likely to form glasses readily would have an internal energy only slightly larger in the amorphous state than when crystalline. (A hypothetical amorphous phase possessing an internal energy considerably higher than the crystal would be prone to rapid devitrification.) The internal energy of a solid is related to its structure (e.g. the Madelung energy of ionic materials). Since it is reasonable to assume that the interatomic interactions in both amorphous and crystalline phases would be very similar, this implies that the atomic structure of the two phases must be similar in certain regards. The structure of amorphous solids is, however, aperiodic, giving rise to broad diffuse haloes in diffraction patterns (Fig. 3.2), instead of sharp spots (or rings) characteristic of crystalline (or polycrystalline) solids. Zachariasen assumed therefore that the oxygen polyhedra found in oxide crystals (triangular, tetrahedral or octahedral) would also occur in glasses, the only difference being the relative orientation of polyhedra should be variable in glasses, giving rise to a non-periodic structure. For example, crystalline forms of SiO_2 contain SiO_4 tetrahedra joined at the corners, and it was assumed that this should also prevail in vitreous SiO_2. The most important aspect of Zachariasen's approach which transcended the importance of his 'rules' for glass formation was that a non-periodic arrangement of atoms could be attained solely as a result of the incorporation of variations in bond angles (for two dimensions) or bond angles and 'dihedral' angles (the relative angle of twist between neighbouring units – see Fig. 3.19) for three dimensions. In this manner, a *continuous random network* (CRN) can be constructed, in which the approach of atoms closer than a normal bond length can be avoided, thereby minimizing repulsive energies and hence the total internal energy. The difference in structures for the glassy and crystalline forms of a hypothetical, two-dimensional oxide, A_2O_3, is shown schematically in Fig. 2.9 in which each oxygen is corner-shared by *two* AO_3

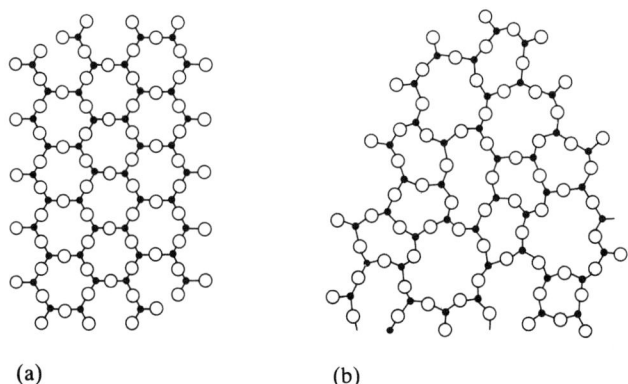

(a) (b)

Fig. 2.9 Schematic two-dimensional representation of the structure of
(a) a hypothetical crystalline compound A_2O_3 and
(b) the Zachariasen model for the glassy form of the same compound.

triangles. (The important concept of a CRN is discussed in considerably more detail in sect. 3.3.2.) Contrast this with the case shown schematically in Fig. 2.10 which is for the hypothetical two-dimensional crystalline oxide AO, where each oxygen is corner-shared by *three* AO_3 triangles; structural disorder cannot be introduced as in Fig. 2.9(a) simply by bond-angle distortions without considerably increasing the internal energy. The amorphous state, as represented by a CRN, is metastable since for crystallization to occur a substantial topological rearrangement of the structural units (having a large activation energy) must occur. Based on these and similar arguments, Zachariasen proposed four rules which an oxide must obey if it is to form a glass:

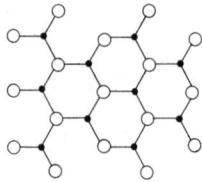

Fig. 2.10 Schematic two-dimensional representation of the structure of the hypothetical crystalline compound AO.

1. An oxygen atom may be linked to not more than two A atoms.
2. The number of oxygen atoms surrounding an A atom must be small (three or four).
3. The oxygen polyhedra share corners with each other, not edges or faces.
4. At least three corners in each oxygen polyhedron must be shared if the network is to be three-dimensional.

Thus, oxides of the type AO and A_2O should not, and indeed do not, form glasses. The rules are satisfied for oxides of the form A_2O_3 provided the oxygens form triangles around each A atom, and for oxides of the form AO_2 and A_2O_5 if the oxygens form tetrahedra. Cooper (1978) has pointed out that Zachariasen's rules are based more on topological than on energetic grounds, since it is found that corner-sharing regular octahedra or cubes can only form crystal-like, periodic arrays, leaving triangular and tetrahedral coordination of oxygens about A atoms the only contenders for glass formation. He has further shown that rule 3 is either redundant or inoperative (depending on the example used), and so should not be included. Consider firstly triangles sharing more than one edge, or tetrahedra sharing more than one face or two edges; these cases violate rule 1, thereby making rule 3 unnecessary. Triangles sharing an edge and a corner, and tetrahedra sharing either two opposite edges or a face and a corner, both form infinite chains (or loops) which can easily be formed into a non-periodic array. Alternatively, tetrahedra sharing one edge and the two opposite corners are equivalent topologically to a single corner-sharing tetrahedron which is therefore able to form a random structure. These two examples show that random structures can result even if rule 3 is violated, and that it should not therefore be considered.

The discussion thus far has centred on the glass-forming ability of simple oxide substances, known as 'network-formers'. Zachariasen also modified the original rules to take account of complex glasses, which contained non-glass-forming oxides in addition. The modified rules are:

1M. A high proportion of (network-forming) cations are surrounded by oxygen tetrahedra or triangles.
2M. The oxygen polyhedra only share corners with each other.
3M. Some oxygen atoms are linked to only two cations, and do not form additional bonds with any other cations.

These rules imply that oxide glasses must contain substantial proportions of glass-forming cations, or other cations which can substitute isomorphously (e.g. Al^{3+} for Si^{4+}). Any other cations (e.g. Na^+) that are present are termed 'network modifiers', since they disrupt the otherwise perfectly connected CRN. Although properly the subject of the next chapter, the question of the structure of modified oxide glasses is mentioned for the sake of continuity. Consider the case of the addition of Na_2O to SiO_2. The action of the modifier is to break up the continuous silica network by introducing 'dangling' or 'non-bridging' oxygens; this process is shown schematically in Fig. 2.11 and is not limited to silicate glasses only, but can

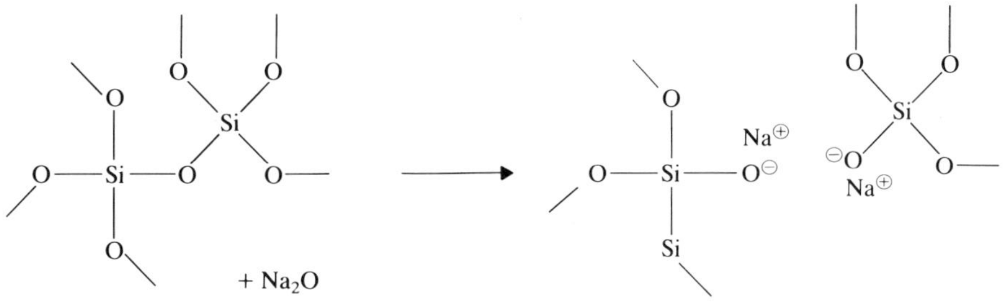

Fig. 2.11 Schematic illustration of the effect of the addition of alkali oxide (e.g. Na_2O) to silica. Each molecule of Na_2O added converts a bridging oxygen to two non-bridging oxygens (negatively charged). Diffusion in the melt would ensure that separation of the non-bridging oxygen atoms would occur.

occur in other oxide systems. The non-bridging oxygens carry a single negative charge and are compensated by a Na^+ ion nearby. The chemical formula of an oxide glass may be written generally as A_mB_nO, where m and n may not be integers; A represents the network-forming cation as before, and B is the network-modifying cation. An interesting intermediate class of oxides, including TeO_2, WO_3, MoO_3, Bi_2O_3, Al_2O_3, Ga_2O_3 and V_2O_5, do not by themselves form glasses, but will do so when mixed with other (modifier) oxides. They are obviously incipient network-formers.

In summary, Zachariasen's rules were formulated for and are only really applicable to oxide glasses, although the concept of the structure of an amorphous covalent solid as described by a continuous random network is of general applicability, as will be seen in the next chapter. The rules are not applicable to other glass-forming systems, such as metallic, organic or ionic glasses, in which directional covalent bonding is not predominant and hence for which the CRN model is inappropriate, nor for some other covalently bonded inorganic systems, such as the chalcogenide glasses, e.g. Se.

2.5.2 Eutectic compositions

For two compounds, A and B, miscible in the liquid state but completely immiscible as solids, the phase diagram will be as shown schematically in Fig. 2.12, provided that no intermediate compound formation occurs. The melting (or liquidus) temperature is depressed and is a minimum at the so-called 'eutectic' composition. It is well known that glass formation is generally favoured for compositions at which there is a deep eutectic. This is as true for metallic glass systems as it is for inorganic materials; an example is given in Fig. 2.13 of the glass-forming region for the system V_2O_5–PbO (an example of an 'incipient' network-former), which is seen to straddle

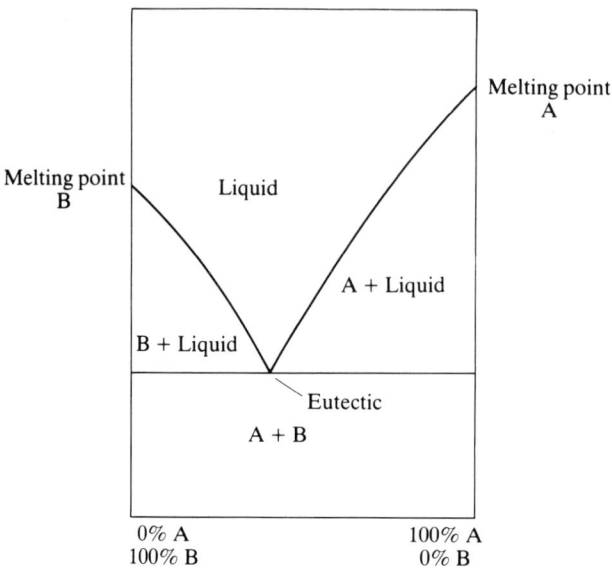

Fig. 2.12 Phase diagram of a simple binary eutectic system.

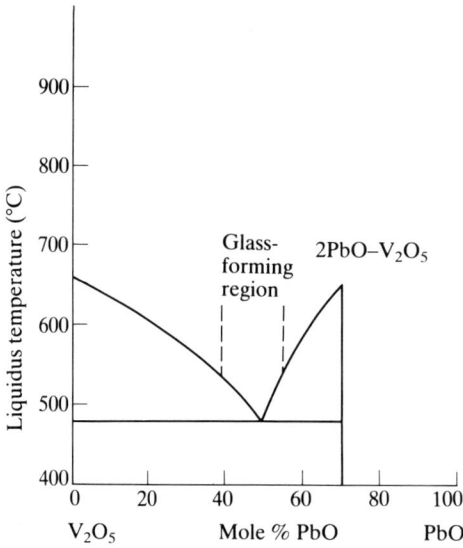

Fig. 2.13 Glass-forming region of the system V_2O_5–PbO in relation to the phase diagram (Rawson 1967).

the eutectic point. Compositions near the eutectic favour glass formation since the melting point is depressed and so the liquid is less supercooled at T_g, thereby reducing the possibility of crystallization (see sect. 1.3.5). Nevertheless, kinetic factors still play an important role in the formation of the glassy state, since if the cooling rate is not fast enough the liquid eutectic alloy crystallizes out into an intimate mixture of, in the simplest case, A and B.

Glass-forming systems

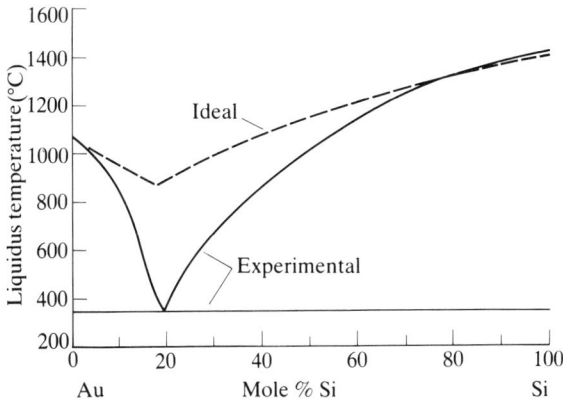

Fig. 2.14 Comparison of Au–Si eutectic phase boundaries determined experimentally and calculated for an ideal liquid (Allen *et al.* 1980).

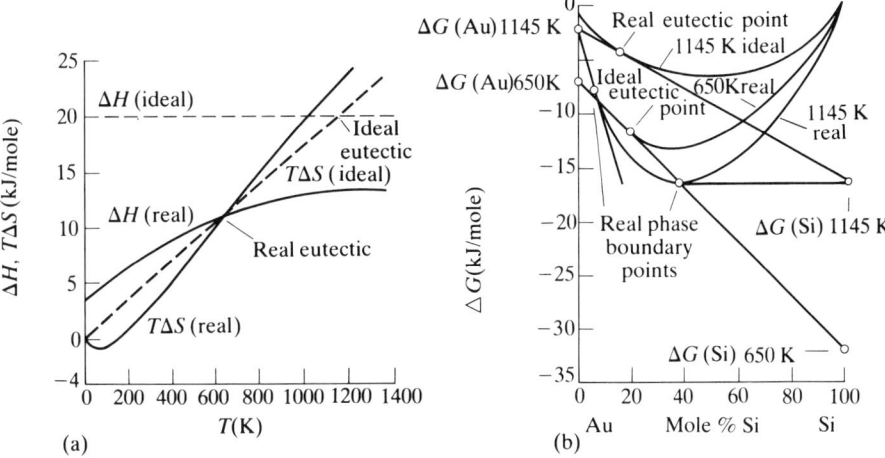

Fig. 2.15 (a) Temperature dependence of the entropy and enthalpy contributions to the free energy balance at the eutectic point of real and ideal liquids:
(b) the common tangent construction leading to the phase boundary shown in Fig. 2.14 (Allen *et al.* 1980).

In a recent study of the glass-forming system $Au_{1-x}Si_x$, which is attractive since it exhibits no intermediate compound formation, Allen *et al.* (1980) have demonstrated that the eutectic temperature is lowered considerably from the value expected for an ideal solution (Fig. 2.14) although the calculated eutectic composition, $x = 0.198$ (and indeed the observed eutectic point $x = 0.186$), is little altered from the ideal value ($x = 0.175$). The large decrease in eutectic temperature is shown to result from a large negative excess free energy of mixing, for which the dominant contribution is the enthalpy, as can be seen from Fig. 2.15(a) where the enthalpy departs much more from ideality than does the entropy. The eutectic point is determined by the condition that the free energies of the liquid and solid are equal, $G^L = G^S$, where

$$G^L = (1-x)G^L_{Au} + xG^L_{Si} + G^L_{AuSi} \qquad [2.12]$$

45

and

$$G^L_{AuSi} = G^I_{AuSi} + G^E_{AuSi} \qquad [2.13]$$

with

$$G^I_{AuSi}(x, T) = -TS^I(x) = RT[x \ln x + (1-x) \ln (1-x)] \qquad [2.14]$$

where G^I and G^E are the ideal and excess mixing terms, respectively. Allen et al. further point out that the eutectic *composition* is determined primarily by the large difference in free energies of solid formation (i.e. $G^S - G^L$) of Si and Au; this is also true for the Ag–Si and Cu–Si systems. This can be seen with reference to Fig. 2.15(b) in which is shown the composition dependence of the free energy of mixing of the liquid for given T. A tangent to the free energy curve at an arbitrary composition intersects the left and right ordinates at the values for the solid–liquid free energy differences (which are themselves temperature dependent) for Au and Si, respectively. For a general case, at some temperature T, the straight line cuts the mixing curve at two compositions; the eutectic point is given by the mixing curve characterized by the eutectic temperature, T_e, for which there is only one intersection, i.e. a tangent, at the eutectic composition, x_e. It can be seen from Fig. 2.15(b) that by this construction, there is little difference in x_e between the ideal and real cases if the solid–liquid free energy differences for the two components differ substantially; if the free energies were more nearly equal, a considerable disparity in values for x_e would occur. The fact that x_e is determined predominantly by the properties of the *pure* components, Au and Si, implies that it is not necessary to invoke solely structural arguments to account for the value of x_e, as had been attempted previously (Polk 1972). Allen et al. also show that in the vicinity of the eutectic point T_e, the excess entropy S^E is negative (see Fig. 2.15(a)), implying a considerable degree of chemical or structural ordering on a local scale, indicative of specific chemical bonding between Au and Si; this point will be discussed more fully in Chapters 3 and 7. Additional evidence for local ordering in metallic glasses comes from a compilation of eutectic compositions for a common metallic glass-forming system, namely transition metals alloyed with 'metalloid' atoms (e.g. B, C, N, Si or P). The frequencies at which eutectic compositions occur for alloys of 23 transition metals is shown in Fig. 2.16 (Gilman 1980), whence it can be seen that, instead of a flat distribution expected if arbitrary compositions produced eutectics, two significant peaks in the distribution are found, corresponding approximately to composition ratios of 6:1 and 5:1. The presence of discrete composition ratios also implies the presence of chemical bonding between metal and metalloid atoms.

A different influence of eutectic formation on glass-forming ability has been proposed by Rawson (1967). Following an earlier suggestion by Sun (1947) that since crystallization involves substantial atomic rearrangement and consequent rupture of bonds, those (oxide) materials having high bond strengths should be less prone to crystallization (and hence better glass-formers), Rawson suggested that the amount of thermal energy available at the melting (or liquidus) temperature might also be a significant factor. The lower the thermal energy (proportional to T) available for breaking bonds, the more difficult should be crystallization, and hence the more easy for a supercooled melt to transform into a glass. More specifically, the parameter relating to ease of glass formation is the ratio of the single bond strength

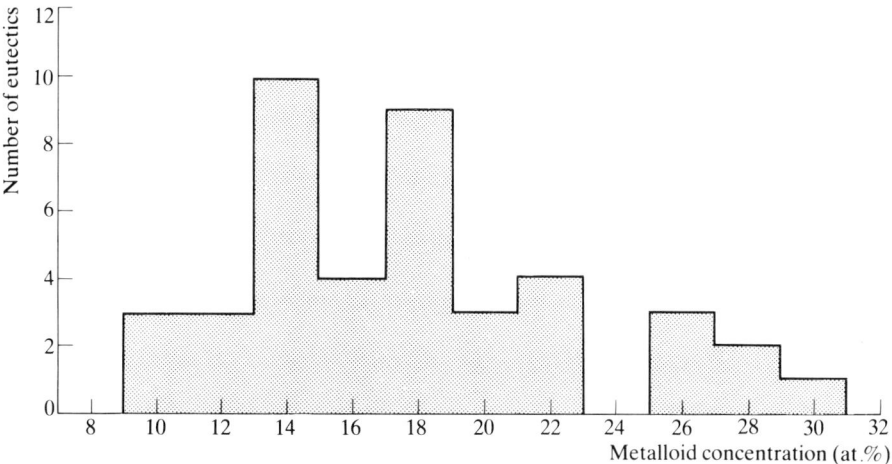

Fig. 2.16 Frequency distribution of eutectic composition for 23 transition-metal–metalloid metallic glasses (Gilman 1980).

to the melting temperature, B/T_m. 'Easy' glass-formers, such as B_2O_3 or SiO_2, have higher values of B/T_m than incipient network-formers (e.g. TiO_2, TeO_2) and considerably higher values than those of network modifiers (e.g. MgO and CaO), seeming to support the suggestion. The influence of eutectic formation in complex alloy glasses is immediately apparent since the melting or liquidus temperature is significantly lowered, thereby increasing the ratio B/T_m even more and hence favouring glass formation. Thus, the reason why the V_2O_5–PbO system should be glass forming in the vicinity of the eutectic composition, although V_2O_5 by itself is not a glass-former, is made understandable. Both Rawson's and Sun's criteria suffer from the limitation that only primary bonds are considered, even though atomic mobility at and above T_g is more likely to be due to rupture and reformation of secondary weaker bonds, such as 'back bonds' or Van der Waals' bonds between molecular units (e.g. between Se chains). The criteria are probably applicable, however, if there is only one type of bonding in the glass, as is probably the case, for example, in non-ionic silicate glasses.

2.5.3 Crystalline polymorphs

It has been pointed out (Goodman 1975, Wang and Merz 1976) that glass formation is often prevalent for those materials which in the crystalline state are found in a variety of polymorphic modifications. An obvious example is silica, the archetypal glass-former, which is found as quartz, cristobalite and tridymite, as well as the high-pressure forms coesite and stishovite. Goodman proposed that glass formation takes place by the generation of clusters of structurally non-related polymorphs which associate on cooling, but cannot act as nucleation centres for the growth of macroscopic regions of one polymorph or another. While being a useful empirical marker for likely glass-forming systems, the structure predicted for the amorphous solid is based on the 'microcrystallite' model, albeit composed of different types of microcrystals, and as such is at variance with the commonly held belief that amorphous *covalent* solids are better described in terms of random networks (sect. 3.3.2). Nevertheless, there is almost certainly some deep connection between the facility of glass formation and the existence of several crystalline polymorphs of the same material. The fact that several structurally unrelated polymorphs of comparable energy exist (e.g. quartz, cristobalite and tridymite) implies that the structural unit involved (in this case the SiO_4 tetrahedron) can occur in several conformations with little difference in strain energy, a circumstance needed if an amorphous solid, construed as a CRN of the structural units, is to exist having a sufficiently low internal (strain) energy that it can be metastable.

2.5.4 Coordination number

It is an interesting empirical observation that most inorganic covalently bonded glasses have relatively low values of atomic coordination number (or average coordination number in the case of alloys). An atom which has all covalent bonds satisfied, obeys the so-called '$8-N$' rule (where N is the number of valence electrons, and the coordination number N_c is given by the number $(8-N)$). Thus Se (a glass-former), has $N_c = 2$, since it is in group VI of the periodic table. For a binary alloy A_xB_{1-x}, the average coordination, m, is given by:

$$m = xN_c(A) + (1-x)N_c(B) \qquad [2.15]$$

In general, of course, m is non-integral; for example, for SiO_2 $m = 2.67$ and for As_2S_3 $m = 2.4$. Thus, m can be regarded as the coordination of a hypothetical 'pseudoatom' forming a structure whose topology is identical to that of the real system.

Phillips (1979) has proposed an interesting connection between glass-forming ability and the average coordination number. He supposes that the glass-forming tendency is maximized when the number of (mechanical) constraints, N_c, experienced by each atom (due to the interatomic forces acting on it) is equal to the number of degrees of freedom, N_d, available to it; the number of constraints per atom is related to the coordination number. A system which has a number of constraints greater than the available degrees of freedom is 'overconstrained' and cannot (easily) form a glass, although an amorphous structure may still be obtained by employing rapid cooling techniques, such as evaporation or sputtering (e.g. as in the case of Ge or Si). A homogeneous, isotropic structure is unlikely to result in this case; instead, 'domains' may preferentially form, with the excess strain relieved to a certain extent within each domain, but being concentrated at the domain boundaries (sect. 3.6.1).

Phillips considered the mechanical constraints experienced by an atom to result from the interatomic forces acting on an atom in the 'valence-force-field' model. In this, the (strain) potential energy can be expressed as a sum of contributions from bond-stretching and from bond-bending forces:

$$U_s = \tfrac{1}{2}\alpha \Delta r^2 + \tfrac{1}{2}\beta r_o^2 \Delta \theta^2 \qquad [2.16]$$

where α and β are the bond-stretching and bond-bending force constants, respectively, and Δr and $\Delta \theta$ represent small deviations in bond length and bond angle from the equilibrium values for the bond length, r_o, and the bond angle, θ_o. For a binary alloy, A_xB_{1-x}, there is only one bond-stretching interaction (α), but there are two bond-bending force constants, $\beta(BAB)$ and $\beta(ABA)$, for bending motions centred on atoms A and B respectively. It is supposed for simplicity that $\beta(BAB) = \beta(ABA)$, and that $\alpha, \beta(BAB)$ and $\beta(ABA)$ all act as rigid mechanical constraints. This is somewhat idealistic (implying, for instance, that the nearest-neighbour bond length r_{AB} and the next-nearest-neighbour bond lengths r_{AA} and r_{BB} are perfectly defined and have zero deviations from equilibrium values), but if the assumption is adopted, then the number of constraints per 'pseudoatom' is given by:

$$N_c = m/2 + m(m-1)/2 = \tfrac{1}{2}m^2 \quad [2.17]$$

where the first term is associated with α interactions (the factor of $\tfrac{1}{2}$ arises because a bond-stretching mode involves two atoms) and the second term is for β interactions. The number of degrees of freedom per atom is simply $N_d = 3$ (for a system in a three-dimensional space), and we have already seen that the condition for an 'ideal' glass which forms most easily can be written as $N_c = N_d$. Therefore, the 'universal' average coordination number most favourable for glass formation is from [2.17] $m_c = \sqrt{6} = 2.45$. Note that this result is interestingly independent of any physical constants such as the melting point, etc.

A somewhat more realistic approach for the case of a binary alloy $A_x B_{1-x}$ is to distinguish the atoms A and B and their constraints, rather than averaging first to yield 'quasi-atomic' parameters. In this manner, for the case of $Ge_x Se_{1-x}$ for example, the total number of constraints is $N_c = 8x + 2(1-x) = 6x + 2$ (two bond-stretching and six bond-bending constraints per atom for Ge, and one bond-stretching and one bond-bending constraint per atom for Se). Equating N_c and $N_D(=3)$ as before gives a value $x = 1/6$ for the composition most favourable for glass formation. The tendency for glass formation in the Ge–Se system has been explored experimentally by finding the minimum quenching rate needed for each composition to avoid crystallization; this was done in a qualitative manner by using three quenching techniques, i.e. quenching in water, air and slow cooling, respectively in order of decreasing quench rate. The data are shown in Fig. 2.17 (Phillips 1979), from which it can be seen that indeed the minimum in glass-forming difficulty (or maximum in glass-forming tendency) occurs near a composition corresponding to $x = 1/6$, seemingly in accord with the predictions of the simple valence-force-field model. The peak at $x = 1/3$ results from compound formation ($GeSe_2$) and represents the position of *chemical* stability, in contrast to the position of *mechanical* stability at the minimum of the curve. A similar analysis for the As–Se alloy system yields $x = 0.4$ for the composition at which glass formation is most favoured; this corresponds of course to the compound $As_2 Se_3$, for which chemical and mechanical stabilities coincide. Unfortunately, no other data relating glass-forming ability (or inverse quench rates) to composition exist for systems other than Ge–Se, and so this appealingly simple model remains to a large extent untested.

2.5.5 Summary

In discussing those factors which favour glass formation, an alternative approach is to consider those factors which *inhibit crystallization* on cooling a melt. Thus, the presence of a deep eutectic (i.e. lowering of the melting temperature) at a given

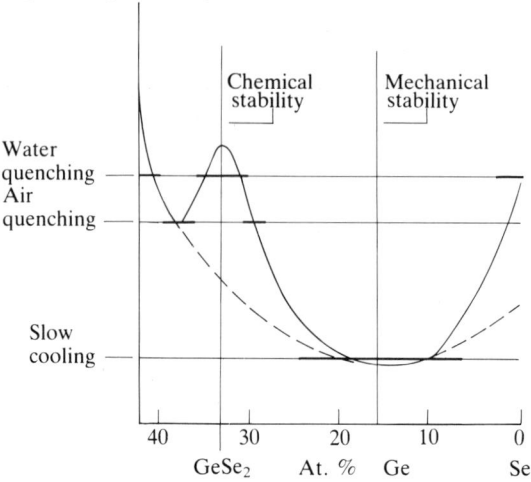

Fig. 2.17 Plot of glass-forming ability (tendency) of the Ge_xSe_{1-x} system. Experimental data are indicated by solid horizontal bars for various rates of cooling (water, air and slow quenching). The solid line is drawn through the data to guide the eye; the prediction of Phillips's mechanical constraint theory is shown by the dashed line.

composition for a liquid alloy will reduce the rate of crystallite nucleation and growth at that composition, since the liquid will be then less supercooled at the glass-transition temperature. In addition, many covalent glass-forming materials are found to have a large number of stable crystalline polymorphs (which have comparable free energies); crystallization into a single polymorph on cooling the melt becomes therefore less likely than glass formation because the structural unit involved can be frozen randomly into a variety of conformations with little extra cost in strain energy. Such energetic considerations also underlie the Zachariasen rules for glass formation in oxide systems. However, another aspect of the Zachariasen rules (for oxides) is the flexibility inherent in a random network, which is a consequence of the particular coordination of the oxygen atoms. In general, it is found that glass formation is easiest for those covalent materials which have a low average coordination number, or in other words have a principal constieuent which itself has a low coordination number. A reason for this may have to do with the fact that the excess free volume characteristic of a liquid, and which is also present in a glass (but not in a crystal), can be distributed in a continuous fashion throughout the bulk in a random network by virtue of the flexibility afforded by the low-coordination atoms; otherwise, an amorphous or polycrystalline aggregate can form, with all the strain and free volume concentrated at the grain boundaries.

Problems

2.1 Show that the Doolittle relation ([2.10]) is equivalent to the Vogel–Tammann–Fulcher law ([2.6]), if the assumption is made that the thermal expansion is linear in temperature. Show further that insertion of the relation for the free-volume temperature dependence into the Doolittle equation gives an equation of the Williams–Landel–Ferry type ([2.11]).

2.2 Deduce the dependence of the glass-transition temperature on the cooling rate (see A. J. Kovacs (1963) *Adv. Polym. Sci.* **3**, 394 for details).

(a) If V is the volume of the glass at time t corresponding to the fictive temperature T_f, and V_∞ is the equilibrium volume at temperature T, then the time dependence of the volume change can be written as:

$$\left(\frac{dV}{dt}\right)_p + \frac{(V-V_\infty)}{\tau} = -\alpha_{Tg} q V_\infty \qquad [2.1P]$$

where τ is the relaxation time, and the term on the right-hand side represents the time-dependent volume change upon cooling at a constant rate $q = -dT/dt$, where $\alpha_{Tg} = (1/V)(dV/dT)$. Show that [2.1P] can be transformed into

$$\frac{d\delta}{dt} + \frac{\delta}{\tau} = \Delta\alpha_T q \qquad [2.2P]$$

where $\delta = (V - V_\infty)/V_\infty$, and $\Delta\alpha_T = (\alpha_T - \alpha_{Tg})$ with $\alpha_T = (1/V_\infty)(dV_\infty/dT)$. The relaxation time can be written (following Tool) as:

$$\tau = A \exp(-K_1 T - K_2 T_f) \qquad [2.3P]$$

where K_1 and K_2 are constants, and the apparent activation energy is (over a small temperature range)

$$E_A = RT^2(K_1 + K_2) \qquad [2.4P]$$

(b) By using the expression for τ in the form $(d \ln \tau)/dT = -\theta$ where $\theta = (K_1 + K_2)$ to give $\tau = (1/\theta q) \exp(\theta q t)$, hence solve [2.2P] to give

$$\delta = -\frac{\Delta\alpha_T}{\theta} e^u E_i(-u) \qquad [2.5P]$$

where $u = \exp(-\theta q t)$, and $E_i(-u) = \int_\infty^u e^{-x} dx/x$ which can be approximated as $\lim_{u \to 0} \{E_i(-u)\} \simeq \ln(\gamma u)$ where γ is the Euler constant ($= 1.781$). By considering the asymptotic linear form for δ, and calling T_g the temperature at which $t = (1/\theta q) \ln q$, hence show that

$$\frac{dT_g}{d \ln q} = \frac{1}{\theta} \qquad [2.6P]$$

Hence derive [2.1] and show that the constant C ($\simeq 3 \times 10^{-5}$) can be obtained using reasonable physical parameters.

Bibliography

C. A. Angell, J. H. R. Clarke and L. V. Woodcock (1981), in *Advances in Chemical Physics*, vol. 48. I. Prigogine and S. A. Rice (eds) (Wiley), p. 397.

G. S. Grest and M. H. Cohen (1981), in *Advances in Chemical Physics*, vol. 48, I. Prigogine and S. A. Rice (eds) (Wiley), p. 455.
J. D. Mackenzie (ed.) (1960), *Modern Aspects of the Vitreous State* (Butterworth).
H. Rawson (1967), *Inorganic Glass-Forming Systems* (Academic).
J. Wong and C. A. Angell (1976), *Glass: Structure by Spectroscopy* (Dekker).

3 Structure

3.1 Introduction

I   MICROSCOPIC STRUCTURE

3.2 Experimental techniques and short-range order
3.2.1 Scattering
3.2.2 Anomalous scattering
3.2.3 Extended X-ray absorption fine structure
3.2.4 Magnetic resonance
3.2.5 Vibrational spectroscopy
3.2.6 Summary

3.3 Structural modelling
3.3.1 Introduction
3.3.2 Continuous random networks
3.3.3 Other structural models
3.3.4 Monte Carlo simulations
3.3.5 Molecular dynamical simulations
3.3.6 Summary

3.4 Medium-range structure
3.4.1 Cluster formation
3.4.2 Chemical ordering

II   MACROSCOPIC STRUCTURE

3.5 Experimental techniques
3.5.1 Microscopy
3.5.2 Small-angle scattering

3.6 Examples of macroscopic structure
3.6.1 Defects in growth morphology
3.6.2 Phase separation
3.6.3 Summary

Problems

Bibliography

3.1 Introduction

Knowledge of the structural arrangement of atoms of a solid substance is an essential prerequisite to a detailed understanding of other physical or chemical properties; this is as true for amorphous solids as for crystalline materials. Determination of the structure of a crystalline solid is made straightforward by the need only to solve the structure within the 'unit cell', containing relatively few atoms in most cases and which is the fundamental 'building unit' of the structure; the structure of the crystal as a whole is then generated by repeating in a periodic fashion the position of the unit cell in space. Such a procedure is impossible for a non-periodic amorphous solid, for which the unit cell may be regarded as being infinite in extent.

The innocent reader at this stage may well then ask: 'why attempt to study the structure of amorphous materials at all (and devote a chapter of this book to it) when, almost by definition, the structure is random and chaotic, and hence structural information, if any, can only be obtained in the form of statistical averages?' The operational definition of an amorphous solid, however, is merely that the structure is *non-periodic* (i.e. non-crystalline), not that it is necessarily truly random in a statistical sense. We shall find in many cases that the structure of many amorphous solids is in fact *non-random*, at least on certain length scales, e.g. there may be a considerable degree of local ordering despite the lack of periodicity. We have seen already in passing an example of this in section 2.5.1 where, for example, the structure of silica and silicates in the glassy state was anticipated as being composed of SiO_4 tetrahedra connected together in some random manner. The presence of discrete SiO_4 tetrahedra, identifiable by a number of techniques as we shall see later, certainly signifies the lack of complete randomness in the structure; a similar occurrence of local order is to be found whenever chemical bonding takes place between atoms in the solid, whether formally as directed covalent bonds or as a weaker component in addition to non-directional forces, such as those involved in metallic or ionic systems. It is readily realized, therefore, that this leads to a very large number of amorphous solids for which the question 'what is the structure?' is very pertinent and important.

This question can only be addressed when the length scale under consideration is stipulated. Quite obviously the determination of atomic positions (relative to an arbitrary origin atom) is generally altogether a different proposition from a discussion of possible phase separation in a glass, and the two problems necessitate the use of different experimental techniques. For this reason, we somewhat arbitrarily divide this chapter into two main sections, the first dealing with 'microscopic' structure (or the disposition of individual atoms or groups of atoms at a local level), and the second concerning itself with the 'macroscopic' structure of amorphous systems (such as phase separation). To a certain degree the distinction, particularly concerning phase separation, is semantic.

We therefore arbitrarily define a length scale which separates *microscopic* structure from *macroscopic* structure, and we take this to be in the region of 100 Å. This division has an added advantage in that completely different and mutually exclusive experimental techniques are used as structural probes in the two domains, and can therefore be discussed separately; e.g. conventional X-ray

diffraction is only sensit[...]
i.e. in the microscopic r[...]
detection of structural i[...]
more, which are in the [...]
subdivide the section o[...]
'short-range order' (of t[...]
'medium-range order' (o[...]
range order' of an amorp[...]
definition, in so far as ag[...]
structure at these two leng[...]
to concentrate first on the [...]
structure of an amorphous s[...]
following it with the section [...]
experimental results in terms of certain modelling studies for covalent solids (sect. 3.3.2). Section 3.4 on medium-range structure should also be read, as should the final section (3.6), dealing with examples of macroscopic structure found in amorphous solids.

We begin with a discussion of the structure of amorphous materials at a microscopic level because this is the more fundamental, and a knowledge of how atoms or groups of atoms are arranged in an amorphous system can often be used to infer the nature of medium-range structure. Note that this chapter employs almost exclusively examples of the structure of *covalent* amorphous solids: the structure of amorphous *metals* forms part of Chapter 7.

I MICROSCOPIC STRUCTURE

We begin this section with a discussion of the techniques that can be employed to obtain structural information for solids when there is no periodicity in atomic positions to aid in solving the structure, as is the case for single crystal work. In so doing, we introduce a variety of concepts and mathematical functions which will help us to describe the structure of a non-crystalline solid. In addition, we will consider first those methods which give information on short-range structure, that which extends over a few ångströms or so, and follow it with a discussion of techniques which probe the structure out to further distances, so-called medium-range order. These have been reviewed by Wright (1974), Wright and Leadbetter (1976), Wagner (1978), Gaskell (1979a) and in a recent book by Waseda (1980).

3.2 Experimental techniques and short-range order

3.2.1 Scattering

It is a fundamental property of all waves that they diffract on meeting an obstacle, and this effect is most pronounced when the size of the obstacle is comparable to the wavelength of the wave. This is as true for electrons and neutrons as it is for, say, X-rays, and since all can have wavelengths comparable to atomic dimensions, it is expected that diffraction from atoms in the condensed state will occur. This, of course, is the basis of the technique of X-ray diffraction used in conventional single crystal studies for which, if constructive interference is to occur among the outgoing beams of X-rays of wavelength λ elastically scattered from regular planes of atoms (Fig. 3.1), the Bragg condition

$$2d \sin \theta = n\lambda \qquad [3.1]$$

is obeyed. Here d is the spacing of planes and n is an integer. Thus diffracted beams of, say, X-rays are observed only for scattering angles 2θ and wavelengths λ for which [3.1] is satisfied; this gives rise to a series of sharp spots being detected in a plane perpendicular to the monochromatic incident beam, similar to the transmission electron micrograph shown in Fig. 3.2(a). We will not discuss further the subtleties of single crystal structure determination, which utilize the intensity variations as well as the position of the spots, since amorphous materials do *not* produce an X-ray or electron diffraction pattern similar to Fig. 3.2(a) (nor even the pattern of concentric *sharp* circles of intensity produced by *randomly* oriented crystals) but instead characteristic *diffuse* haloes are observed, as shown in Fig. 3.2(b). In this case, where there is no periodic structure, Bragg's law ([3.1]) is inappropriate, and we must derive afresh the conditions for diffraction to occur from a random structure.

We consider first, for simplicity, the case of X-ray diffraction by an amorphous sample; there are several good reviews of the application of scattering techniques (particularly by X-rays) in the literature (Warren 1969, Wright 1974, Wright and Leadbetter 1976, Wagner 1978), and we will primarily follow the treatment of Warren here. X-rays are scattered by the electrons present in the atoms of the sample (an important point to which we will return later), and although a full quantum-mechanical calculation should be used, a simpler classical treatment gives equivalent results.

The scattering from an electron, regarded as a point charge, is described by the (classical) Thomson equation, for which the differential scattering cross-section in SI units is given by:

$$\frac{d\sigma}{d\Omega} = \left(\frac{e^2}{4\pi\varepsilon_0 m_e c^2}\right)^2 \frac{(1+\cos^2\theta)}{2} \qquad [3.2]$$

Experimental techniques and short-range order

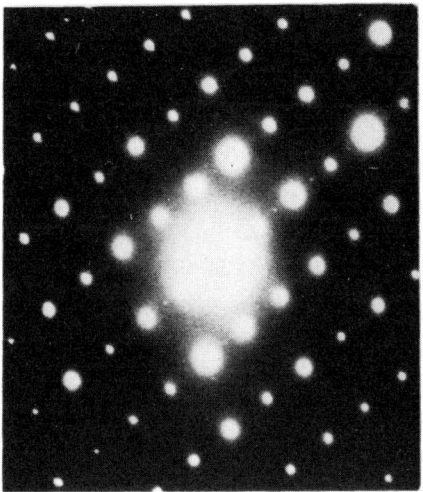
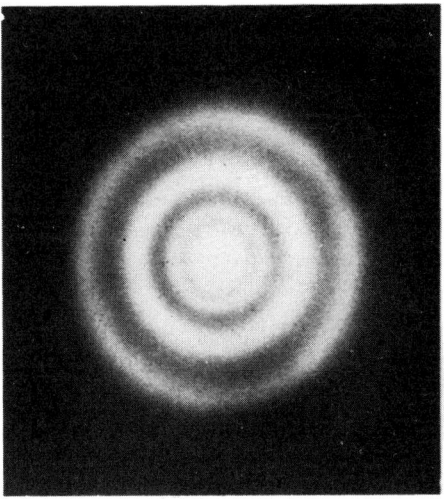

Fig. 3.2 Transmission electron diffraction micrographs of thin films of As_2Se_3 in (a) crystalline and (b) amorphous phases. (N. F. Mott and E. A. Davis, *Electronic Processes in Non-Crystalline Materials*, first edn (OUP: 1971).

This describes the 'unmodified' or 'coherent' scattering where the photon momentum (and hence the wavelength) is conserved (in contrast to Compton or 'modified' scattering where momentum is exchanged with the electron); the factor $\frac{1}{2}(1+\cos^2\theta)$ arises if the incident beam is unpolarized. Thus, incident radiation of intensity I_o falling on a point electron at O and scattered through an angle 2θ (Fig. 3.3(a)) has an intensity at a point P, distance R from O, given by

$$I = \frac{I_o}{R^2}\left(\frac{e^2}{4\pi\varepsilon_o mc^2}\right)^2 \frac{(1+\cos^2\theta)}{2} \qquad [3.3]$$

Scattered intensities in units of the right-hand side of [3.3] are termed 'electron units' of intensity, and these units will be adopted in the following discussion.

We next consider the scattering from an electron whose charge is distributed in a certain volume dV as $\rho \, dV$, rather than being confined to a point; the scattering is then $\rho \, dV$ times the amplitude of the classical scattering from a single electron. The path differences between the various scattered rays from different parts of the charge distribution at **r** from an arbitrary origin must be considered, giving the following expression for the scattering intensity in e.u. as:

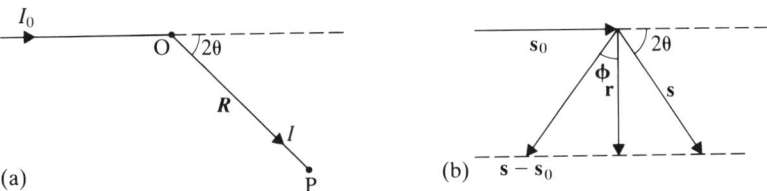

Fig. 3.3 (a) Scattering geometry for a point charge; (b) relation between scattering vectors.

$$I_{eu} = f_e f_e^* \quad [3.4]$$

where the 'scattering factor' f_e is given by:

$$f_e = \int \exp\left[\frac{2\pi i}{\lambda}(s-s_o)\cdot r\right]\rho(r)\, dV \quad [3.5]$$

The relation of the scattering vectors s and s_o to the vector r is shown in Fig. 3.3(b), from which it can readily be deduced that $(s-s_o)\cdot r = 2r\sin\theta\cos\phi$. We also introduce the widely used abbreviation (termed the scattering vector) given by:

$$k = \frac{4\pi \sin\theta}{\lambda} \quad [3.6]$$

If we assume spherical symmetry (true for an eventual summation over closed shells of electrons) then $\rho(r) = \rho(r)$, and the scattering factor per electron can be written as

$$f_e = \int_{r=0}^{\infty}\int_{\phi=0}^{\pi} \exp[ikr\cos\phi]\rho(r)2\pi r^2 \sin\phi\, d\phi\, dr$$

Integration over ϕ gives:

$$f_e = \int_0^{\infty} 4\pi r^2 \rho(r) \frac{\sin kr}{kr}\, dr \quad [3.7]$$

For an atom containing several electrons, the 'atomic scattering factor' or 'form factor' is simply the sum of the individual amplitudes:

$$f = \sum_n f_{en} = \sum_n \int_0^{\infty} 4\pi r^2 \rho_n(r) \frac{\sin kr}{kr}\, dr \quad [3.8]$$

and the total scattered intensity is $I_{eu} = ff^*$. Clearly, the quantity $\sum_n \int_0^{\infty} 4\pi r^2 \rho_n(r)\, dr$ is simply the total number of electrons in the atom (i.e. the atomic number Z); and since $\sin kr/kr \to 1$ for small kr, f tends to the value Z as k (or $\sin\theta/\lambda$) tends to zero. The dependence of the atomic scattering factor f on k is shown schematically in Fig. 3.4 and is tabulated in the literature for all atomic species.

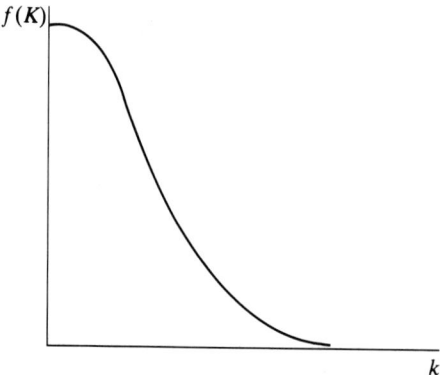

Fig. 3.4 Schematic illustration of the dependence of the X-ray atomic form factor on scattering vector.

The treatment thus far has depended on three assumptions: that the scattered waves can be treated as plane waves (true if $|R| \gg |r|$), that the electron distribution has spherical symmetry (true for heavier atoms where the fraction of valence electrons utilized in bonding is small), and finally that the X-ray wavelength used is far removed from the wavelengths corresponding to absorption edges of the atom. The latter case can be allowed for by the introduction of 'dispersion corrections' to the atomic scattering factor for wavelengths near an absorption edge, $f = f_o + \Delta f' + i\Delta f''$. These corrections act as the basis of the technique utilizing 'anomalous dispersion' which will be discussed in section 3.2.2.

For an assembly of atoms, the scattered intensity (in electron units) is given by the summation of the scattered amplitude from each atom m, having position r_m, multiplied by its complex conjugate:

$$I_{eu} = \sum_m f_m \exp\left[\frac{2\pi i}{\lambda}(s-s_o)\cdot r_m\right] \sum_n f_n \exp\left[-\frac{2\pi i}{\lambda}(s-s_o)\cdot r_n\right] \quad [3.9]$$

where we have assumed the atomic scattering factors to be real. This expression is true for *all* materials, whether crystalline or amorphous. In the case of scattering from crystals, where the atoms are arranged in a periodic manner such that the position vectors of the atoms are simply related through a lattice vector $r_o (= r_m - r_n)$, the summations in [3.9] form a geometric progression which can be evaluated. For amorphous materials, however, no such simple relationship exists between the position vectors, and the summations must be left in the analysis. Equation 3.9 can be written more succinctly by introducing the difference vector $r_{mn} = r_m - r_n$, such that

$$I_{eu} = \sum_m \sum_n f_m f_n \exp\left[\frac{2\pi i}{\lambda}(s-s_o)\cdot r_{mn}\right] \quad [3.10]$$

A simplifying assumption that is usually valid for amorphous solids is that the material is isotropic, so that the vector r_{mn} may adopt all orientations with equal probability, or equivalently can take all positions on the surface of the sphere shown in Fig. 3.5. In this case the average of the exponential term in [3.10] may be written as:

$$\left\langle \exp\left[\frac{2\pi i}{\lambda}(s-s_o)\cdot r_{mn}\right]\right\rangle = \frac{1}{4\pi r_{mn}^2} \int_{\phi=0}^{\pi} \exp[ikr_{mn}\cos\phi]2\pi r_{mn}^2 \sin\phi \, d\phi$$

$$= \frac{\sin kr_{mn}}{kr_{mn}} \quad [3.11]$$

Substituting this result in [3.10] yields the 'Debye equation' for a random array of scattering atoms

$$I_{eu} = \sum_m \sum_n f_m f_n \frac{\sin kr_{mn}}{kr_{mn}} \quad [3.12]$$

To proceed further, one must take into account the type of atom within the sample; the analysis is considerably simplified if it is assumed, as we do at first, that the material is monatomic. We further define a quantity which is extremely

Structure

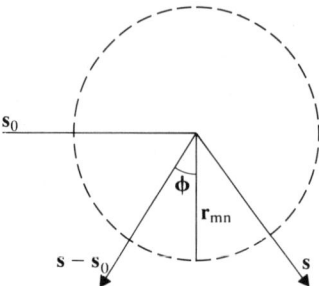

Fig. 3.5 Illustration of the positions that r_{mn} may take for an isotropic material.

important in the field of structure determination of amorphous materials, namely the 'radial distribution function', $4\pi r^2 \rho(r)$. This is defined such that the average number of atom centres lying between r and $r + dr$ from the centre of an arbitrary origin atom is given by $4\pi r^2 \rho(r)\,dr$. The function $\rho(r)$ is thus essentially a pair correlation function, being large at those distances for which there are on average many atoms from a given origin atom, and small otherwise. The relationship of the density function $\rho(r)$ to the actual atomic structure is shown schematically in Fig. 3.6 in a two-dimensional representation of an amorphous structure. It can be seen that the first peak in $\rho(r)$ corresponds to the first shell of atoms, the second peak to the second shell, and so on. It is important to note that in *three* dimensions, features in

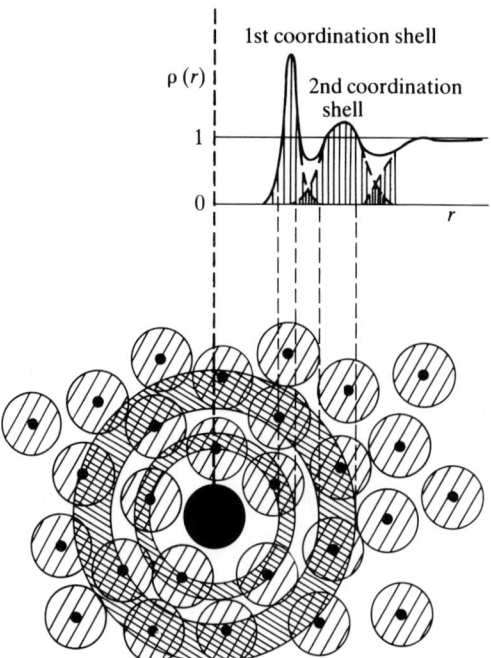

Fig. 3.6 Schematic illustration of the structural origin of certain features in the density function $\rho(r)$ for an amorphous solid (after Ziman 1979).

$\rho(r)$ beyond the first peak have contributions from higher order shells as well. It is important to note as well that the radial distribution function (RDF) is a *one-dimensional representation of a three-dimensional structure* and as such is spatially averaged (strictly only valid if the material is isotropic, as we have already assumed), and that as a result of this the RDF can only carry a limited amount of structural information. In addition, the RDF is clearly only a useful concept if the atoms making up the amorphous solid have well-defined sizes and rather well-defined closest distances of approach or bond lengths.

Rewriting [3.12] as

$$I_{eu} = \sum_m f^2 + \sum_m \sum_{n \neq m} f^2 \frac{\sin kr_{mn}}{kr_{mn}}$$

and introducing the density function $\rho_m(r_{mn})$ (for the origin atom m) gives

$$I_{eu} = \sum_m f^2 + \sum_m f^2 \int \rho_m(r_{mn}) \frac{\sin kr_{mn}}{kr_{mn}} dV_n$$

where the summation over n has been replaced by the integral over the sample volume and spherical symmetry has been assumed as before. The (microscopic) density function averaged over all atoms, m, in the sample can be written as $\rho(r) = \langle \rho_m(r_{mn}) \rangle$, and if the (macroscopic) average density is written as ρ°, then adding and subtracting a term in ρ° yields:

$$I_{eu} = \sum_m f^2 + \sum_m f^2 \int 4\pi r^2 [\rho(r) - \rho^\circ] \frac{\sin kr}{kr} dr + \sum_m f^2 \int 4\pi r^2 \rho^\circ \frac{\sin kr}{kr} dr \quad [3.13]$$

The summation over m yields simply N, the total number of atoms in the sample.

Amorphous materials possess no long-range order, and an equivalent way of expressing this is to note that the density function $\rho(r)$ tends to ρ° at large r (see Fig. 3.6). Thus, the quantity $[\rho(r) - \rho^\circ]$ tends to zero for distances greater than a few primary atom separations, and hence the integral in the second term of [3.13] is dominated by scattering predominantly from close scattering centres, and therefore the Fraunhofer approximation assumed so far holds. This is not so for the third term in [3.13] which involves interactions between distant points in the sample, and Fresnel diffraction theory must be used. It can be shown (see e.g. Warren 1969) that the third term represents small-angle scattering due to the finite sample size, which is generally limited to scattering vectors $k \lesssim 2\pi/R$ where R is the specimen size. For R of the order of millimetres and λ a few ångströms, this corresponds to scattering at such extremely small angles that the primary beam completely obscures it. For sample geometries such that this scattering is observed at finite, but small angles, it can be calculated and subtracted from the total observed intensity to leave only that scattering intensity resulting from atomic correlations.

These considerations transform [3.13] into:

$$I_{eu} = Nf^2 + Nf^2 \int_0^\infty 4\pi r^2 [\rho(r) - \rho^\circ] \frac{\sin kr}{kr} dr \quad [3.14]$$

where the upper limit to the integration is valid since the size of the sample is much greater than atomic dimensions. Two further functions can now be defined in order to simplify [3.14]. The 'reduced scattering intensity', $F(k)$, is defined as:

$$F(k) = k\left[\frac{I_{eu}/N - f^2}{f^2}\right] \qquad [3.15]$$

and the 'reduced radial distribution function', $G(r)$, is defined as:

$$G(r) = 4\pi r[\rho(r) - \rho^\circ] \qquad [3.16]$$

Both functions are termed 'reduced' because they oscillate about zero instead of being an increasing or decreasing function of k or r. Thus [3.14] can be written in the form:

$$F(k) = \int_0^\infty G(r) \sin kr \, dr \qquad [3.17a]$$

This equation is important since it links a quantity which is directly obtainable from experiment, namely $F(k)$, with a function which describes the real-space structure of the amorphous solid, i.e. $G(r)$. The form of [3.17a] is immediately suggestive of a Fourier transform, and thus the equation can be inverted:

$$G(r) = \frac{2}{\pi} \int_0^\infty F(k) \sin kr \, dk \qquad [3.17b]$$

Note that the sine function which arises in [3.17a] results solely from the analysis for the scattering intensity (cf. the Debye equation, [3.12]), whereas that which appears in [3.17b] arises from the Fourier transform procedure. Note too that the equation by which the real-space correlation function is obtained from experimental data ([3.17b]) requires the data for a very large range of values of k (ideally infinite in extent); the problems that arise in practice when this condition is not met will be discussed later.

For convenience, various real-space distribution functions and their reciprocal-space analogues are shown for a-Ge in Fig. 3.7, together with their interrelationships and definitions. It should be noted that these curves strictly refer to X-ray scattering; this is particularly true for the total intensity curve ($I(k)$), where the background curve ($f^2(k)$) is k-dependent, in contrast to (non-magnetic) neutron diffraction as will be seen later. Various points of interest pertaining to each function are given below:

1. *Density function* $\rho(r)$

The features of $\rho(r)$ at large r have been discussed above. There can be no *atomic* correlations for distances less than the first peak (if the atomic size is well defined), and so $\rho(r)$ should be zero in this region. However, since X-rays are in fact scattered from *electrons* and not atoms, those electrons forming covalent bonds between atoms can act as scattering centres and correlations other than interatomic spacings can be registered, particularly for light atoms for which a significant fraction of the valence electrons are involved in bonding.

2. *Radial distribution function* $J(r)$

This is one of the most common forms of representing real-space structural

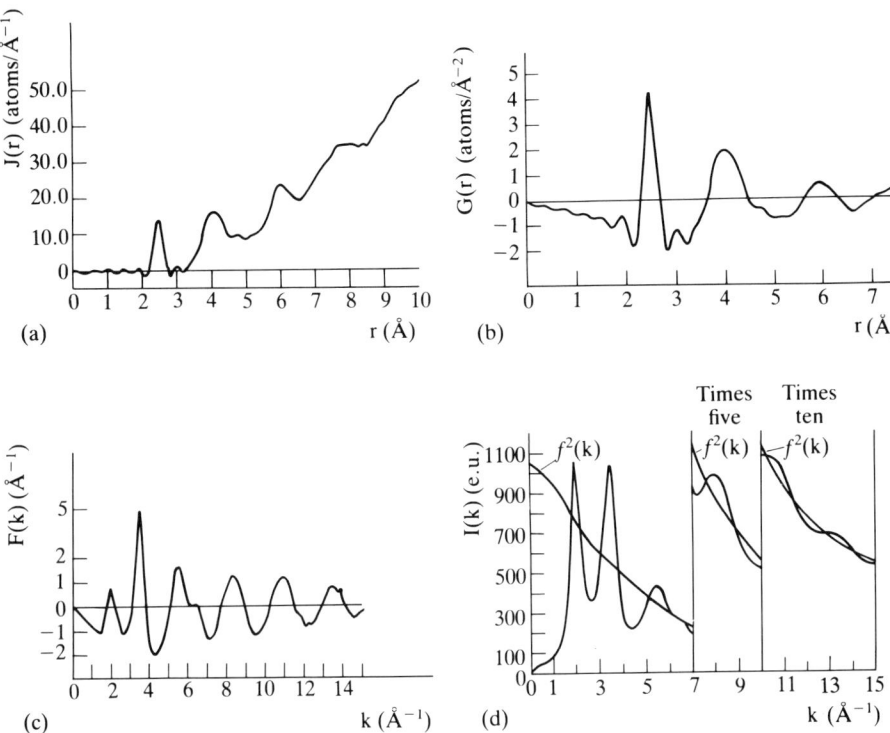

Fig. 3.7 Various structural correlation functions in real and reciprocal space for sputtered a-Ge films deposited on to substrates at 150°C (Temkin *et al.* 1973).
(a) RDF, $J(r) = 4\pi r^2 \rho(r)$;
(b) reduced RDF, $G(r) = (J(r)/r - 4\pi r \rho^0)$;
(c) reduced scattering intensity, $F(k) = [(I/N - f^2)/f^2]k$;
(d) scattering intensity, $I(k)$.

information, and it has the typical form shown schematically in Fig. 3.7(a). The characteristic features are a series of peaks becoming broader and less well defined with increasing r, superimposed upon the 'average density parabola' given by $4\pi r^2 \rho^0$, and to which the RDF tends at large r when there are no further significant atomic correlations; note that the RDF is ideally zero below the first peak for the reasons given in the previous section. The importance of the RDF lies principally in the fact that the area under a given peak gives the effective coordination number of that particular shell. The first peak is generally sharp; its mean square width, σ^2, is caused ideally only by the thermal vibrations for the case of crystals, whereas amorphous materials may have an additional broadening contribution from the presence of static disorder in the bond length. The position of the first peak gives a value for the nearest-neighbour bond length, r_1, and similarly the position of the second peak gives the next-nearest-neighbour distance, r_2; a knowledge of both immediately gives a value for the bond angle θ:

$$\theta = 2 \sin^{-1}\left(\frac{r_2}{2r_1}\right) \qquad [3.18]$$

Note that the width of the second peak is generally wider than the first for covalent amorphous materials; this is a reflection of the presence of a static variation in the bond angles (which is typically of the order of $\pm 10\%$) in addition to the thermal disorder. In crystals, on the other hand, no static bond-angle distortion exists in general, and so the widths of the first two peaks in the RDF should be equal. An important point to note is that the RDF yields only a limited amount of information by inspection, restricted essentially to the local structure around a given atom, i.e. bond lengths and bond angles. The structural origin of more distant correlations corresponding to higher order peaks in the RDF cannot be obtained directly, but only in conjunction with additional information as may be obtained for example from modelling studies (see sect. 3.3).

3. Reduced radial distribution function $G(r)$

This function is of importance since it is obtained directly by Fourier transformation of experimental data. It oscillates about zero since the average density background curve has been subtracted, and because it is a difference curve and weighted by a factor of r, in many cases it displays structural correlations more clearly than the RDF itself (see Fig. 3.7(b)). Note that in the region below the first peak where the RDF, $J(r)$, is zero, $G(r)$ decreases linearly as $-4\pi r\rho^0$ and hence the slope gives a value for the average (atomic) density directly (in units of atoms Å^{-2}). In this manner ρ^0 can be experimentally determined and the RDF constructed from $G(r)$.

4. Reduced scattering intensity $F(k)$

This function is important as the precursor to obtaining the RDF (Fig. 3.7(c)), but is often informative in its own right since subtle structural differences between samples (say, prepared under different conditions) may often be more readily discerned in a comparison of the respective $F(k)$ curves than by comparing the RDFs. Further, the curves are obtained directly from the measured data and hence do not suffer from spurious features that can be introduced by the Fourier transform process (see later). A major disadvantage in the examination of correlation curves in reciprocal space, however, is that it is difficult to identify real-space structural correlations simply by inspection of the curves. This can be seen more clearly by a consideration of how the $F(k)$ curve is built up.

Consider first the case of a random system in which only nearest-neighbour correlations are well defined; beyond the first shell there are no correlations. This would correspond to a single peak in $G(r)$ at r_1, having a width σ_1^2 resulting from both thermal and static disorder contributions. Assuming this can be approximated by a Gaussian, the Fourier transform of $G(r)$ gives a damped, simple sinusoidal form for $F(k)$. The degree of damping is determined by σ_1^2, and the periodicity k_1 of the sinusoid by the nearest-neighbour separation r_1, ($k_1 = 2\pi/r_1$). If it is now assumed that a second well-defined shell of atoms of radius r_2 exists, then this would by itself give rise to a damped sinusoid, but having a *different* damping parameter and periodicity. Thus, at low values of k, there will be considerable mixing of these two contributions giving rise to a complex form for $F(k)$. However, as we have already seen, higher order shells usually have widths considerably larger than that of the first shell and so the attenuation of the contribution to $F(k)$ of the second and higher shells will be considerably greater than that of the first (since $\sigma_2^2 > \sigma_1^2$, etc.). Hence the behaviour of $F(k)$ at large values of k should solely be determined by the first peak in $G(r)$ and should be a simple sinusoid of period $2\pi/r_1$. Thus, if data can be taken to sufficiently large values of k and $F(k)$ is sinusoidal in this region, then a limited amount of structural information (namely the value for the bond length r_1) can be obtained by inspection. It should be obvious to the reader that the first major peak in $F(k)$ does *not* correspond in a simple manner in general to any real-space structural correlation, and it is incorrect to assume (as is sometimes done) that its position is uniquely determined by the nearest-neighbour separation. Exceptions to this caveat are the small, sharp peaks in $F(k)$ often observed in chalcogenide glasses at around $k \simeq 1$ Å^{-1}, which are believed to result from Fourier components of spatial frequency $\simeq 6$ Å, perhaps residual layer–layer correlations which are predominant in the crystalline forms of these materials, but which cannot be identified directly in the RDF.

5. Total scattering intensity $I(k)$

This is the curve directly obtained from experiment after appropriate subtraction of incoherent contributions to the scattering such as Compton scattering. For the case of X-ray diffraction, $I(k)$ oscillates about the curve of the square of the atomic scattering amplitude f^2, as can be seen by inspection of [3.14] and Fig. 3.7(d), where f^2 varies strongly with scattering vector. This is in marked contrast to the case of neutron diffraction, where the scattering amplitude is *independent* of k for closed shells, i.e. in the absence of *magnetic* scattering by unpaired electrons (see sect. 7.2.1).

The experimental arrangement employed in an X-ray diffraction study is shown schematically in Fig. 3.8. X-radiation from a source (e.g. MoK_α, $\lambda = 0.71$ Å or CuK_α, $\lambda = 1.54$ Å) is collimated by the first slit L_1 and may then be

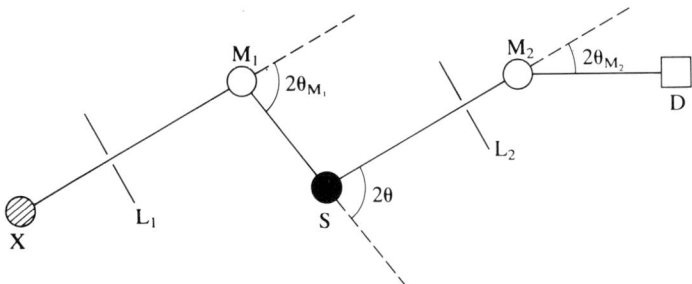

Fig. 3.8 Schematic illustration of the experimental scattering geometry used in diffraction experiments. X = source of radiation; L_1, L_2 collimating slits; M_1, M_2 monochromators; S sample; D detector.

monochromatized by the first single crystal (or filter) monochromator M_1 and diffracted so as to strike the sample. The diffracted radiation from the sample, scattered through 2θ, is then further collimated by L_2 and monochromatized by M_2 before entering a detector (e.g. a scintillation counter). The scattered intensity is measured as a function of 2θ which can be converted via [3.6] into wavevector k.

The scattered intensity as measured contains contributions from the substrate of the sample (if present), noise and includes a polarization factor. These can be computed and eliminated, leaving the total unnormalized scattered intensity which is a sum of coherently and incoherently (Compton) scattered intensities, $I'(k) = I'_{coh}(k) + I'_{incoh}(k)$, of which we are only concerned with $I_{coh}(k)$. The Compton contribution is difficult to calculate and hence to correct for; a better procedure is to attempt to eliminate it directly. An early method developed by Warren and Mavel (1965) is based on fluorescence; the diffracted beam strikes a thin foil which has an X-ray absorption edge at an energy only very slightly smaller than the photon energy of the incident beam, in which case coherently scattered radiation excites fluorescence which can be detected, while the modified, Compton down-scattered radiation cannot do so. Nowadays, energy dispersive detectors are used, which are capable of measuring only the elastically scattered intensity. The resulting coherent scattered intensity can be converted from arbitrary units to electron units by arranging that $I'(k)$ oscillates equally above and below the curve of $f^2(k)$ at large values of k, whereupon it is normalized automatically, or by using the more complicated Krogh-Moe–Norman method (see Warren 1969).

A considerable problem exists in the conversion of reciprocal-space data to real-space correlation functions via Fourier transforms, since the range of values of k accessible to experiment is strictly limited. This may be seen by reference to the defining equation ([3.6]) for $k = (4\pi \sin \theta)/\lambda$; the maximum value possible for k is when $\sin \theta = 1$. For MoK_α radiation ($\lambda = 0.71$ Å) this gives $k_{max} = 17.7$ Å^{-1}, but for CuK_α ($\lambda = 1.54$ Å), the maximum value of k is only $k_{max} = 8.2$ Å^{-1}. Transformation between reciprocal and real space is only perfect if $F(k)$ is known for an *infinite* range of k values (cf. [3.17b]); in practice, this is obviously not the case and the data are terminated at some finite k_{max}. This is equivalent to $F(k)$ being multiplied by a 'modification function' $M(k)$, where $M(k) = 1$ for $k \leqslant k_{max}$ and 0 for $k > k_{max}$; i.e. the effective result is

$$G'(r) = \frac{2}{\pi} \int_0^{k_{max}} F(k) \sin kr \, dk = \frac{2}{\pi} \int_0^\infty F(k) M(k) \sin kr \, dk.$$

This leads to the introduction of 'termination errors' in the Fourier transform (i.e. in $G(r)$), which for the step-function form of $M(k)$ given above is equivalent to peaks in $G(r)$ being convoluted with the cosine transform of $M(k)$, i.e. the sinc function $(\sin rk_{max})/\pi r$. The effect of this is twofold: a loss in resolution is incurred since wavelengths in real space less than $2\pi/k_{max}$ are lost, resulting in a peak width $\simeq \pi/k_{max}$, and spurious 'termination' ripples at values $rk_{max} \simeq \pm((4n+1)\pi)/2$ are introduced (where n is an integer), which decay with increasing r.

The problem can be alleviated to a certain extent experimentally by taking data to high values of k_{max} (i.e. using smaller wavelengths), a situation more easily achievable in neutron scattering than in X-ray diffraction; this approach is particularly successful if, due to sufficiently broad peaks in the real-space correlation functions, the oscillations in $F(k)$ have died away at $k = k_{max}$, there being therefore no discontinuity in the experimental data at k_{max}. If this cannot be achieved satisfactorily, a variety of procedures may be used to minimize termination error. The usual method employs a damping (or convergence) factor which multiplies the finitely ranged $F(k)$ in the Fourier transform, replacing a sharp discontinuity at k_{max} by a smoothly varying function. Two kinds of function are commonly used; the artificial temperature factor due to Bragg and West (1930):

$$M(k) = \exp\left(-\frac{2.3k^2}{k_{max}^2}\right) \quad [3.19]$$

and that due to Lorch (1969), which is more favourable since it reduces the ripples more rapidly:

$$M(k) = \frac{k_{max}}{\pi k} \sin\left(\frac{\pi k}{k_{max}}\right) \quad [3.20]$$

Both functions reduce the termination ripples but at the expense of worsening the real-space resolution.

Thus far, we have considered only scattering by X-rays. While this technique has many advantages including the ready availability of sources and ancillary equipment in the laboratory, there are a number of distinct disadvantages and drawbacks, among which number the low values of k_{max} achievable using common X-ray sources, the fact that intensity variations (in $I(k)$) are superimposed on a rapidly decreasing background of $f^2(k)$ (so data at high values of k are subject to considerable error) and that *electron* correlations are determined which are not the same as atomic correlations for low Z atoms. An alternative technique, which resolves some of the drawbacks of X-ray diffraction but at the cost of necessitating a large facility as a source and needing large volumes (~ 1 cm^3) of sample, is neutron scattering. Neutrons emerging from a nuclear reactor pile have wavelengths typically in the range $\lambda \sim 0.1$ to 1 Å or larger, corresponding to energies ranging from 8 eV to 80 meV, the latter being 'thermal' neutrons produced by the use of a moderator. In practice, the minimum wavelengths for which useful fluxes occur are ~ 0.7 Å (for an ambient source) and ~ 0.5 Å (for a 'hot' source). In any scattering event, there is generally an energy transfer $\hbar\omega = E_o - E$ and a momentum transfer $\hbar Q = \hbar k_o - \hbar k$ between the scattered wave and the scattering object. For the case of thermal neutrons, their characteristic energy is of the same order of magnitude as that of lattice vibrations in solids and so considerable energy transfers (inelastic events) are possible. X-rays have much higher characteristic energies and so the fractional energy changes due to phonon processes are very small (and $k \simeq k_o$), although different inelastic (Compton) processes are possible. Thus, neutron scattering may be used in two completely different ways, since *elastic* scattering ($\hbar\omega = 0$; $Q = Q_o$) is related to the spatial distribution of centres (atoms) and *inelastic* scattering ($\hbar\omega \neq 0$) is related to the dynamics of the centres (see sect. 4.4.4).

The scattering law $S(Q, \omega)$ is a function of both momentum and energy transfer, and is related to a generalized pair correlation function $G(r, t)$ by a double Fourier transform

$$S(Q, \omega) = \frac{1}{2\pi} \int_0^\infty \int_{-\infty}^\infty G(r, t) \, e^{i(Q \cdot r - \omega t)} \, dr \, dt \quad [3.21]$$

and

$$G(r, t) = \frac{1}{(2\pi)^3} \int_0^\infty \int_{-\infty}^\infty S(Q, \omega) \, e^{i(\omega t - Q \cdot r)} \, dQ \, d\omega \quad [3.22]$$

In the experimental arrangement conventionally used in neutron diffraction, namely a 'twin-axis spectrometer' (identical to the configuration shown in Fig. 3.8 with the omission of the second monochromator M_2), *all* neutrons are detected after scattering at a given angle irrespective of their energy; this is termed 'total scattering' and is the same for X-ray diffraction. Thus, this involves an integration over all values of ω, which by reciprocity is equivalent to taking $t=0$. Thus in this case, the so-called 'static approximation', [3.22] reduces to

$$G(r, 0) = \frac{1}{(2\pi)^3} \int_0^\infty S(\mathbf{Q}) e^{-i\mathbf{Q}\cdot\mathbf{r}} \, d\mathbf{Q}$$

where $S(\mathbf{Q}) = \int_{-\infty}^{\infty} S(\mathbf{Q}, \omega) \, d\omega$ and $G(r, 0)$ is the instantaneous pair correlation function. A full quantum-mechanical calculation of the scattering (see, e.g., Wright 1974) leads to the following expression for the total neutron scattering intensity from a monatomic isotropic system in the *static* approximation:

$$I_N^T(Q) = N\overline{b^2} + N\overline{b}^2 \int_0^\infty 4\pi r^2 \rho(r) \frac{\sin Qr}{Qr} \, dr \qquad [3.24]$$

where $\overline{b}$ is the neutron 'scattering length' averaged over all isotopes of the particular element concerned and N is the number of atoms. Defining

$$i_N(Q) = I_N^T(Q) - N\overline{b^2} - I_N^0(Q) \qquad [3.25]$$

where $I_N^0(Q)$ is the experimentally inaccessible contribution to the scattering at $Q=0$ from the average density (cf. the last term of [3.13]), then:

$$Q i_N(Q) = N\overline{b}^2 \int_0^\infty G(r) \sin Qr \, dr \qquad [3.26]$$

This equation is entirely equivalent to that derived for coherent X-ray scattering ([3.14]) if the following equivalents in terminology between neutron and X-ray scattering are made, $\mathbf{Q} \equiv \mathbf{k}$, $Q\,i(Q) \equiv F(k)$; the scattering length b is analogous to the atomic form factor $f(k)$. In practice deviations from the static approximation lead to [3.24] being no longer obeyed, and the so-called Placzek corrections have to be applied to the first term in [3.24] (see Wright 1974).

The static approximation is avoided by detecting the *elastically* scattered neutrons by incorporating a second monochromator, i.e. using a *triple* axis spectrometer which passes only those neutrons having the same energy as the incident beam (exactly as shown in Fig. 3.8). In this case the scattering function which is relevant is $S(\mathbf{Q}, 0)$ which is related to $G(r, \infty)$, or the time-averaged correlation function when the averaging time is much longer than the period of an atomic vibration. The elastically scattered intensity can be written (Leadbetter and Wright 1972) as:

$$I_N^{el}(Q) = N\overline{b^2} e^{-2W} + N\overline{b}^2 \int_0^\infty 4\pi r^2 \rho^{eq}(r) e^{-2W} \frac{\sin Qr}{Qr} \, dr \qquad [3.27]$$

where e^{-2W} is the so-called 'Debye–Waller' factor, and W is a function of the mean square atomic displacement $\overline{u^2}$ due solely to thermal vibrations:

$$W = \tfrac{1}{6} Q^2 \overline{u^2} \qquad [3.28]$$

By comparison, the expression for the total diffracted intensity can also be rewritten in terms of a separation of atomic positions into equilibrium and time-dependent thermal displacement terms:

$$I_N^T(Q) = N\overline{b^2} + N\overline{b}^2 \int_0^\infty 4\pi r^2 \rho^{eq}(r) e^{-2W} \frac{\sin Qr}{Qr} dr \quad [3.29]$$

A comparison of the two sets of data for a given material therefore yields an estimate for atomic displacements. Note, however, that only atomic displacements of purely thermal origin are obtained in this way; any static distortions in atomic positions will not be detected, since for this case both elastic and total diffracted intensities take the form of [3.29].

Both experimental arrangements discussed so far, namely double and triple axis spectrometers, use a fixed wavelength incident beam and vary 2θ in order to trace out the Q (or k) dependence of the scattering. An alternative approach is to use the so-called 'time-of-flight' method in which white, unmonochromatized radiation containing all wavelengths is incident upon the sample, and the scattered intensity is monitored at fixed scattering angles. In practice, the incident neutron beam is pulsed, either by employing a mechanical chopper with a steady-state reactor or by using a pulsed neutron source such as that produced by electrons from an electron linear accelerator impinging on to a uranium target ('spallation' source). The experimental configuration is shown schematically in Fig. 3.9; the two flight paths l_0 and l_1, are of the order of metres in length with $l_1 \ll l_0$. The elastic momentum transfer is given by the expression:

$$Q_{el} = \left[\frac{2m_n(l_0 + l_1)}{\hbar}\right] \frac{\sin \theta}{\tau} \quad [3.30]$$

where τ is the flight time and m_n is the neutron mass. The technique offers the advantages over conventional fixed λ, variable 2θ experiments of much higher values of Q_{max} (values of ~ 60 Å^{-1} are attainable) resulting in an increase in resolution, higher count rates (at least for pulsed sources) and since all values of Q are measured simultaneously, the possibility of performing time-dependent experiments, e.g. monitoring kinetic changes. Similar experiments are possible for X-rays using the broad-band radiation emitted by synchrotrons. However, the problem of the

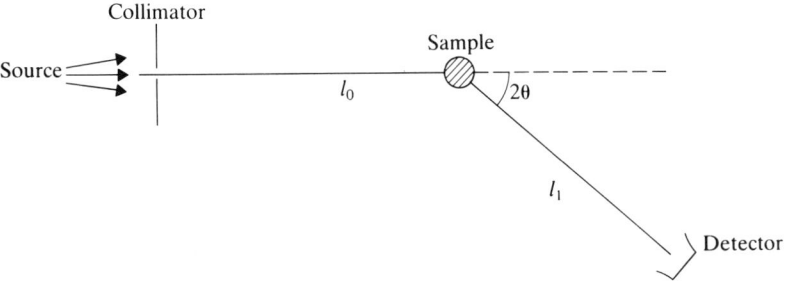

Fig. 3.9 Schematic illustration of the geometry employed in time-of-flight diffraction experiments.

corrections needed to eliminate the Compton scattering at high k-values make the technique potentially less attractive than for neutrons.

A very important feature of neutron scattering is the behaviour of the scattering length, $\bar{b}$. This is independent of Q (or k) for closed shells, in marked contrast to the equivalent entity in X-ray scattering, the atomic form factor $f(k)$, which is strongly dependent on k. In addition, $\bar{b}$ varies in a random fashion between the different elements, whereas $f(k)$ varies in a systematic way from element to element since $f(0)$ is equal to Z, the atomic number. Furthermore, $\bar{b}$ is different for different isotopes of the *same* element, again varying in a random fashion, although $f(k)$ of course is the same for all isotopes, being just dependent on the electronic structure of the atom. These features will prove to be of importance in the discussion of multicomponent systems later.

The final type of scattering probe that has been used in the study of the structure of amorphous materials is the electron; a review of the subject has been given by Dove (1973). The experiment is performed in the electron microscope with typically 50–100 keV electrons, which correspond to wavelengths of the order of 0.05 Å; again λ is kept constant and the scattering angle varied by deflecting the electrons with a scanning coil. The electrons are scattered both by the electrons and the nuclei of the atoms in a sample, and as such the technique formally lies between X-ray and neutron scattering. The form factor for electron scattering is given by

$$f^e(k) = \frac{2m_e e^2}{\hbar^2} [(Z - f^x(k))/k^2]. \qquad [3.31]$$

Because of the coulombic interaction the scattering is much stronger than for X-rays, and multiple scattering becomes important for film thicknesses beyond a few hundreds of ångströms. In addition, there is a large background due to inelastic loss processes such as plasmon formation, etc. The necessity of using very thin films raises the doubt, furthermore, of whether the structure of such thin films is truly representative of the structure of bulk specimens.

The advantages and disadvantages of each of the three scattering techniques are summarized in Table 3.1.

Thus far we have confined our discussion exclusively to monatomic systems, and the equations that have been derived or stated are valid for that case alone. Far more common, of course, are multicomponent alloy systems for which matters are somewhat more complicated. Take, for example, the simplest case, that of a binary alloy AB. In this case, a single real-space correlation function (say the RDF) is insufficient to describe fully the structure; instead *three* partial functions are required, dealing respectively with A–A, A–B and B–B type pair correlations. Thus, three different diffraction experiments are required in order to extract each pair function unambiguously. A combination of X-ray, neutron and electron scattering studies on the same sample, or isotopically substituted samples in conjunction with neutron diffraction are possible ways of achieving the desired goal of three scattering experiments each with different scattering factors ($f(k)$ or b) in order that the partial correlation functions may be extracted. While this may be possible for a binary alloy it is not feasible in general, since for an n-component system $n(n+1)/2$ partial functions are required. For these complicated systems a chemically specific structural probe (e.g. EXAFS – see sect. 3.2.3) is required. A simple diffraction experiment obviously yields a gross average, both spatially and chemically, and although the nearest-neighbour bond length may be well defined, it is not always possible to obtain unambiguously other structural parameters (even including the nearest-neighbour coordination number) from a Fourier transform of the intensity.

For a polyatomic system, the equation previously derived for X-ray scattering from a monatomic material can be generalized to give:

$$\frac{I_{eu}}{N} = \sum_{i=1}^{n} x_i f_i^2 + \sum_{i=1}^{n} \sum_{j=1}^{n} x_i f_i f_j \int_0^\infty 4\pi r^2 \rho_{ij}(r) \frac{\sin kr}{kr} dr$$

$$- \left[\sum_{i=1}^{n} x_i f_i\right]^2 \int_0^\infty 4\pi r^2 \rho^0 \frac{\sin kr}{kr} dr \quad [3.32]$$

where x_i is the atomic fraction of element i, f_i is the scattering factor of element i and $\rho_{ij}(r)$ is the average number of j atoms per unit volume at distance r from any i atom. Defining the following functions $\langle f \rangle = \sum_{i=1}^{n} x_i f_i$, $\langle f^2 \rangle = \sum_{i=1}^{n} x_i f_i^2$, and

$$\rho(r) = \sum_{i,j} W_{ij} \rho_{ij}(r) \quad [3.33]$$

where the weighting factors are given by

$$W_{ij} = x_i f_i f_j / \langle f \rangle^2 \quad [3.34]$$

leads to a reformulation of [3.32]

$$\frac{I_{eu}}{N} = \langle f^2 \rangle + \langle f \rangle^2 \int_0^\infty 4\pi r^2 [\rho(r) - \rho^0] \frac{\sin kr}{kr} dr \quad [3.35]$$

The reduced scattering intensity is now:

$$F(k) = k \left[\frac{I_{eu}/N - \langle f^2 \rangle}{\langle f \rangle^2}\right] = \int_0^\infty 4\pi r [\rho(r) - \rho^0] \sin kr \, dr \quad [3.36]$$

This expression is superficially similar to that for monatomic systems ([3.15] and [3.16]) except that weighted values for the scattering factors are used instead, and that $\rho(r)$ is a function of the scattering factors through the term in W_{ij}.

The equation for the reduced scattering intensity can only be inverted simply and straightforwardly without approximation to give real-space information if W_{ij}, and hence f_i and f_j, are *independent of k* or have the same k-dependence. This is the case for neutrons ($f_i = b_i$) but not in general for X-rays ($f_i = f_i(k)$); thus for neutrons, the reduced RDF is given by the same equation as derived previously ([3.17b]), but with $\rho(r) = \sum_{i,j} W_{ij} \rho_{ij}(r)$. Furthermore, since different isotopes of the same elements often have markedly different scattering lengths, it is feasible in principle to design materials which have particular combinations of scattering length and concentrations such that the partial correlation functions may be extracted. A suitable choice for this is the Ni–Ti system, for which the natural abundance values of the scattering lengths are $b_{Ni} = +1.03 \times 10^{-12}$ cm and $b_{Ti} = -0.33 \times 10^{-12}$ cm (see Table 7.2). For example, if by suitable admixture of isotopes of a particular constituent of a binary alloy it is possible to make the net coherent scattering length

Table 3.1 Advantages and disadvantages of various scattering techniques

	Advantages	Disadvantages
X-rays (fixed λ, variable θ)	Convenient laboratory sources and equipment. Powdered or thick film samples can be used. Atomic form factor can be calculated.	Small k_{max} for many X-ray sources. Significant Compton contribution to scattering at large k. Partial correlation functions difficult to extract for multicomponent systems. Atomic form factor strongly decreasing function of k.
Neutrons (fixed λ, variable θ, double or triple axis)	Scattering length independent of k. Partial correlation functions easier to extract for multicomponent systems. Isotopic substitution possible to determine partial correlation functions.	Reactor source required. Placzek corrections required for double-axis experiments (static approximation). Large volume of material required. b must be determined experimentally for each isotope. Low count rates.
(fixed θ, variable λ, time of flight)	(In addition to the above) Large values of k_{max} attainable. Fixed geometry simplifies experimental set-up, e.g. for pressure studies. Simultaneous collection of data for all k-values facilitates time-resolved experiments. Higher count rates (if LINAC used).	Pulsed source required (chopper + steady-state reactor or LINAC source).
Electrons (fixed λ, variable θ)	Uses scanning electron microscope. In situ grown thin films can be examined. High count rates.	Scattering very strong and multiple scattering important for thicknesses greater than 100 Å. Thin films may not be representative of the bulk structure. Large inelastic scattering background (due to plasmons, etc.).

b_2 of element 2 zero, then the partial interference function $I_{11}(k)$ is obtained immediately in a scattering experiment, and so forth (see problem 3.1 and sect. 7.2.1).

We conclude this section on scattering techniques and their application to structural determination of amorphous materials with a collection of structural parameters obtained experimentally in this way for a variety of amorphous solids. We have seen already that scattering experiments only yield *local* structural information unambiguously, and this is limited to nearest-neighbour and next-nearest-neighbour separations r_1 and r_2, together with their respective coordination numbers. For the case of those materials with directed (i.e. covalent) bonding, knowledge of r_1 and r_2 enables the bond angle to be calculated ([3.18]); this obviously is not appropriate for those materials which have predominantly non-directional bonding such as is the case for metallic and ionic glasses. We therefore give, primarily for reference purposes, in Table 3.2 experimental structural data (r_1, r_2, θ and $\rho°$) for a selection of covalently bonded amorphous materials. The structure of metallic glasses forms the subject of section 7.2.

3.2.2 Anomalous scattering

It was pointed out in the previous section that even for the simplest multicomponent system, a binary alloy AB, a single diffraction experiment is not sufficient to yield all the structural information about the system, namely the three distribution functions $\rho_{AA}(r)$, $\rho_{BB}(r)$ and $\rho_{AB}(r)$. To achieve this goal requires three separate scattering experiments in each of which the scattering factors are varied in some manner. For the case of neutron diffraction this can be effected by means of isotopic substitution as discussed in the previous section. This avenue is not available, however, for X-ray scattering.

Nevertheless, a possibility for systematically varying the atomic scattering factors in X-ray diffraction does exist in the phenomenon of anomalous scattering (or dispersion). In general, the atomic scattering factor is a function of both k, the change in scattering wavevector and ω, the frequency of the X-rays:

$$f(k,\omega) = f°(k) + f'(k,\omega) + if''(k,\omega) \qquad [3.37]$$

The first term in this equation, $f°(k)$, is the Fourier transform of the electronic charge density ([3.8]) and is independent of ω; this is the quantity we have used previously. The other factors, $f'(k,\omega)$ and $f''(k,\omega)$, are the so-called 'dispersion corrections' and are only significant when ω lies very close to an absorption threshold of the atom. Their form is shown for the case of Ge in Fig. 3.10 as a function of energy (or ω); the imaginary part of the scattering factor, $f''(k,\omega)$, is related to the absorption coefficient $\alpha(\omega)$ by

$$f''(\omega) = \frac{\omega \alpha(\omega)}{4\pi c} \qquad [3.38]$$

where c is the speed of light, and f' and f'' are related by a Kramers–Krönig transformation

$$f'(\omega_0) = \frac{2}{\pi} \int_0^\infty \frac{\omega f''(\omega)\, d\omega}{(\omega^2 - \omega_0^2)} \qquad [3.39]$$

In principle, therefore, it should be possible to extract partial correlation functions for, say, a binary alloy, by measuring the scattering intensity for three X-ray wavelengths including the regions near an absorption edge such that f', f'' and hence f vary widely between each experiment. The weighting factors $W_{ij}(k)$ can then be written as

$$W_{ij}(k) = x_i \frac{[(f_i° + f_i')(f_j° + f_j') + f_i'' f_j'']}{\langle f \rangle^2} \qquad [3.40]$$

where $\langle f \rangle^2 = [\sum x_i(f_i° + f_i')]^2 + [\sum x_i f_i'']^2$.

Table 3.2 Structural parameters for some amorphous materials

Material	Preparation method	r_1 (Å)	r_2 (Å)	θ	ρ (g cm^{-3})	Radiation type	Comments	Reference
Elements								
C	TD	1.425	2.45	119	0.923	n	Prepared from polyfurfuryl alcohol resin	1
P	TD	2.18	3.48	106	2.16	X	Prepared from white P	2
As	VT	2.49	3.76	98	4.7–5.1	X	Hydrogen transport	3
Se	LQ	2.33	3.70	105	4.25	X	—	4
Si	E	2.35	3.86	109	2.10	e		5
Ge	E	2.463	3.997	109	4.79	n	Electron beam evaporated	6
Tetrahedral materials								
SiO$_2$	TD	1.61	2.62/3.13	109(Si)/152(O)	2.20	n	Spectrosil	7
GeO$_2$	LQ	1.74	3.18	109(Si)/132(O)	3.65	X		8
BeF$_2$	LQ	1.54	2.56	109(Be)/156(F)	2.00	n		9
ZnCl$_2$	LQ	2.29	3.72	109(Zn)/110(Cl)	2.71	n		10
Chalcogenide materials								
As$_2$S$_3$	LQ	2.28	3.48	102(As)/99.5(S)	3.17	X		11
As$_2$Se$_3$	LQ	2.43	3.70	100(As)/95(Se)	4.56	X		11
As$_2$Te$_3$	LQ	2.65	4.00	98(As)/95–98(Te)	5.53	X		11
Sb$_2$S$_3$	LQ	2.50	3.85	90(Sb)/100.7(S)	4.21	X		11
Sb$_2$O$_3$	LQ	1.99	2.80	92(Sb)	5.07	X	5% B$_2$O$_3$ added to stop crystallization	12
B$_2$O$_3$	LQ	1.37	2.38	120(B)/130(O)	1.84	n		13

Key: TD = thermal decomposition; VT = vapour transport; LQ = liquid quenched; E = evaporated; X = X-rays; n = neutrons; e = electrons.
References: (1) D. F. R. Mildner and J. M. Carpenter (1982), *J. Non-Cryst. Sol.* **47**, 391; (2) H. Krebs and H. U. Grüber (1967), *Z. Naturf.* **22a**, 96; (3) H. Krebs and R. Steffen (1964), *Z. anorg. allg. Chem.* **327**, 224; (4) P. Andonov (1982), *J. Non-Cryst. Sol.* **47**, 297; (5) S. C. Moss and J. F. Graczyk (1969), *Phys. Rev. Lett.* **23**, 1167; (6) G. Etherington, A. C. Wright, J. T. Wenzel, J. C. Dore, J. H. Clarke and R. N. Sinclair (1982), *J. Non-Cryst. Sol.* **48**, 265; (7) A. C. Wright, G. Etherington, J. A. Erwin Desa, R. N. Sinclair, G. A. N. Connell and J. C. Mikkelsen (1982), *J. Non-Cryst. Sol.* **49**, 63; (8) A. J. Leadbetter and A. C. Wright (1972), *J. Non-Cryst. Sol.* **7**, 37; (9) A. J. Leadbetter and A. C. Wright (1972), *J. Non-Cryst. Sol.* **7**, 156; (10) J. A. Erwin Desa, A. C. Wright, J. Wong and R. N. Sinclair (1982), *J. Non-Cryst. Sol.* **51**, 57; (11) L. Cervinka and A. Hruby (1982), *J. Non-Cryst. Sol.* **48**, 231; (12) H. Hasegawa, M. Sone and M. Imaoka (1978), *J. Phys. Chem. Glasses* **19**, 28; (13) P. A. V. Johnson, A. C. Wright and R. N. Sinclair (1982), *J. Non-Cryst. Sol.* **50**, 281.

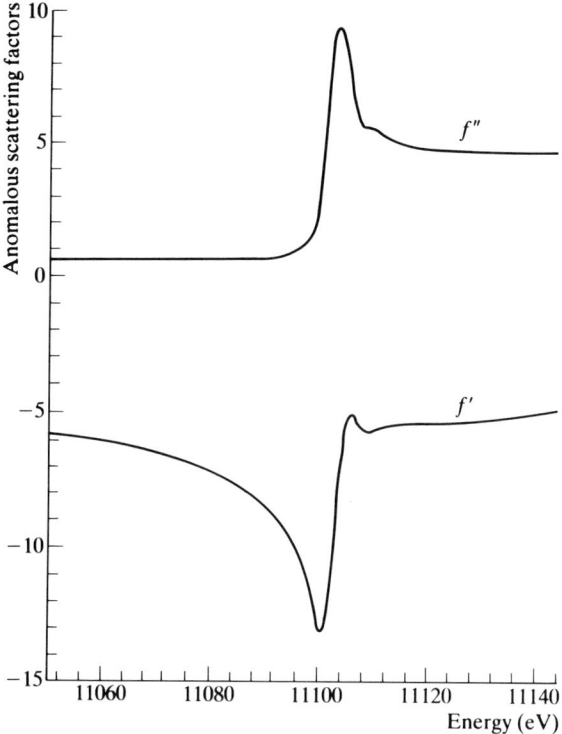

Fig. 3.10 The real and imaginary parts of the scattering factor $f = f' + if''$ for Ge (Fuoss 1980).

In practice, however, the changes in f' and f'' are perhaps only several tens of per cent of f° even in the most favourable situation, giving rise to large errors in the derived partial correlation functions. A more promising approach is to make use of *differential* anomalous scattering, since the changes in the *derivatives* of f' and f'' with energy are much larger than is the case for f' or f'' alone (Shevchik 1977). The derivative of the scattering intensity for a binary alloy AB with respect to photon energy near the absorption edge of atom A is:

$$\left[\frac{\delta I(k)}{\delta E}\right]_k = x_A \frac{\delta}{\delta E}[f_A(k)f_B^*(k) + f_A^*(k)f_B(k)]F_{AB}(k) \qquad [3.41]$$

where $x_A F_{AB} = x_B F_{BA}$ has been used, and F_{AB} is related to the partial real-space correlations function by:

$$F_{AB}(k) = \int_0^\infty 4\pi r[\rho_{AB}(r) - \rho_B^\circ]\sin kr\,dr \qquad [3.42]$$

Thus Fourier transformation of the intensity derivative ([3.41]) gives information about the environment about atom A (i.e. $\rho_{AB}(r)$), and in the general case a weighted sum of the partial functions $\rho_{A\beta}(r)$ over all atom types β other than A in the multicomponent system. In practice $I(k)$ is measured for several wavelengths in the region of each edge, and the difference taken between various wavelengths.

The technique obviously depends for success on having available a wide range of X-ray wavelengths to encompass all absorption edges of interest. For this purpose, synchrotron radiation sources which produce broad-band, high-intensity radiation are well suited. The advantages of the (differential) anomalous scattering technique in extracting structural information for multicomponent systems lie in the simplicity of the approach; once an appropriate radiation source is available no special procedures in sample preparation are needed (such as isotopic or isomorphous substitution) and large volumes of sample are not required as for neutron diffraction. Furthermore, data can be measured over a range of k values extending to $k=0$; contrast this with the case for EXAFS (see sect. 3.2.3) in which data at low values of k are not easily interpretable using simple theories, and thus structural information at large separations cannot readily be ascertained. The disadvantages lie in the fact that the dispersion corrections, f' and f'', must be known with reasonable accuracy if appreciable errors in the final partial RDFs are not to be incurred; errors are inherent in the differential technique in any case since to form the differential two quantities, comparable in size, are subtracted from each other. A further drawback is that the maximum value of k achievable is determined by the lowest energy absorption edge used, and this automatically limits the spatial resolution in the transform of the diffracted intensity because of the small k_{max}. In addition, an experimental complication is the large fluorescent background for photon energies near an absorption edge, which can be 10–100 times the desired coherent intensity, and which must be subtracted before the data can be manipulated. Nevertheless, like EXAFS to be discussed next, differential anomalous scattering is chemically specific in that with a tunable radiation source, absorption edges for different elements in the sample can be selected and hence the structure about that atom investigated. (Note in passing that anomalous dispersion can also occur in neutron scattering, where the nucleus (e.g. Sm) undergoes a nuclear reaction on absorption of a neutron; this possibility is limited, however, to the very few nuclei which react with low-energy thermal neutrons, and therefore does not have the general applicability of anomalous X-ray scattering.)

An example of the application of the technique is for amorphous GeSe (Fuoss et al. 1981). There has been a considerable controversy over the coordination in this material; both 3:3 coordination, characteristic of *crystalline* GeSe (which need not involve 'wrong' or homopolar bonds), and 4:2 coordination as found in amorphous or crystalline GeSe$_2$ (i.e. silica-like, and which must involve Ge—Ge bonds to satisfy stoichiometry in GeSe) have been proposed. The scattering intensity was measured at several photon energies within 100 eV of the Ge K-edge (~ 11.1 keV) and the Se K-edge (~ 12.65 keV), and by means of the differential method discussed previously, the distribution of atoms around Ge and Se, respectively were determined. These are shown in Fig. 3.11(a), from which it can be seen that the areas of the first peaks are nearly identical (3.1 atoms for Ge and 2.5 atoms for Se) and the second peaks are coincident in position, lending support to the 3:3 model, for which it is probable that the bond angles would be similar. This can be contrasted with similar measurements taken on a-GeSe$_2$ (Fig. 3.11(b)), which yield very different coordinations of nearest neighbours (3.8 for Ge and 2.3 for Se) and the bond angle about the Ge atom is larger than about the Se atom, consistent with the notion of sp^3 hybridization of the fourfold coordinated Ge ($\theta \simeq 109°$) and p-bonding of the twofold coordinated Se ($\theta \simeq 90°$).

Experimental techniques and short-range order

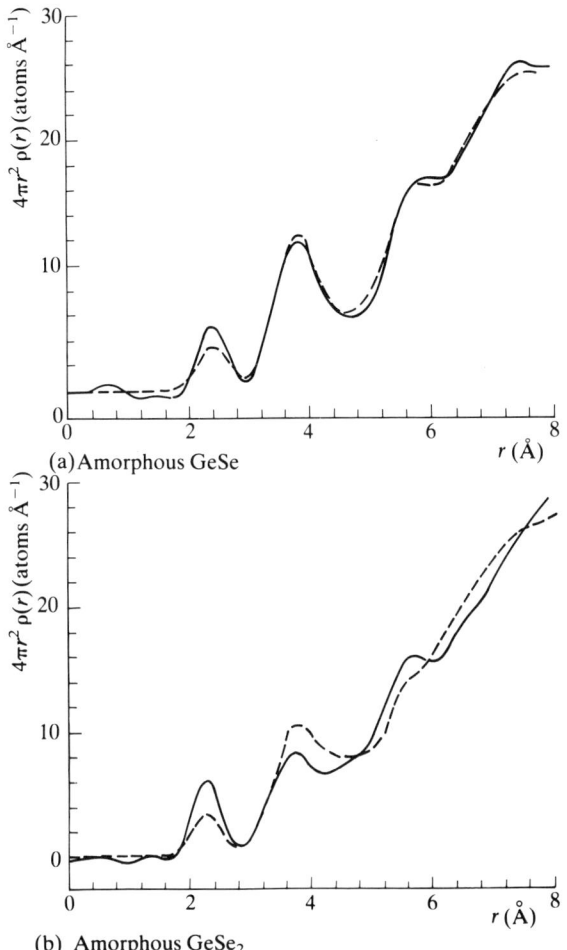

Fig. 3.11 Partial radial distribution functions for (a) a-GeSe and (b) a-GeSe$_2$ deduced from anomalous scattering experiments. The solid line refers to data taken about the Ge edge and the dashed line refers to data taken about the Se edge (Fuoss *et al.* 1981).

3.2.3 Extended X-ray absorption fine structure

The acronym for 'Extended X-ray Absorption Fine-Structure' is EXAFS and like anomalous scattering it is a chemically specific structural probe. It differs from the methods cited in sections 3.2.1 and 3.2.2 in that it is not a direct scattering technique. The X-ray absorption coefficient of atoms is generally a decreasing function of energy, except at certain discrete energies at which there are discontinuous increases in absorption, i.e. absorption edges. Consider what happens when a photon of energy slightly higher than threshold is absorbed by an atom, and a photo-electron is ejected. The photo-electron wave propagates outwards to infinity if the absorbing atom is isolated (e.g. in a rare gas); in this case the absorption coefficient decreases smoothly from the absorption edge. However, if other atoms surround the absorbing atom, as in a gaseous molecule or in a condensed phase, the outgoing

77

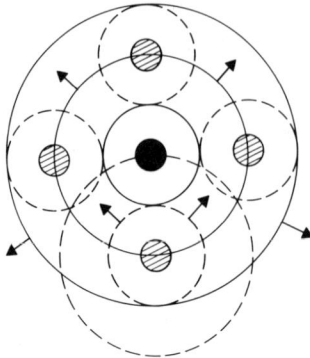

Fig. 3.12 Schematic illustration of the EXAFS process. An atom (solid circle) absorbs an X-ray creating an outgoing photo-electron wave which back-scatters off neighbouring atoms (hatched circles). The solid circles represent the outgoing electron wave and the dashed circles represent the back-scattered electron waves; interference can take place where these overlap.

photo-electron wave will be back-scattered, and the back-scattered waves will interfere with the outgoing waves (Fig. 3.12). This in turn influences the matrix elements for the absorption process itself. Thus, as the X-ray photon energy is increased above threshold, the energy of the photo-electron also increases and consequently the electron wavelength decreases, and constructive (destructive) interference occurs when the interatomic spacing is an integral (half-integral) number of wavelengths. Thus one expects to observe a periodic oscillation of the X-ray absorption above the absorption edge – the so-called 'fine structure'. This is indeed observed, and an example is shown in Fig. 3.13 for the case of a-As_2Se_3, where fine structure above both the As and Se edges is observed.

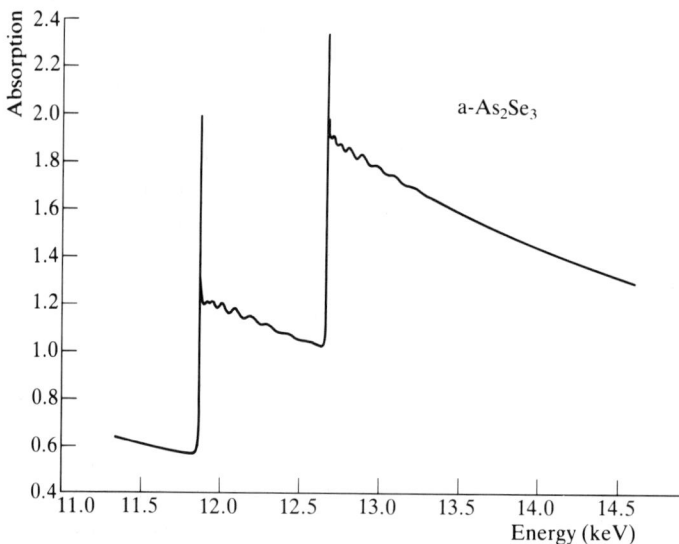

Fig. 3.13 Extended X-ray absorption fine structure for a-As_2Se_3 showing the fine structure beyond the As and Se K edges (R. Pettifer – private communication).

The photo-electric absorption process can be treated semi-classically by regarding the photon as a classical electromagnetic field but treating the electron quantum-mechanically. In the dipole approximation, where the wavelength of the photon is large compared with the spatial extent of the excited core state, the absorption can be treated to first order in perturbation theory and Fermi's golden rule then yields for the absorption coefficient:

$$\mu_X = 4N_0\pi^2 e^2 \frac{\omega}{c} |\langle f|z|i\rangle|^2 \rho(E^F) \qquad [3.43]$$

where $|i\rangle$ and $|f\rangle$ are the initial core state and final photo-electron wavefunction, respectively, ω is the X-ray frequency, $\rho(E^F)$ is the density of final states and N_o is the number of atoms of one type in the sample. Since in the dipole approximation the photon field can be regarded as being spatially uniform, it can be approximated by a scalar potential proportional to the distance z if the X-ray polarization is in the z-direction. Hence, only the matrix element or the density of states in [3.43] could give rise to oscillatory behaviour of μ_X. For photon energies well above threshold, $\rho(E^F)$ varies monotonically since it is well described by the free electron value. Therefore, the matrix element alone must be responsible for the oscillatory behaviour, and this is because the final state wavefunction $|f\rangle$ is made up of contributions from both the outgoing wave and the back-scattered wave, and interference between the two modulates the matrix element and hence μ_X (Stern 1978).

EXAFS is conventionally measured using a transmission geometry in which the incident (I_o) and transmitted (I) X-ray intensities passing through a thin foil of thickness d are measured as a function of photon energy; they are related to the absorption coefficient μ_X by $I = I_o \exp(-\mu_X d)$. Two sources of broad-band X-rays may be used. Either the *brehmstrahlung* spectrum from conventional X-ray tubes or the radiation emitted by electron synchrotrons may be used, synchrotron radiation offering the advantage of a factor of $\sim 10^4$ increase in photon flux and concomitant reduction in counting time; in both cases the beam is monochromatized by a crystal. For dilute samples, the desired EXAFS is swamped by background intensity in a transmission mode, and the more sensitive technique of fluorescence detection must be employed. In this, the K-shell hole left by the photo-electron is filled by a p-electron from the L-shell, emitting an X-ray photon of characteristic energy less than that of the exciting X-rays. Thus by tuning to the fluorescent wavelength, only those atoms which are excited are monitored, with a consequent dramatic increase in sensitivity.

The magnitude of the EXAFS signal is obtained from the measured value of $\mu_X(E)$ by subtracting the background μ_o and normalizing:

$$\chi(E) = \frac{\mu_X(E) - \mu_o}{\mu_o} \qquad [3.44]$$

E is the photo-electron energy which, however, is not known exactly since the threshold energy (i.e. the zero of energy E_0) cannot be positioned precisely; instead it is left as an adjustable parameter. The background absorption μ_o is also difficult to measure, and it is usually fitted by means of a polynomial.

The formula describing the EXAFS can be calculated on the basis of the model above and in atomic units is (Sayers et al. 1971):

$$\chi(k) = -\sum_j \frac{N_j}{R_j^2} \frac{|f_j(\pi)|}{k} \exp\left(-\frac{2R_j}{\lambda_e}\right) \exp(-2\sigma_j^2 k^2) \sin(2kR_j + 2\delta(k) + \eta_j(k))$$
[3.45]

where a sum over all shells of atoms, j, is taken, each containing N_j atoms at a distance R_j from the absorbing atom. The magnitude of the EXAFS is proportional to N_j, inversely proportional to R_j^2 (since both the outgoing and back-scattered waves are assumed to be spherical, decreasing in amplitude as $1/R$) and proportional to the back-scattering amplitude $|f_j(\pi)|$ from the atoms in the jth shell. The amplitude is attenuated because of the finite mean free path λ_e of the electrons in the material and by the Debye–Waller term involving r.m.s. displacements σ_j (static or thermal) about the equilibrium position. Finally, the amplitude is sinusoidally modulated by a function involving the phase shift of the electron; the additional phase shifts $\delta(k)$ and $\eta(k)$ arise because the photo-electron is emitted and back-scattered, respectively from atomic potentials. The wavenumber of the photo-electron, k, is given by $k = 2\pi/\lambda = [2m_e(E-E_0)]^{1/2}/\hbar$; note that this is *half* the *scattering* wavevector Q or k ([3.6]). Thus, from [3.45] we expect the EXAFS to be a complicated oscillatory function, each shell contributing a sinusoid of a different period which mix together; note, however, that it is only the first few shells which contribute strongly because of the limited mean free path of the photo-electrons, and the large widths, σ, of higher-lying shells.

It is important to emphasize at this juncture that the simplicity of [3.45] is the result of several approximations. Firstly, multiple scattering is neglected, i.e. it is assumed that the electrons are back-scattered only once by a particular shell before interfering at the origin atom, and are not scattered by intervening atoms. Furthermore, it is assumed that the electron waves can be regarded as planar at the scattering shells. These two approximations are valid for energies above ~ 80 eV from the edge. In the energy range $40 < E < 80$ eV from the edge, the curvature of the waves cannot be neglected and the simple [3.45] is no longer appropriate and the full curved wave formula due to Lee and Pendry (1975) must be used (which is still a single scattering theory). For energies close to the edge, multiple scattering is very important and EXAFS theory is inappropriate. The phenomenon has been given the name XANES (X-ray absorption near-edge structure) and a theory based on low-energy electron diffraction (LEED) has been given by Durham et al. 1981; since scattering by atoms between the back-scattering shell and the origin atom is explicitly considered, *orientational* information (i.e. three-body correlations) in principle is obtainable, in contrast to simple EXAFS which essentially affords pair correlations.

The importance of EXAFS for the study of the structure of amorphous materials lies in the fact that the theories describing it are equally valid for ordered or disordered structures, and perhaps more importantly that the local structure around a given atom may be ascertained simply by measuring the absorption near the edge of that particular atom. If the absorption edges of the various elements in a multicomponent material are sufficiently separated in energy (say by a separation of ~ 1 keV) that the EXAFS beyond one edge does not run into the next edge (see e.g. Fig. 3.13), the local structure around *each* atom type may be determined. This makes possible the detailed study of complex systems (say common mixed alkali silicate glasses), a circumstance completely impracticable using conventional scattering techniques. Furthermore, the substantial increases in sensitivity that detection of EXAFS by X-ray fluorescence affords means that the structure around atoms in low concentrations (say less than 1 at. %) can be studied.

The structural information available from EXAFS experiments can be seen by reference to [3.45]. In principle, the coordination number, N_j, the interatomic spacing, R_j, and the mean square deviation, σ_j^2, for each shell, j, of atoms surrounding the absorbing atom is obtainable; in practice, only the first and second neighbour shells are significant. Two methods of extracting the structural information are commonly employed. One approach is to treat all the structural parameters in [3.45] as adjustable variables and to vary these (perhaps with crystalline values as input parameters) until the best fit between the calculated EXAFS spectrum and the experimental data is achieved. The major difficulty in this, as in most analyses of EXAFS, is what values to take for the phase shifts δ and η. There are two possible remedies to this problem: either values may be calculated theoretically by solving Schrödinger's equation taking account of the perturbations to the potential caused by neighbouring atoms, or else comparison is made with a standard (usually a crystal) for which the structure is known and by fitting parameters to the EXAFS data of the standard, values of δ and η are thereby obtained.

The second method of analysis of EXAFS data is to take the Fourier transform:

$$\phi(r) = \frac{1}{(2\pi)^{1/2}} \int_{k_{min}}^{k_{max}} \chi(k) M(k) k^n \exp(2ikr)\, dk \qquad [3.46]$$

where $M(k)$ is a window function, and n can be 1, 2 or 3, the larger values weighting the data more at high k-values. This procedure gives rise to a series of peaks in real space corresponding to the first, second and perhaps higher shells (see Fig. 3.14): since EXAFS retains phase information, in contrast to X-ray diffraction, the Fourier transform of $\chi(k)$ consists of positive and negative excursions (Fig. 3.14(a)). The peaks do *not* lie at the correct distances if, as in [3.46], the Fourier transform is taken with respect to the argument $2ikr$ only, ignoring the phase-shift term

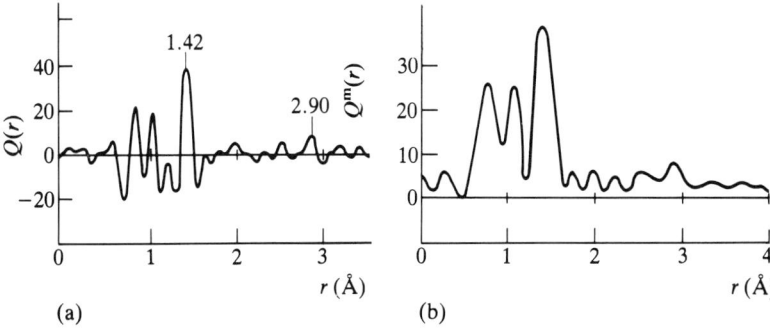

Fig. 3.14 The real part (Q) and the magnitude (Q^m) of the Fourier transform of the Ge edge EXAFS for amorphous GeO$_2$ transformed over the range 2.2–17.5 Å^{-1} (Sayers et al. 1975).

$2\delta(k) + \eta_j(k) = \alpha_j(k)$. This is because the k-dependence of $\alpha_j(k)$ also contributes to the frequency of oscillation of $\chi(k)$; if $\alpha_j(k)$ is expanded in powers of k as $\alpha_j(k) = \alpha_o + (\partial\alpha/\partial k)k$, then the frequency of $\chi(k)$ would be proportional to $R_j + \partial\alpha/\partial k$ and conversely the positions of the peaks in the simple Fourier transform, [3.46], would appear at distances $R_j - \partial\alpha/\partial k$. Both techniques depend for success on the assumption of transferability of phase shifts and back-scattering amplitudes; i.e. δ, η and f are invariant between systems which are chemically similar. Although interatomic spacings obtained in this fashion can be very accurate (± 0.02 Å), the values of the coordination number are not so precise and can be $\sim 50\%$ low due to changes in the emitter potential on ejection of the photo-electron, but the errors can be corrected for in comparative studies employing standards.

Information about relative displacements σ_j of atoms is relatively straightforwardly obtained from EXAFS measurements of the unknown (amorphous) and standard (crystalline) samples. If transferability of phase shifts and back-scattering amplitudes is assumed, and it is further assumed that the atomic separations R_j are the same (generally valid at least for the first two shells of covalently bonded solids), then from [3.45] the ratio of the EXAFS for the *same* shell for amorphous and crystalline samples is:

$$\ln\left[\frac{\chi_j(A)}{\chi_j(C)}\right] = \ln\left[\frac{N_j(A)}{N_j(C)}\right] - 2[\sigma_j^2(A) - \sigma_j^2(C)]k^2 \qquad [3.47]$$

In practice, the total measured EXAFS of both unknown and standard is Fourier transformed into real space, the appropriate peak (first or second) is selected and back-transformed into reciprocal space. If this peak can be represented by a simple Gaussian, then the resulting back-transform is a simple damped sinusoid for both unknown and standard, having the same period if R_j, η and δ are the same, but different damping factors. The ratio of these two curves taken at each k value and plotted versus k^2 then yields as slope the difference of the relative displacements ([3.47]). The result of this analysis for EXAFS above the K edge of Ge in GeO_2 is shown in Fig. 3.15, whence it can be seen that the curve for the first shell has zero slope and intercept implying, from [3.47], that σ^2 is the same for both amorphous and crystalline forms (and is thermal in origin), and that each material has the same

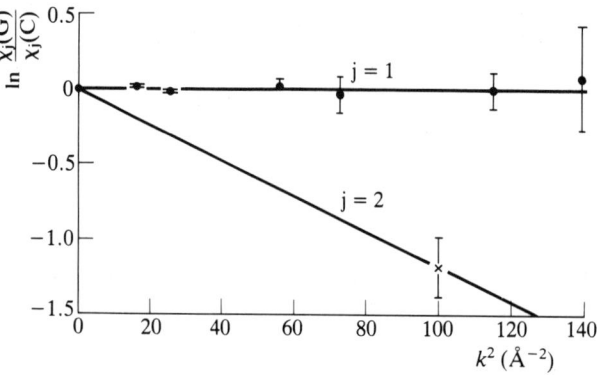

Fig. 3.15 Plot of the logarithm of the ratio of the EXAFS for a given shell (j) versus k^2 for glassy and crystalline GeO_2 (Sayers et al. 1975).

coordination number. Although the signal-to-noise ratio was significantly worse for the second peak, a similar analysis gives a line of non-zero slope; the larger value of σ^2 for a-GeO_2 was found to be due to a static distortion of the bond angle of $\pm 6.5°$ about the (crystalline) Ge—O—Ge bond angle of $130°$.

The advantages of EXAFS lie principally in the fact that it is a chemically specific, local-structure probe which is very suitable for the study of complex multicomponent systems, such as comprise many amorphous materials of interest, and for which conventional scattering techniques only give limited information. It is an easy experiment to perform, and with the high photon fluxes available from synchrotron radiation sources, spectra can be taken in a matter of minutes. Fine structure above about 50 eV from the edge can be simply analysed using either the plane wave approximation formula ([3.45]) or the full curved wave equation, either by parameter fitting or Fourier transformation. The disadvantage of the latter technique is in the finite value of k_{min} to be used in the Fourier transform, since the data below ~ 50 eV ($k \lesssim 3$ Å^{-1}) are only properly described by multiple scattering (XANES) theory. Unlike the case for conventional diffraction, the data cannot be analytically continued to $k_{min} = 0$; however, unless another absorption edge intercedes, only signal-to-noise limits the upper value of k_{max}. The finite value of k_{min} means that much information for structural correlations at large distances is lost, and information is essentially confined to the first two shells only. This should be contrasted with the case for anomalous scattering, where no such problem exists.

3.2.4 Magnetic resonance

The phenomenon of magnetic resonance can be detected in those systems which possess a magnetic moment μ as a consequence of having angular momentum $\hbar G$, where G may be either a nuclear (I) or electron spin (S). These are related by:

$$\mu = \gamma_g \hbar G \qquad [3.48]$$

where the constant γ_g is the gyromagnetic ratio given by:

$$\gamma_g = g\left(\frac{e}{2m_o}\right) \qquad [3.49]$$

m_o is the electron mass m_e or the proton mass m_p depending whether electron spin resonance (ESR) or nuclear magnetic resonance (NMR) is being considered; the quantity ($e\hbar/2m_o$) is known as the Bohr magneton (β) or the nuclear magneton (μ_N) (e being negative for ESR and positive for NMR). The so-called g-factor can be calculated for the case of ESR, and for a free spin is given by:

$$g_s \simeq 2\left(1 + \frac{\mu_o c e^2}{4\pi h}\right) = 2.0023 \qquad [3.50]$$

When both orbital and spin moments are present, g becomes a function of the coupling between them and can also be calculated; for the case of NMR, g must be obtained from experiment through [3.51].

In the presence of a strong magnetic field, H, the Zeeman interaction between the field and the magnetic dipole moment μ is given by:

$$E = -\boldsymbol{\mu} \cdot \boldsymbol{H} = -g\left(\frac{e\hbar}{2m_o}\right) \boldsymbol{G} \cdot \boldsymbol{H} = -g\left(\frac{e\hbar}{2m_o}\right) mH \qquad [3.51]$$

and m is the appropriate magnetic quantum number. Resonance occurs when an oscillating magnetic field of an electromagnetic field at right angles to the static field and of (circular) frequency:

$$\omega = g\left(\frac{e\hbar}{2m_o}\right)\frac{H}{\hbar} \qquad [3.52]$$

induces transitions between the Zeeman levels. For $g=2$ and $H=0.33$ T, the resonance frequency is $\nu = 9.3$ GHz for ESR, which lies in the so-called X-band range (9.2–12.4 GHz) of microwaves; in contrast, resonant frequencies for NMR lie in the radio-frequency range (MHz) because of the ratio of masses m_p/m_e ($=1836$) and hence the gyromagnetic ratios.

Magnetic resonance is often measured by detection of the derivative signal obtained by the application of a small amplitude, oscillating magnetic field in addition to the static Zeeman field. The resonance condition may be probed by either sweeping the frequency or the magnetic field (cf. [3.52]); in practice for ESR, it is generally the field that is varied. However, many modern NMR spectrometers now monitor the time evolution of the signal, the so-called 'free-induction decay', after the application of a pulse or series of pulses of r.f. power; this is then Fourier transformed to give a conventional spectrum in the frequency domain.

Magnetic resonance is therefore only observed in systems which have a non-zero angular momentum. For NMR, this is limited to those isotopes which have non-zero nuclear spin, and the following examples have been studied in the amorphous state (I given in units of $\hbar$ in parentheses): ^{1}H($\frac{1}{2}$), ^{10}B(3), ^{11}B(3/2), ^{17}O(5/2), ^{27}Al(5/2), ^{29}Si(1/2), ^{31}P(1/2), ^{51}V(7/2), ^{75}As(3/2), ^{77}Se(1/2), ^{125}Te(1/2) and ^{205}Tl(1/2). On the other hand, ESR results when unpaired electrons exist, and this can be either when transition metals or rare earth elements (having an incomplete d or f shell) are present, or when unpaired electrons exist at defects, such as broken bonds; since in this chapter we are concerned with the structure of (perfect) amorphous solids, we will only consider the first possibility here and leave the second for discussion in the chapter on defects (Ch. 6).

The structural information that can be obtained from magnetic resonance spectra arises from the perturbations due to neighbouring atoms on the energy levels of the resonant species, whether nucleus or electron. For the case of ESR, values of the g-factor as measured using [3.52] often differ from the free spin value g_s ([3.50]). For spin–orbit coupling between the electron spin and the magnetic field resulting from the motion of other charges (for the case of nuclei with even spin), second-order perturbation theory gives for the g-tensor:

$$g_{ij} = g_s(\delta_{ij} - \lambda \Lambda_{ij}) \qquad [3.53]$$

where δ_{ij} is the Kronecker delta, λ is the spin–orbit coupling constant, and $\boldsymbol{\Lambda}$ is the tensor formed from a combination of the ground-state wavefunction $|0\rangle$ and excited states $|n\rangle$ which are mixed by the spin–orbit interaction, and the energy difference

$(E_n - E_o)$ between the states:

$$\Lambda_{ij} = \sum_{n \neq 0} \frac{\langle 0|L_i|n\rangle\langle n|L_j|0\rangle}{(E_n - E_o)} \qquad [3.54]$$

where L_i is the angular momentum operator. The g-shift is simply given by $(g_{ij} - g_s) = -g_s \lambda \Lambda_{ij}$. It is always possible to find a coordinate system whereby the g-tensor is rendered diagonal and so the principal components g_{11}, g_{22} and g_{33} are generally recorded. Amorphous solids are generally isotropic, and so all orientations are taken by the paramagnetic centres; therefore resonance no longer occurs at the three discrete frequencies determined by g_{11}, g_{22} and g_{33} (if all different) but instead the absorption spectrum is broadened out into a 'powder pattern' (Taylor et al. 1975) in which the frequencies associated with the principal values of the g-tensor occur at shoulders or peaks in the spectrum. The directly measured derivative spectrum may therefore exhibit a split peak in both the positive and negative going parts, but if the (dipolar) broadening is too large, this resolved structure disappears and a broad, asymmetric line results (Fig. 3.16). The structural information obtainable from the g-tensor devolves on [3.54]. It is often possible to determine the coordination of a transition-metal ion in a glass (tetrahedral or octahedral, say) from a consideration of crystal–field splittings due to the ligands and the effect these have on the g-shifts; in addition, distortions from symmetry can often be distinguished. Wong and Angell (1976) give a large number of examples of the use of ESR in the study of paramagnetic ions in amorphous solids.

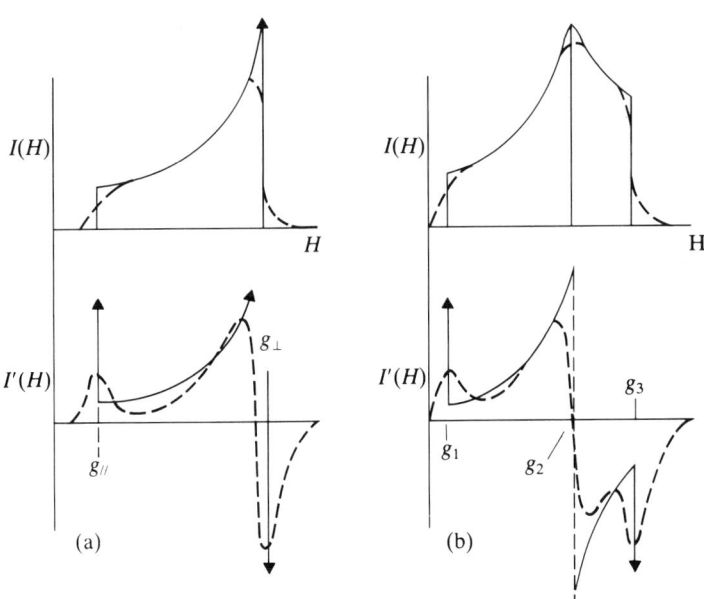

Fig. 3.16 Schematic illustration of the absorption (powder pattern) $I(H)$ and first derivative $I'(H)$ for paramagnetic centres which have a g-tensor that is
(a) axially symmetric, and
(b) anisotropic (P. W. Atkins and M. C. R. Symons, *The Structure of Inorganic Radicals* (Elsevier: 1967)).

In general, the atomic nucleus (or nuclei) may have a non-integral spin, and the so-called 'hyperfine' interaction between the electron and nuclear spin occurs, resulting in hyperfine structure in the ESR line. The effective spin Hamiltonian representing ESR can thus be written:

$$\mathcal{H}_{ESR} = \beta \mathbf{S} \cdot \mathbf{g} \cdot \mathbf{H} + \mathbf{S} \cdot \mathbf{D} \cdot \mathbf{S} + \mathbf{S} \cdot \mathbf{A} \cdot \mathbf{I} + \mathbf{I} \cdot \mathbf{Q} \cdot \mathbf{I} - \gamma_g \hbar \mathbf{I} \cdot \mathbf{H} \qquad [3.55]$$

The first term describes the Zeeman splitting of the lowest energy levels, the second is the fine structure interaction produced by ligand field or spin–spin interactions in zero field, the third is the hyperfine interaction between the electron and the nucleus mentioned previously, the fourth is the nuclear quadrupolar interaction and the final term is the direct interaction between the nucleus and the magnetic field $\mathbf{H}$. The quantities $\mathbf{g}$, $\mathbf{D}$, $\mathbf{A}$ and $\mathbf{Q}$ are all tensors.

The Hamiltonian describing NMR can be written as:

$$\mathcal{H}_{NMR} = \gamma_g \hbar \mathbf{I} \cdot \boldsymbol{\sigma} \cdot \mathbf{H} - \gamma_g \hbar \mathbf{I} \cdot \mathbf{H} + \mathbf{I} \cdot \mathbf{Q} \cdot \mathbf{I} + \mathbf{I} \cdot \mathbf{D} \cdot \mathbf{I} \qquad [3.56]$$

where $\boldsymbol{\sigma}$ is the magnetic shift tensor, and represents a 'chemical shift' for diamagnetic or slightly paramagnetic solids or a 'Knight shift' for metals. Comparison of [3.55] and [3.56] shows that a formal similarity between ESR and NMR exists in that the electronic Zeeman term ($\beta \mathbf{S} \cdot \mathbf{g} \cdot \mathbf{H}$) has the same form as the nuclear Zeeman plus the magnetic shift term ($\gamma_g \hbar \mathbf{I} \cdot (\boldsymbol{\sigma} - 1) \cdot \mathbf{H}$) and the ESR fine-structure interaction ($\mathbf{S} \cdot \mathbf{D} \cdot \mathbf{S}$) is formally the same as the quadrupolar interaction in NMR ($\mathbf{I} \cdot \mathbf{Q} \cdot \mathbf{I}$); there the analogy ends, however, since there are no terms in the NMR Hamiltonian corresponding to the nuclear terms in ESR.

A general characteristic of NMR spectra obtained from solids is the broadness of the lines, particularly compared with liquids, and the consequent lack of resolution of the spectra. This broadening is a result of any 'anisotropic' terms (i.e. having different values of the tensor components associated with various directions in the coordinate frame) in the interactions comprising the NMR Hamiltonian ([3.56]) and arises as shown schematically in Fig. 3.16. It can be dramatically reduced to give high-resolution NMR in the solid state by adopting one or both of two techniques, 'magic-angle spinning' or 'multiple-pulse sequencing'. The dipolar interaction is often the principal cause of line broadening and the Hamiltonian $\mathbf{I} \cdot \mathbf{D} \cdot \mathbf{I}$ can be rewritten in the form of a sum over all pairs of nuclei i, j (Andrew 1981):

$$\mathcal{H}_D = \sum_{i<j} \tfrac{1}{2} \gamma_i \gamma_j \hbar^2 r_{ij}^{-3} (\mathbf{I}_i \cdot \mathbf{I}_j - 3 I_{iz} I_{jz})(3 \cos^2 \theta_{ij} - 1) \qquad [3.57]$$

where r_{ij} is the internuclear displacement and θ_{ij} the angle between r_{ij} and the Zeeman field H_o which is in the z-direction in the laboratory frame. It is immediately apparent from [3.57] that the thermal motion of atoms that occurs in fluids reduces the dipolar interaction to zero, since the isotropic average of $\cos^2 \theta_{ij} = 1/3$. But what of solids? If the sample is rotated at an angle ϕ relative to the magnetic field direction, each vector r_{ij} rotates about the axis of rotation with the result that θ_{ij} becomes time dependent. In fact, the dipolar Hamiltonian can be separated into a constant term and a time-dependent part which has zero mean value; the constant term involves the factor $(3 \cos^2 \phi - 1)$ and so if the sample is rotated at an angle of $\cos^{-1}(1/\sqrt{3}) = 54° \, 44'$ (the 'magic' angle) to the magnetic field direction, the dipolar interaction is removed (Andrew 1981). In fact, all the anisotropic tensors in [3.56] contain a factor $(3\cos^2\theta_{ij}-1)$ and so magic-angle spinning can in principle remove all such contributions to the line broadening, revealing high-resolution spectra from which structural information, such as coordination, can be obtained. The drawback of the technique is that the sample rotation frequency must at least equal the dipolar line width for line narrowing to occur; this is of the order of 1–10 kHz, depending on the nuclear species. An alternative approach to reduce the homonuclear dipolar line width is to manipulate the dipolar interaction spin operators rather than the spatial coordinates by means of a series of short r.f. pulses each rotating the spins by 90°, together with various phase shifts and pulse separations which have the effect of 'decoupling' the spins – a full discussion of this multiple-pulse sequence technique is beyond the scope of this book and the interested reader should refer to the article by Mansfield (1981). The technique of high-resolution NMR in solids is relatively new however, and has been little applied to the study of amorphous materials as yet.

An example of the application of NMR to the study of the structure of amorphous solids concerns glassy B_2O_3, and the structural changes that occur upon the addition of alkali oxides. There is much evidence from e.g. X-ray and neutron diffraction studies, that in pure glassy B_2O_3, each B atom is coordinated by three O atoms in the form of a planar triangle, as in the hexagonal crystalline modification, but how the triangles themselves are connected together is still a matter of some dispute. NMR studies are of use principally in this system through the quadrupolar interaction. The quadrupolar interaction arises for those nuclei having an electric quadrupole moment (only for $I \geqslant 1$) which can interact with the electric field gradient at the nucleus resulting from the distribution of electrons around the nuclei; this distribution is a sensitive function of the local structure and bonding. The quadrupolar interaction shifts the Zeeman levels, splits the resonance line into $2I$ lines and broadens the central line ($m=1/2 \to m=-1/2$, e.g. for $^{11}B(3/2)$) by an amount dependent on the quadrupolar coupling constant $Q_{cc}=(e^2qQ)/h$, where eQ is the nuclear quadrupole moment and eq is the component of maximum magnitude of the electric field gradient tensor. The ^{11}B NMR spectra for both glassy and hexagonal crystalline B_2O_3 are very similar, and both are characterized by a large quadrupolar coupling constant of 2.76 MHz, a value which is theoretically predicted for a planar —BO_3 configuration. However, on addition of up to about 33% of Na_2O, a very sharp central line grows continuously in the spectrum (Fig. 3.17). This has been ascribed to the creation of tetrahedral —BO_4 units on the addition of alkali oxide. Certain anomalies in the UV absorption spectra are also observed in glassy B_2O_3 upon the addition of alkali oxides and this effect has been termed the 'borate anomaly'. Originally it was thought that the anomalies in the optical spectra coincided exactly with the creation of tetrahedral BO_4 units, but this correlation does not appear so clear when quantitative NMR data are examined. Q_{cc} for ^{11}B in a near-tetrahedral configuration is very small (~ 50 kHz) because of symmetry, and hence the line is narrower. Analysis of the area under the NMR absorption curves confirmed the validity of the theoretical relation linking the number of tetrahedral B atoms (N_4) to the mole percentage (x) of modifier: $N_4 = x/(100-x)$ (each oxide ion added converts two B atoms to tetrahedral coordination). This would be extremely difficult to detect using scattering techniques, and conventional EXAFS would be unsuitable since very soft X-rays are required to

Structure

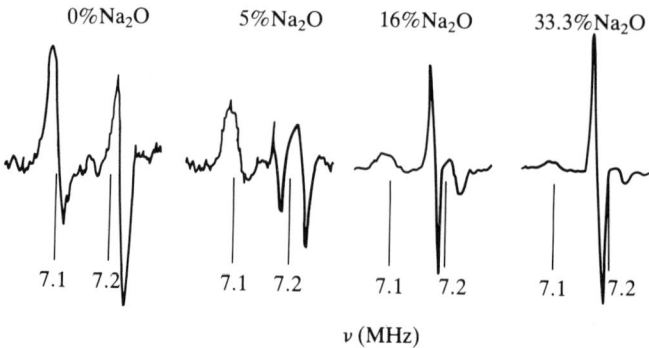

Fig. 3.17 ^{11}B NMR spectra for a series of Na$_2$O–B$_2$O$_3$ glasses (Silver and Bray 1958).

probe the B K-edge. A review of the application of NMR to the study of borate glasses has been given by Bray (1978).

A related technique which utilizes the quadrupolar interaction in a different way is *nuclear quadrupole resonance* (NQR). In the absence of a magnetic field, the quadrupolar interaction produces two doubly degenerate energy levels (for $I = 3/2$) for which resonance occurs at a frequency:

$$v = \frac{Q_{cc}}{2}\left(1 + \frac{\eta_a^2}{3}\right)^{1/2}$$

[3.58]

where η_a is the asymmetry parameter of the electric field gradient tensor and is defined by $((V_{xx} - V_{yy})/V_{zz})$, where $V_{xx} = \partial^2 V/\partial x^2$, etc. and V is the potential. Conventional NMR experiments for ^{75}As ($I = 3/2$)-containing glasses are impossible because of the large value of Q_{cc} and so NQR is the only option available. The NQR spectrum has been recorded for a-As, and the line shape is shown in Fig. 3.18; it is very broad and asymmetric compared with the orthorhombic crystalline modification. The asymmetry in the NQR signal is a direct consequence of the asymmetry in the local bonding and reflects the distribution of atoms around the resonant atom. A convenient way of describing the local distribution of atoms is in terms of the 'dihedral angle', ϕ, defined as the angle of rotation about a common bond required to bring two neighbouring structural units into coincidence; the dihedral angle for the case of As (or any threefold coordinated atom) is shown in Fig. 3.19. A completely *random* distribution of dihedral angles, $P(\phi)$ (i.e. $P(\phi)$ flat), produces a *symmetric* NQR line shape; an asymmetry in $P(\phi)$ produces a concomitant asymmetry in the NQR spectrum. A continuous random network model constructed by Greaves and Davis (1974) to simulate the structure of a-As (see sect. 3.3) was found to have an asymmetric $P(\phi)$, the 'staggered' configuration ($\phi = 180°$) being approximately twice as probable as the 'eclipsed' configuration ($\phi = 0°$). The NQR line shape calculated for this $P(\phi)$ is also shown in Fig. 3.18; the lack of agreement in line widths may be ascribed to the use of hydrogen-like orbitals rather than more realistic wavefunctions and the lack of precision in the experimental value of the nuclear quadrupole moment for ^{75}As.

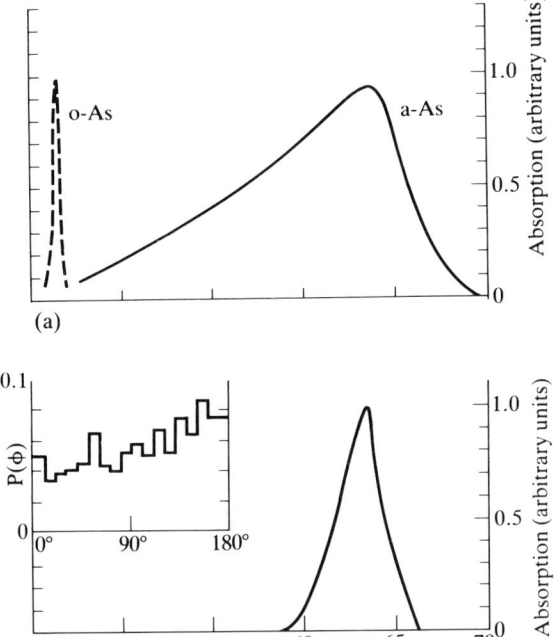

Fig. 3.18 (a) Nuclear quadrupole resonance spectrum for amorphous As (solid curve) compared with that for orthorhombic As (dashed line)
(b) Calculated NQR spectrum for a-As obtained using the dihedral-angle distribution shown in the inset (Jellison *et al.* 1980).

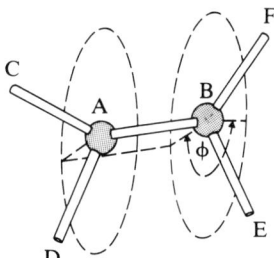

Fig. 3.19 Schematic illustration of the dihedral angle ϕ in a triply-coordinated network (e.g. As). ϕ is defined as the angle between the bisectors of the pairs of bonds AC, AD and BE, BF when projected on to a plane perpendicular to the bond AB (Greaves *et al.* 1979).

3.2.5 Vibrational spectroscopy

Optical excitation of vibrational modes of an amorphous solid, as in the techniques of infra-red (IR) or Raman spectroscopy, can be a most sensitive, albeit indirect, probe of the structure of the material. In fact, the structure can be explored at several different levels; both microscopic and macroscopic details can be ascertained from the relevant vibrational spectra, but we will content ourselves with a discussion only of the application of IR and Raman techniques to the study of short-range order, leaving a more general account of these techniques to the chapter dealing with lattice vibrations (Ch. 4). Modes are IR-active if there is a change in *dipole moment* associated with the vibration; covalent homopolar materials do not normally possess a static dipole moment, and so their activity arises from a *dynamic* mechanism in which charge is transferred from extended to compressed bonds (see Fig. 4.7) (Alben *et al.* 1974). Raman activity, on the other hand, arises from a change in *polarizability*, a second rank tensor quantity. As will be seen in the next chapter, the intensity of the Raman scattering depends on the polarization of the electric field vector for both the incident and scattered radiation. Infra-red and Raman spectra do not generally provide direct structural information, unlike scattering techniques, but instead calculated vibrational characteristics of assumed structures may be compared with experiment.

Vibrational spectroscopy is valuable as a *short-range* structural probe when there are structural units present which are capable of being vibrationally excited *independently* of the surrounding amorphous matrix. This might occur in one of two ways: (i) 'terminator' atoms may be present which have a different bonding coordination to the majority of atoms forming the amorphous network and where, if the terminator atoms are considerably lighter than the other atoms, high frequency localized 'impurity' modes will result (an example of this is the hydrogen found in glow-discharge deposited amorphous silicon); (ii) if the structural units are vibrationally 'decoupled' from each other, which happens when the units are connected together by atoms at which the bond angle is close to $90°$ (Sen and Thorpe 1977). In cases other than these, the vibrations are not confined to a small local region and can only properly be described in terms of network dynamics in which all the atoms present contribute to the modes, as will be discussed in the next chapter.

As an example of the application of IR spectroscopy as a structural probe, we will first consider the hydrogenated amorphous silicon system. As indicated in Chapter 1, the material may be prepared in a variety of ways, including the glow-discharge (GD) decomposition of SiH_4, reactive sputtering in Ar/H_2 and chemical vapour deposition from SiH_4 (possibly with post-hydrogenation in an H-plasma). The presence of (5–10 at. %) H is vital for the desirable electronic properties of this material (e.g. its dopability), yet preparation even by the same technique under different conditions can produce material with widely differing hydrogen contents, which for the GD method can even include polysilanes, a-SiH_x. A technique which is capable of measuring the concentration of H in a sample and which can also determine the bonding *configuration* of the H is obviously of great value, and in certain circumstances IR absorption can fulfil such a need. Samples can be deposited on single crystal Si wafers (polished on both sides and wedged in thickness to eliminate interference fringes), and the IR transmittance measured as a function of wavelength in a double-beam spectrophotometer with an identical uncoated Si substrate in the reference beam to correct for absorption, if any, of the substrate. Crystalline Si wafers are suitable as substrates for a-Si:H films because of the good match in refractive indices which minimizes interference fringes from multiple reflections within the film. The absorption coefficient α can be obtained from the equation

$$T = (1-R)^2 \, e^{-\alpha d}/(1 - R^2 \, e^{-2\alpha d}) \qquad [3.59]$$

where T is the percentage transmission, R the reflectivity and d the thickness of the film – [3.59] is strictly valid only for a freely supported film, but is a sufficiently good approximation in this case. Figure 3.20(a) shows the IR transmission of a series of films prepared by GD decomposition for different applied r.f. powers, for which the H content increases with r.f. power from ~ 9 at. % to ~ 15 at. % (see Lucovsky and Hayes 1979, Lucovsky et al. 1979). The spectra can be analysed by noting that since $m_H/m_{Si} \sim 0.03$, the vibrational modes can be regarded as arising solely from the vibration of the H atoms; these are assumed to be attached tetrahedrally to Si and can occur in $\mathord{>}$Si—H, $\mathord{>}$Si—H$_2$ or —Si—H$_3$ configurations. The local vibrations for these groups are shown in Fig. 3.20(b), where the frequencies are assigned from a knowledge of analogous vibrations in the respective polysilane molecules and from the fact that stretching vibrations lie higher in energy than bending modes, in turn higher than rocking vibrations. There is still some debate about the precise identification of the bending modes, however; in particular whether the mode at 845 cm^{-1} is due to —SiH$_3$ groups or to interacting $\mathord{>}$SiH$_2$ groups (see e.g. Brodsky et al. 1977). The effect of r.f. power on the microstructure of a-Si:H films is apparent from Fig. 3.20(b); low power samples (which have the lowest H content) contain primarily monohydride configurations, with $\mathord{>}$Si—H$_2$ configurations becoming more prevalent in higher power samples. The differences characteristic of —Si—H$_3$ groups (higher values for stretch and bend modes) are only observed in material deposited on substrates at temperatures less than 200 °C but are not shown in this figure. If the feature of ~ 2100 cm^{-1} were solely due to $\mathord{>}$SiH$_2$ groups, then the intensities of this and the bending modes at ~ 850–890 cm^{-1} should scale together with changes in preparation. However, it is found experimentally that this is not true, and another contribution to the feature of 2100 cm^{-1} must be present, perhaps due to $\mathord{>}$Si—H groups (Shanks et al. 1980).

In addition to providing local structural information, the IR absorption can also give quantitative estimates of the overall H content of the films (Brodsky et al. 1977). The contribution of an IR-active vibration of a Si—H bond of frequency ω_o to the dielectric constant can be shown to be:

$$\Delta\varepsilon(\omega) = \frac{4\pi N \varepsilon_s^{*2}}{\mu(\omega_o^2 - \omega^2 - i\gamma\omega)} \quad [3.60]$$

where N is the density of such bonds, μ the reduced mass, γ the linear width factor and ε_s^* the appropriate effective charge for Si—H in an a-Si matrix. Integration of $\omega\,\Delta\varepsilon$ over the given absorption band yields the sum rule:

$$N = \frac{\mu}{2\pi^2 \varepsilon_s^{*2}} \int \omega \mathrm{Im}(\Delta\varepsilon)\,d\omega = \frac{cn\mu}{2\pi^2 \varepsilon_s^{*2}} \int \alpha(\omega)\,d\omega \quad [3.61]$$

or

$$N = \frac{cn\omega_o \mu}{2\pi^2 \varepsilon_s^{*2}} \int \frac{\alpha(\omega)}{\omega}\,d\omega \quad [3.62]$$

where c is the speed of light and n is the refractive index. Thus, the integrated absorption $\int (\alpha(\omega)/\omega)\,d\omega$ should be directly proportional to the concentration of H (in the particular configuration producing the absorption band under study). The

Structure

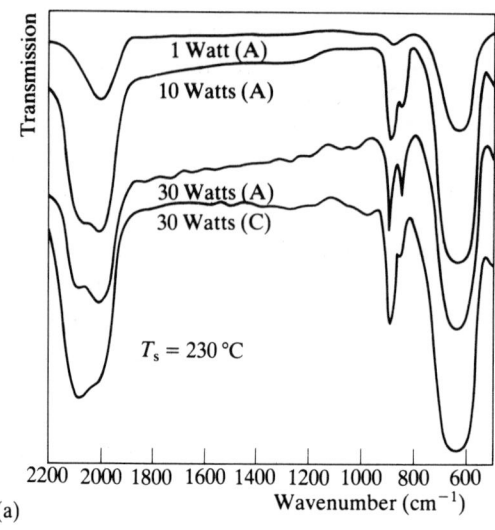

(a)

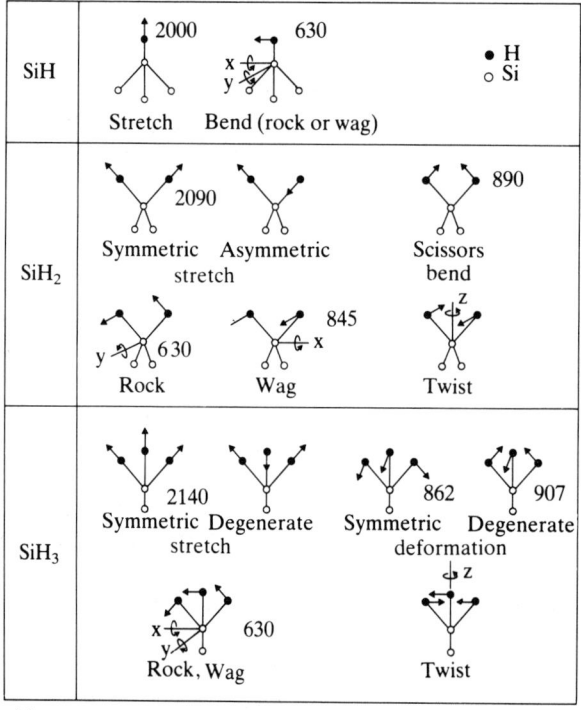

(b)

Fig. 3.20 (a) Infra-red transmission of films of a-Si:H deposited on to substrates at 230 °C as a function of r.f. power in the discharge. (A refers to material deposited on the anode, C to material deposited on the cathode.)
(b) Local Si:H vibrations for SiH, SiH_2 and SiH_3 groups, with the frequencies (in cm^{-1}) shown for each mode (Lucovsky *et al.* 1979).

problem with this approach is what to take for ε_s^*, since in a solid the bonds experience a *local* field rather than the applied external electromagnetic field; both Clausius–Mosotti and Maxwell–Garnett models have been used, the former is unsatisfactory because it is valid only for a homogeneous medium, and the latter perhaps dubious because it is a macroscopic model used on a microscopic scale. Shanks *et al.* (1980) discuss in detail the effects on the local field of a variety of models and shapes of cavities surrounding the vibrating bond.

Our second example of the application of vibrational spectroscopy to the study of short-range order is the series of amorphous materials having the general formula AX_2, where A is the cation and X is the anion. Examples of these are SiO_2 and the Ge-chalcogenides (S and Se); in all cases the structure is believed to be comprised of tetrahedral AX_4 units linked either through the apices at the chalcogen atom or through shared edges. While the tetrahedral nature of these units is well established from a variety of experiments (e.g. diffraction), the angle subtended at the linking chalcogen atom is not in general known precisely, for the reasons referred to in section 3.2.1. However, the oxide glasses are believed to have higher A—X—A bond angles than other chalcogenide materials (e.g. for SiO_2, $\theta \sim 140°$ and for GeS_2, $\theta \sim 105°$), and this feature plays a crucial role in the interpretation of the vibrational spectra, and indeed in the appearance of such spectra. Sen and Thorpe (1977) have shown that for a simple model employing central (bond-stretching) and much weaker non-central (bond-bending) forces, the vibrational modes change from being 'band-like' (i.e. phonon modes involving all the atoms in the network in a cooperative motion) to 'molecular-like' (i.e. motion of isolated tetrahedra) as the chalcogen bond angle is reduced. Molecular modes are characterized by sharp, narrow features in the vibrational density of states, and therefore in the IR and Raman spectra, whereas band-like modes are typically broad and featureless. Thus, as the A—X—A bond angle is reduced, the coupling between molecular modes of different tetrahedra diminishes, and in fact is zero for an angle of 90° for a model in which there are central forces only (see Problem 4.1). This is the basis of this aspect of the use of vibrational spectroscopy as a structural probe, for if the signature of molecular-like modes is observed in the spectrum, the implication is that the chalcogen bond angle is near 90° or, as the calculations of Sen and Thorpe (1977) show, less than the critical angle $\theta_c = \cos^{-1}(-2m_X/3m_A)$, where m is the atomic mass; for SiO_2, θ_c is 117° and for GeS_2 it is 107°. Although detailed interpretation of IR and Raman spectra is complicated by matrix element effects, which mean that the vibrational spectra do not necessarily reflect accurately the shape of the vibrational density of states (see Ch. 4), nevertheless the independent molecular model has been used quite successfully to account for the vibrational spectra of Ge and As chalcogenide glasses (see e.g., Lucovsky *et al.* 1974). For materials like SiO_2 or GeO_2 for which the chalcogen bond angle $\theta > \theta_c$, the molecular model is completely insufficient and the lattice dynamics of an entire (CRN) cluster must be calculated. Note also that the simplest molecular approach, i.e. using AX_4 tetrahedra is only strictly valid for stoichiometric compositions, and for other compositions, like-bonds (e.g. Ge—Ge or S—S) as well as larger molecular units must be considered (see sect. 3.4.1).

3.2.6 Summary

The aim of this section has been to introduce a variety of experimental techniques which may be used to gain an understanding of the short-range structure of amorphous materials, and these have been illustrated, for the most part, by examples of covalently bonded amorphous solids. The structure of amorphous metals, and those experimental techniques which are particularly suited for metallic systems (such as 'magnetic' neutron scattering) are left for discussion in Chapter 7.

The somewhat depressing conclusion of this section is that the determination of the structure of amorphous solids is an exceptionally difficult task, and no single experimental technique is wholly sufficient. The situation is easiest for monatomic systems and for these, conventional diffraction techniques using X-rays, neutrons or electrons are the most suitable, which can yield structural information about the local order and possibly, with the help of modelling studies (see sect. 3.3), medium-range order. For multicomponent systems which, after all, form the majority of amorphous materials, the situation is much worse. Diffraction techniques and their variants (anomalous scattering, isomorphous and isotopic substitution or magnetic neutron scattering) are really only suitable for binary compounds which are described by three partial pair correlation functions (and hence need three separate diffraction experiments to determine them). For systems containing more than two components, only EXAFS is really suitable as a direct structural probe. It has the virtue of being chemically specific, so the average local order around the atomic species which absorbs the X-ray photon is determined. Its disadvantages are that it is insensitive to atomic correlations much beyond the second coordination shell and it is difficult to deduce coordination numbers quantitatively. Other techniques such as magnetic resonance and vibrational spectroscopy can supply indirect structural information, usually as an adjunct to more direct structural probes.

3.3 Structural modelling

3.3.1 Introduction

We have seen in the last section that a variety of experimental techniques can give information about the microscopic structure of amorphous solids. In most cases, and for diffraction experiments in particular, this information is limited almost entirely to the first two coordination shells; i.e. the bond lengths and angles of nearest-neighbour atoms comprising the basic structural unit (e.g. SiO_4 tetrahedra in a-SiO_2) can be determined, but the relative disposition of such units cannot be ascertained with certainty. For the case of RDFs derived from scattering experiments on monatomic systems, for example, the problem lies in the fact that peaks other than the first and second cannot be uniquely associated with a particular interatomic correlation, but are made up from a variety of contributions from higher-lying shells (see Fig. 3.6). Matters are complicated considerably for multicomponent materials for which a single diffraction experiment does not identify the origin of any peak in the RDF in terms of specific atomic pair correlations.

One solution to these difficulties is the construction of *models* which purport to simulate the structure. Such structural information as is known can be used as a basis for the generation of a random structure. The form of randomness involved depends to a great extent on the type of bonding which is prevalent in the material. The structure of amorphous metals, ionic solids and molecular organic solids (and also some liquids), which are held together predominantly by *non-directional* forces, can be described in terms of a 'dense random packing (DRP) of hard spheres', and these models will be discussed more fully in the chapter on metallic glasses (sect. 7.2.3). Non-metallic materials (excluding ionic solids) having predominantly covalent *directed* bonds can often be described in terms of a 'continuous random network' (CRN) – these form the subject of much of the next section. In either case, structural parameters such as the RDF, density etc. may be computed for the structural model, and the theoretical predictions compared with experiment. The importance of structural modelling lies in the fact that a detailed understanding of the structure may thereby be gained. For instance, the structural origin of features in the (computed) RDF may readily be ascertained with the use of a model, information often impossible to ascertain in any other way.

3.3.2 Continuous random networks

The term 'random network' is somewhat of a misnomer in that such networks are not truly random in a statistical sense, but have a considerable degree of local order conferred upon them by the presence of directed bonding. Thus, covalently bonded amorphous solids generally have a well-defined bond length, bond angles and nearest-neighbour coordination, often very similar to those found in corresponding crystalline polymorphs. The fundamental polyhedra formed by the nearest neighbours (e.g. —SiO_4 in SiO_2, —BO_3 in B_2O_3 or —As_4 in As) are connected together in the network, and their relative orientation is characterized by the dihedral angle, ϕ (see Fig. 3.19). If the units were completely randomly arranged, the dihedral angle distribution would be flat; in practice, it is found for most CRNs that although all values of dihedral angle are represented, certain values are more probable than others. Thus, although a CRN may not be truly *random*, it is certainly *non-crystalline*, and this is achieved by allowing some degree of bond-angle distortion and more or less complete freedom for values of the dihedral angle. Continuous random network models can be built in this manner, seemingly without limit in size and without accruing undue strain, utilizing values for the bond lengths and angles taken from experimental RDFs; typically bond-angle distortions are found to be about 10% (less for intra-unit angles in e.g. —SiO_4 or —BO_3) and bond-length distortions are 1% or less.

There are two ways of constructing structural models. The first involves the physical building of a model, which for a covalently bonded material consists of a 'ball-and-stick' arrangement, in which tubes of the appropriate length represent covalent bonds, and atomic sites are represented by either balls with holes drilled at the appropriate orientations or units with spokes at the correct relative angles (Fig. 3.21 shows two CRNs of these types for the oxide glasses SiO_2 and B_2O_3); dense random packed (DRP) models can simply be constructed by mixing balls of the same (or different) diameter and gluing the resulting non-periodic assembly to form

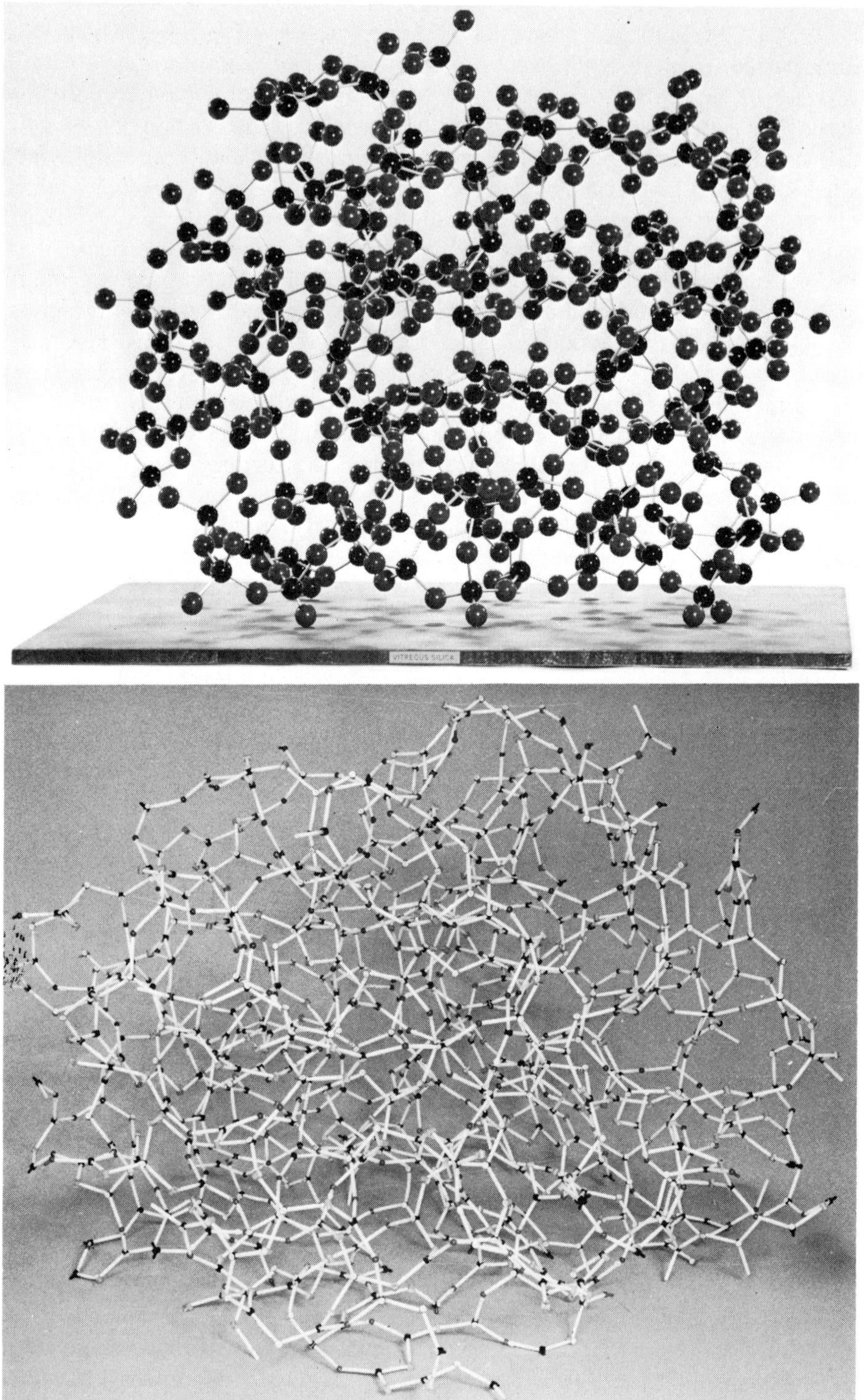

Fig. 3.21 (a) Continuous random network model simulating the structure of amorphous SiO_2 (Bell and Dean 1966) (Crown copyright).
(b) Continuous random network model simulating the structure of vitreous B_2O_3 (no boroxol rings) (Williams and Elliott 1982).

the model. In both cases, each atom is numbered and a connectivity table constructed; the spatial coordinates of all the atoms are measured either by stereo-photography or by some form of surveying. The other way of constructing structural models is by using a computer and this technique in itself can take several forms. The method which comes closest to physical construction is the serial addition of atoms, adatom by adatom, to a starting seed, usually chosen to be a polyhedron not found in any crystalline polymorphs (e.g. containing fivefold rings for the case of Ge or Si) so as to preclude possible periodic growth of the cluster. The algorithm used must search among all the possible empty bonding sites to find those most favourable, i.e. those which minimize strain and the number of non-bonded atoms (dangling bonds) in the case of covalent materials, or which maximize the density of random packings of spheres for metals. An alternative way of using computational methods to generate structural models is the *modification* of existing models usually by altering the energy of the structure by appropriate algorithms; for the case of CRN models this means minimizing the local strain energy resulting from bond-length and bond-angle distortions (e.g. using [4.11]) and for DRP models minimizing the total energy assuming the potentials to be of the Lennard–Jones type or similar ([7.6]). Note that for the former case the topology (i.e. local geometry) is unaltered since only deviations from the mean are reduced, but the topology is not necessarily invariant for the latter case. The last type of computer-generated model is that which is obtained by *transformation* of existing models in which a model simulating the structure of one material is converted into a different network simulating the structure of another material, i.e. Ge to H_2O (Boutron and Alben 1975) and As to B_2O_3 (Elliott 1978). The former case (Ge $\rightarrow$ H_2O) is an example of a direct type of transformation, where the atomic sites in the original model (Ge) are replaced by a new species (e.g. O) and the other atomic species (H) are inserted relative to these; the latter case (As $\rightarrow$ B_2O_3) is an example of an indirect transformation in which the atomic sites of the original network serve to define a local geometry within which the new structural unit is constructed (e.g. the mid-points of the three As bonds in a pyramidal unit become O sites and B atoms are inserted at the centroid of the resulting planar O_3 triangle). A further interesting form of topological transformation is the interconversion of a tetrahedral CRN and a DRP model. A random network containing only even-membered rings can be subdivided into two interpenetrating networks and the removal of one of these leads to a structure which is randomly close-packed with ~ 12 neighbours per atom (Connell 1975); Wright *et al.* (1980) review the subject of geometrical transformations of random networks in detail.

The principal advantage of simulating the structure of amorphous materials using physical models is that many structural features may be identified directly by inspection. As an example may be cited the case of the $\sim 20\%$ density deficit between the crystalline and amorphous forms of arsenic which was not explained before the construction of a threefold coordinated CRN demonstrated that voids of sufficient size to account for the density difference arose naturally as a result of the asymmetric As_4 building unit forming the basis of a random network (Greaves and Davis 1974). In addition, it is a relatively straightforward matter to construct a CRN (or DRP) model of sufficient size (say ~ 500 atoms) to give a reasonably small amount of statistical noise in the calculated RDFs. The disadvantage of physical model building lies mainly in the tedium of measuring the coordinates of large numbers of atomic sites (although it is not necessary to measure them precisely if energy relaxation procedures are to be applied subsequently), and also in the possibility that a bias may be introduced into the network by the model builder.

The advantage of computer generation of models is that large numbers of atoms can be generated in this way without bias and that the model can be energy relaxed as it is formed. Transformation of networks would be virtually impossible to achieve without using a computer. The disadvantage of computer-generated models is that they cannot be examined visually, and that it is extremely difficult to produce algorithms that ensure perfect connectivity in the interior of a network for a covalent material.

A list of the CRNs built to simulate the structure of a variety of amorphous solids is given in Table 3.3, together with a summary of their relevant details.

The principal structural parameter that can be calculated for a structural model and compared with experiment is of course the RDF. Knowledge of the atomic coordinates of the model enables the RDF to be calculated by simply noting the frequency with which interatomic correlations occur as a function of radial distance, taking in turn each atom as origin; the resulting RDF in histogram form must be corrected for the finite size of the model. It should be noted that *direct* comparison between experiment and theory in this way is not possible; the RDF calculated directly from the model coordinates can only be compared with data obtained indirectly by means of a Fourier transform from experimental scattering measurements (and is therefore susceptible to termination errors because of the finite k_{max}) and, vice versa, the raw (corrected) experimental data can only be compared with the calculated scattering intensity resulting from a Fourier transform of the model RDF, itself subject to error because of the finite model size. The influence of factors such as bond-angle and dihedral-angle distributions and the distribution of ring sizes on the shape and positions of peaks in the model RDF can be investigated which, if the model RDF agrees with experiment, is important since these quantities cannot be obtained directly from experimental data but only inferred from such modelling studies. The distribution of ring sizes is a significant parameter since it influences the shape of features in the electronic density of states in the valence band for certain amorphous semiconductors (e.g. Ge or Si) (see sect. 5.2.1); Greaves (1976) and Temkin (1978) have shown how the RDF can be synthesized from a knowledge of the ring statistics and the bond-length and dihedral-angle distributions for threefold (e.g. As) and fourfold (e.g. Ge) coordinated networks, respectively. We show in Fig. 3.22 an experimental RDF derived from a recent high-resolution neutron scattering study of a-Ge, compared with a calculated RDF for a CRN model; also shown is another structural parameter ($P(\phi)$) for the network. It can be seen that the agreement is reasonable, but this worsens if the relative proportion of fivefold to sixfold rings is changed from the value of 0.34:1.5. It must be admitted however that the position of the first two peaks is fixed for a tetrahedral material such as a-Ge, since the first peak of the model RDF is scaled to the experimental value for the nearest-neighbour distance, and the average bond angle (and hence the position of the second peak) is determined by symmetry to be the tetrahedral angle, 109° 28'. It is the shape and position of higher-lying peaks

Table 3.3 CRN models for amorphous solids

Material	No. of atoms	Construction	Comments	Reference
Tetrahedral materials				
Si, Ge	440	Ball and spoke	Coordinates not measured	1
Si, Ge	519	Ball and spoke	Expanded Polk model; bond-length deviations minimized by computer	2
Si, Ge	500	Computer generation	Energy relaxed	3
Si, Ge	201	Computer generated	Energy relaxed	4
Si, Ge	238	Ball and spoke	Even-membered rings only – suitable for conversion to III–V structures	5
Si, Ge	238	Computer modification	Odd-membered rings introduced into Connell–Temkin model by Monte Carlo methods	6
Si, Ge	1000	Computer generation	Simulated real growth mechanism – 5% dangling bonds in interior	7
Si, Ge	61	Computer modification	Random repositioning of original b.c.c. lattice; periodic boundary conditions	8
Si	651	Computer transformation	O atoms removed from SiO_2 model of Evans and King	9
Si	54	Computer generation	Random repositioning of original f.c.c. lattice; periodic boundary conditions	10
Si:H	314(Si):83(H)	Ball and stick	5 SiH_2 groups, 73 SiH groups	11
Si:H	54(Si):6 or 8(H)	Computer transformation	Periodic boundary conditions	12
Oxides				
SiO_2	651(Si):1302(O)	Ball and stick	—	13
SiO_2	188(Si):426(O)	Ball and stick	—	14
B_2O_3	I 533(B):866(O) II 513(B):831(O)	Computer transformation	No boroxol rings: I derived from odd/even ring As model (Greaves and Davis); II from even-ring As model (Matthews *et al.*)	15
B_2O_3	414(B):600(O)	Ball and stick	No boroxol rings	16
B_2O_3	500(B):762(O)	Ball and stick	Equal number of B atoms in boroxol rings and BO_3 triangles	17
H_2O	892(H):519(O)	Computer transformation	O positions derived from Polk–Boudreaux model; H positioned according to ice rules	18
H_2O	442(H):266(O)	Ball and stick	14 large cages included (as in clathrates)	19
As_2O_3	263(As):372(O)	Ball and stick	Energy relaxed to $\theta(As)=100°$, $\theta(O)=126°$	20
Pnictides and chalcogenides				
As	533	Ball and stick	Odd/even rings	21
As	513	Ball and stick	Even rings only	22
P, Sb	533	Computer modified	Energy relaxed Greaves–Davis model to give $\theta(P)=102°$ and $\theta(Sb)=96°$	23

Table 3.3 (continued)

Material	No. of atoms	Construction	Comments	Reference
Se, Te	539	Ball and stick	Polymeric (no rings); energy relaxed with Lennard–Jones potential	24
As_xSe_{1-x}		Computer generation	Monte Carlo positioning of atoms to fit RDF	25
$CdGeAs_2$	39(Cd):39(Ge):78(As)	Computer transformation	Derived from even-ring Connell–Temkin-type tetrahedral CRN	26

References: (1) Polk (1971); (2) Polk and Boudreaux (1973); (3) Duffy et al. (1974); (4) Steinhardt et al. (1974); (5) Connell and Temkin (1974); (6) Beeman and Bobbs (1975); (7) Shevchik and Paul (1972); (8) Henderson and Herman (1972); (9) Evans et al. (1974); (10) Guttman (1976); (11) Weaire and Wooten (1980); (12) Guttman (1981); (13) Evans and King (1966); (14) Bell and Dean (1972); (15) Elliott (1978); (16) Williams and Elliott (1982); (17) Wright et al. (1982); (18) Alben and Boutron (1975); (19) Boutron and Alben (1975); (20) Beeman et al. (1980); (21) Greaves and Davis (1974); (22) Matthews et al. (to be published); (23) Davis et al. (1977); (24) Long et al. (1976); (25) Renninger et al. (1974); (26) Popescu et al. (1977).

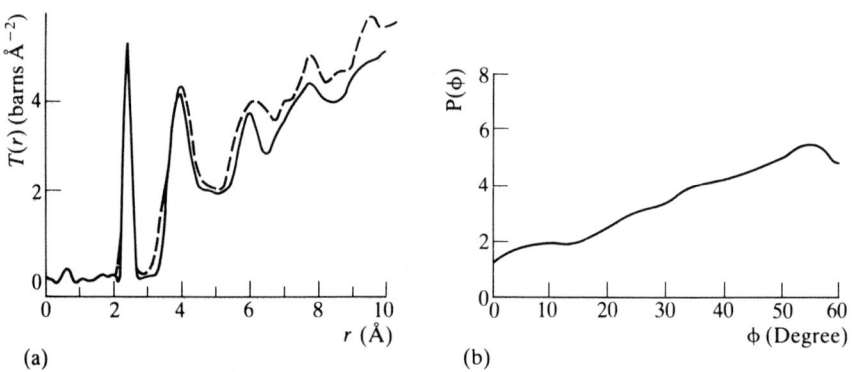

Fig. 3.22 (a) Comparison of the Beeman–Bobbs (1975) model (H5) (dashed line) with the experimental RDF (solid line) ($T(r) = J(r)/r$) of a-Ge (Etherington et al. 1982).
(b) Dihedral angle distribution $P(\phi)$ (solid line) for the Beeman–Bobbs H5 model.

which are characteristic of a particular model (through $P(\phi)$, etc.) and whose agreement with experiment is necessary for the model to be valid.

For the case of models containing more than one component, all the structural parameters discussed above can of course be obtained, but in addition and most importantly, the individual partial correlation functions can be computed and the relative contribution of each to the total weighted RDF ($\sum W_{ij}\rho_{ij}(r)$) can be ascertained. As can be seen with reference to Table 3.3, several binary systems have been modelled but as yet only one ternary system has been simulated. As an example of the power of structural modelling applied to multicomponent systems, we show in Fig. 3.23 data for the archetypal glassy system, SiO_2, and the structural information about the individual pair correlations that can be obtained; again agreement between theory and experiment is good.

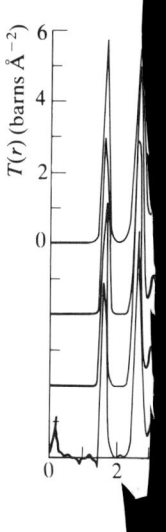

Fig. 3.23 Co[...]
energy relaxed [...]
Wright and Si[...]

At this point it must be emphasized that a model is an idealization of the real amorphous structure; ideal in that most CRN models are perfectly connected in the interior of the model, admitting no dangling bonds except at the surface, even though it is known from other studies (e.g. ESR) that appreciable densities ($\simeq 1\%$) of bonding defects can occur, particularly in vapour-deposited materials. In addition, no model of an amorphous structure can be unique even though a good fit with experimental scattering data may be achieved; this is because the RDF is relatively insensitive to small variations in e.g. the dihedral-angle distribution. As a corollary, it must be admitted that there is no ideal structure for a real amorphous solid either, unlike the crystalline case, since variations in preparation technique can dramatically affect the structure on both microscopic and macroscopic scales. An example of this is given by the arsenic chalcogenide systems (e.g. As_xS_{1-x}). The conventional modelling approach would dictate that all arsenic atoms be threefold coordinated and all sulphur atoms twofold coordinated, satisfying their normal valence in accordance with the '$8-N$' rule, and having no homopolar ('wrong') bonds at the stoichiometric composition As_2S_3, and with the atoms forming a CRN. However, neutron scattering experiments (Daniel *et al.* 1979) show clear evidence for the presence of As_4S_4 molecular units in as-deposited evaporated films of amorphous As_xS_{1-x}, presumably held together by Van der Waals' forces; As_4S_4 molecules form the basis of the molecular crystal realgar and are the predominant vapour species (together with sulphur) above As–S alloy melts (see sect. 1.3).

Furthermore, CRN (and DRP) models are idealistic in that they simulate the amorphous structure in a dense state; that is voids, or other such growth defects, are not included unless they arise naturally in the construction of the model (e.g. when low-dimensional structural units are involved, as in the As or B_2O_3 CRNs). Thus, a tetrahedral CRN simulating a-Ge or a-Si and containing no voids has a density 99% of the diamond cubic form (Polk and Boudreaux 1973). This demonstrates the

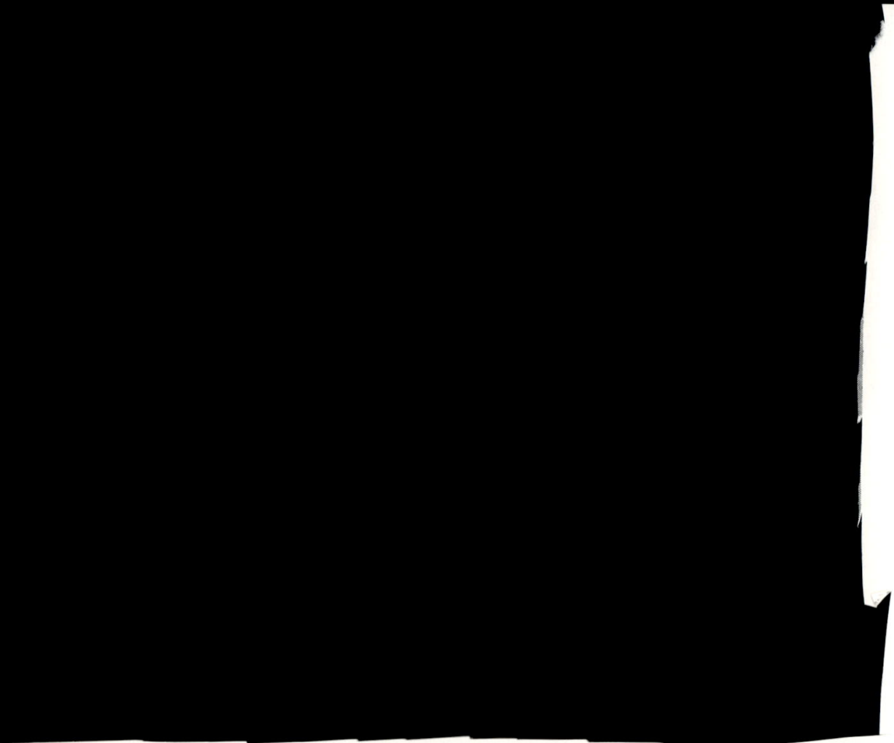

Fig. 3.24 (a) Bright-field electron micrograph of a-Ge film produced at a substrate temperature $T_s = 25\,°C$. Note that the void network observed is completely absent in a film prepared at $T_s = 250\,°C$ (Donovan and Heinemann 1971).

important point that model amorphous structures can be built, without the supposed stabilizing influences of the presence of voids or dangling bonds, such that the inbuilt strain (e.g. in bond-angle distortions) does not increase as the size of the model increases. However, real amorphous structures are often *not* as homogeneous on the macroscopic scale as these models may suggest. Figure 3.24 shows two such examples of growth inhomogeneity found in amorphous semiconductors. The first (Fig. 3.24(a)) shows an electron micrograph of an a-Ge film, evaporated on to a substrate at 25°C, having a morphology of interconnecting voids. The presence of such voids has a profound effect on the electro-optic properties of the material (see Fig. 6.13); it is found that films produced on substrates held at 250 °C do not show any evidence for void structure, surface atom mobility presumably being higher in this case, preventing nucleation at certain points only and resultant void formation where two such growth islands coalesce. The second illustration is hydrogenated amorphous silicon produced by the glow-discharge decomposition of silane. Samples produced at low r.f. power and with high ratios of $SiH_4:Ar$ are characteristically homogeneous as monitored by electron microscopy, particularly if high substrate temperatures are used, while samples produced under other conditions appear to have a columnar growth morphology (Fig. 3.24(b)). Material prepared under these conditions and having a columnar structure is found to have a large density of point defects which adversely affect possible device performance. Furthermore, there is evidence from proton NMR studies (Reimer *et al.* 1980) that the H atoms are not homogeneously distributed throughout the bulk of the material (as had previously been believed on the basis of the CRN hypothesis, in which

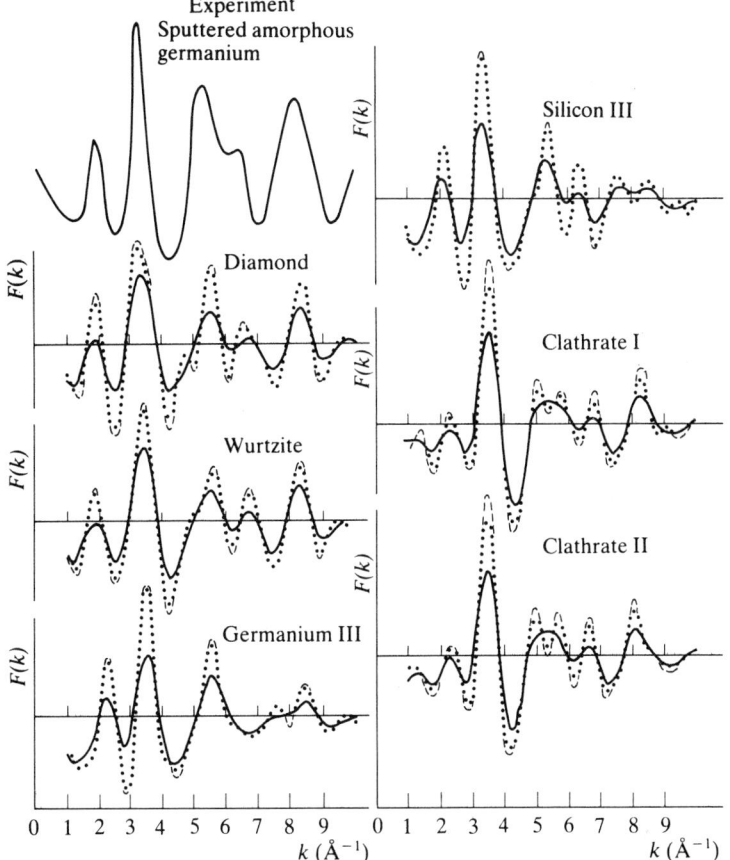

Fig. 3.25 (a) Reduced scattering intensity ($F(k)$) for a-Ge compared with calculated curves for microcrystallites of various structures. In each case, the solid line refers to crystallites of small radii ($\simeq 4$ Å), whereas the dotted and dashed lines refer to crystallites of medium and large ($\simeq 6$ Å) radii (Weinstein and Davis 1973/4).

a random network model consisting essentially of a mixture of diamond-like units and so-called 'amorphons', regular dodecahedra consisting of planar pentagonal rings (Fig. 3.26(a)). The fivefold rings result from the presence of eclipsed bonds (i.e. $\phi = 0°$) in contrast with the staggered configuration ($\phi = 60°$) characterizing the diamond cubic structure. However, since the interior angle of a regular planar pentagonal ring is 108°, there is an angular misfit of 1° 28' from the ideal tetrahedral angle, implying either that if regular amorphons are used the structure cannot be completely space filling or that distortions must be accommodated, increasing the strain energy of the model. Note that such a model is not a true CRN in the sense we have previously been discussing, since only two values of dihedral angle, 0° and 60°, are allowed. Another approach, similar in spirit but different in practice, has been suggested by Gaskell (1975) and is based on the 'polytetrahedral' units shown in Fig. 3.26(b). Each polytetrahedral unit consists of five tetrahedra in an ordered, diamond cubic structure and the four $\{111\}$ faces contact other units by eclipsed bonds. Again,

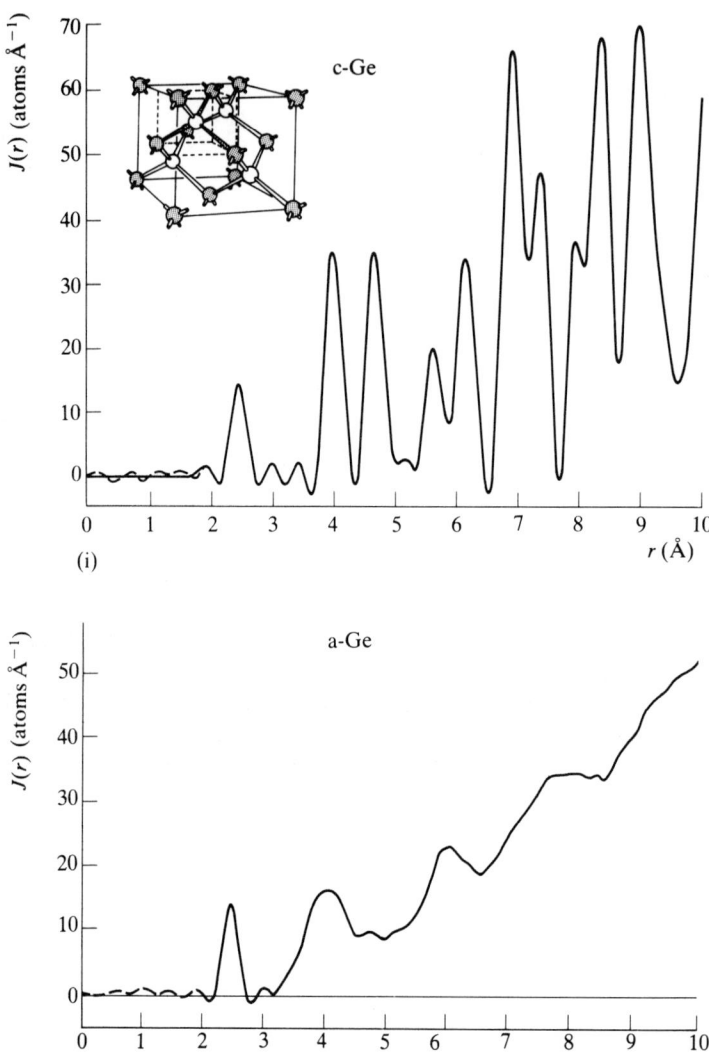

Fig. 3.25 (b) Radial distribution functions for (i) crystalline Ge (with inset of diamond cubic lattice, and (ii) amorphous Ge (sputtered on to a substrate at 150°C) (Temkin *et al.* 1973).

angular mismatch ensures that the models are not completely space filling, but energy relaxation of models containing several hundred atoms does produce structures whose calculated RDFs match qualitatively, but not quantitatively, experimental RDFs (Gaskell *et al.* 1977).

3.3.4 Monte Carlo simulations

One of the few modelling studies on an alloy system with a *range* of compositions is that by Renninger *et al.* (1974) on the As_xSe_{1-x} system ($0 < x < \frac{1}{2}$) using a Monte Carlo approach. Random starting structures, containing clusters of about 150

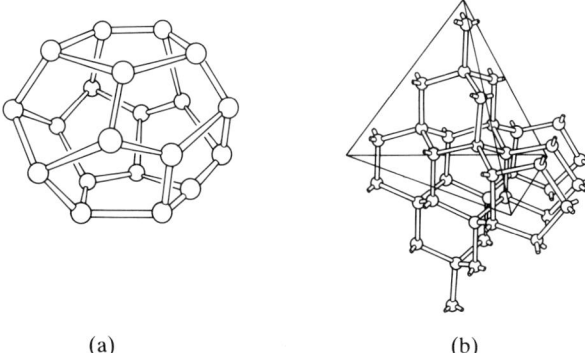

Fig. 3.26 Two forms of cluster which have been proposed to exist in amorphous tetrahedral solids: (a) amorphon; (b) polytetrahedral unit.

atoms, were generated using structural data known from experiment, e.g. bond lengths, coordination number (in accord with the $(8-N)$ rule) and bond angle. The Monte Carlo approach consisted of choosing atoms at random and moving them in an arbitrary direction and only if the move resulted in an improvement to the fit of the calculated RDF to the experimental curve was the move accepted; this process was continued until the r.m.s. deviation was less than a certain amount. As the percentage of As was increased it was found that the tendency to form large clusters of atoms increased, the Se chains becoming progressively cross-linked; a trend from chains to rings was therefore predicted. This is demonstrated in Figs. 3.27(a) and (b) which are cross-sections of portions of the models, for the cases of Se ($x=0$) and stoichiometric As_2Se_3 ($x=0.4$). The agreement between experimental and theoretical RDFs is good; Fig. 3.27(c) shows relevant curves for the composition As_2Se_3.

Another application of Monte Carlo methods has been the conversion of the even-ring Connell–Temkin (1974) CRN into ones containing varying fractions of odd-membered rings by the random repositioning of atoms with subsequent energy relaxation (Beeman and Bobbs 1975).

3.3.5 Molecular dynamical simulations

The final method of computer simulation differs in spirit from all the preceding approaches. Molecular dynamics, as its name implies, involves the study of the time evolution of the classical motion of a collection of atoms (or molecules), interacting by means of a given interatomic potential, as a function of the change in temperature (or some other thermodynamic variable) of the system. In this manner, the course of solidification on reducing the temperature of the computer 'liquid' may be followed, and importantly, the thermal history (trajectories, momenta) of all the atoms in the cluster are known. In practice, a small cluster of atoms, 100–1000 in number, is confined to a cube whose dimensions accord with the known density of the liquid or solid under investigation, and each atom is assigned to a site (usually random). The

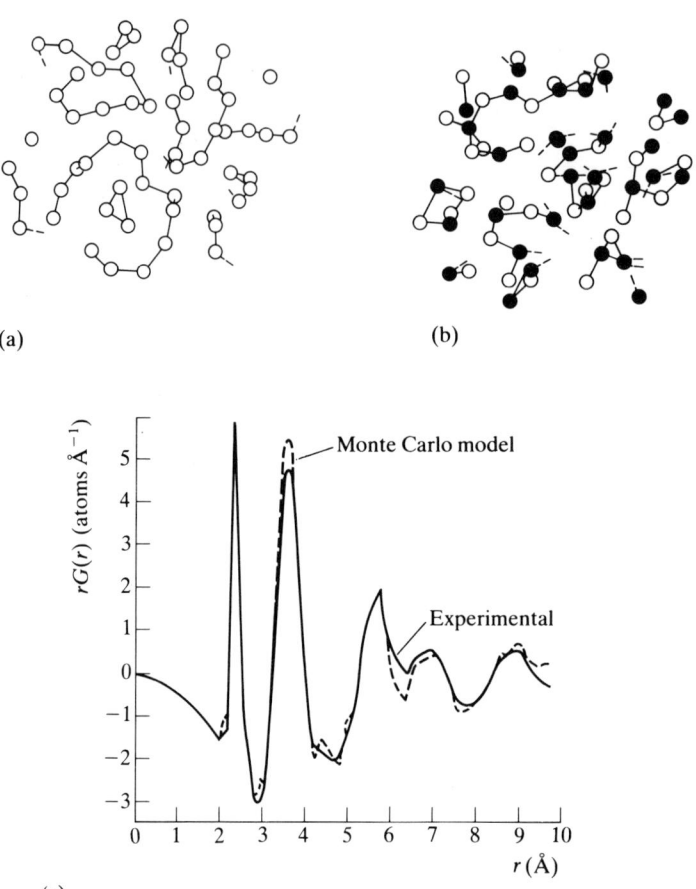

Fig. 3.27 Monte Carlo generated models of the structure of a-Se (a) and a-As$_2$Se$_3$ (b). The calculated reduced RDF is compared with the experimental curve for a-As$_2$Se$_3$ in (c) (Renninger *et al.* 1974).

atoms are then given a random Boltzmann distribution of velocities corresponding, typically, to a temperature of several thousand Kelvin. The Newtonian equations of motion are then solved for a series of time steps each of typical length $\sim 10^{-15}$ s. Cooling of the system is achieved by a step-wise reduction in temperature followed by an isothermal molecular dynamics run for say $\sim 10^{-10}$ s in total to ensure thermal equilibration. This is possible in these time-scales for liquids, but not for *glasses* where the time taken to equilibrate with respect to time-averaged atomic positions becomes much longer than the 'experimental' time-scale; the transition between these two situations is of course the glass transition (see Ch. 2). The structure (e.g. the RDF) of the system can be calculated at the end of each temperature decrement.

The advantages of the molecular dynamics (MD) approach include a detailed knowledge of the time evolution of the atomic positions and momenta as the liquid forms a glass, and the ability to impose instantaneous changes of thermodynamic variables, such as pressure and temperature, on the system and to monitor the concomitant changes, if any, on the glass transition and the final structure of the glass. The disadvantages of the method, which are considerable, are that the effective rate of cooling in such computer simulations is very high, of the order of 10^{11} K s^{-1} or higher, and the loss of internal equilibrium during cooling, marking the transition to the glass (i.e. the glass-transition temperature), occurs often at temperatures much higher than those encountered in the laboratory. The very high rate of cooling is simply a result of computer economics; an isothermal run can seldom exceed $\sim 10^{-10}$ s (i.e. $\sim 10^5$ time steps) since it takes several hours of computer time, and if 10 K temperature decrements are chosen, the minimum cooling rate is consequently $\sim 10^{11}$ K s^{-1}. Such rates are well in excess of any achievable in the laboratory, exceeding probably even the effective cooling rates encountered in thermal evaporation (Table 1.1). The other major drawback of the MD approach is the fact that as yet only non-directional (e.g. ionic) interatomic potentials have been used; the attractive part of the potential is simply of coulombic form or similar for ionic systems or Van der Waals' for rare gas systems, whereas the repulsive part may be either an exponential function or an inverse power ($1/r^n$) of the interatomic separation.

Glasses have been simulated for Lennard–Jones systems (e.g. Ar), simple ionic glasses (e.g. KCl) (which cannot normally be vitrified in the laboratory), BeF_2 and $ZnCl_2$ (which are among the few halides to be relatively easy glass-formers) and also SiO_2 (Angell *et al.* 1977). The latter system is interesting because tetrahedral coordination of Si by O (i.e. —SiO_4 tetrahedra) was found, which is surprising in view of the complete neglect of directional covalent forces in the calculation, a purely ionic potential being used. However, the tetrahedra were more distorted than appears to be the case experimentally from the X-ray studies of Mozzi and Warren (1969), perhaps resulting from the neglect of covalent interactions. Nevertheless, it is remarkable that a MD simulation using only central forces should predict tetrahedral coordination for this system. What is perhaps even more remarkable are the recent MD simulations for B_2O_3 (Soules 1980 and Amini *et al.* 1981) in which, again using ionic potentials, —BO_3 planar triangular units connected at the apices are found for the computer 'glass' in agreement with experimental structural studies. However, it appears that the choice of ionic radii (which appear in the expression for the potential) is quite crucial in determining the planar trigonal configuration (or the tetrahedral configuration of SiO_2). The empirical choice of these parameters casts some doubt as to whether the MD approach really can predict the coordination geometry of such systems using only ionic potentials, or whether such a result is an artefact of the particular values of parameters used in the simulation.

In conclusion, MD simulations can be very informative about the *dynamics* of liquids and the behaviour on cooling to form a glass. However, the cooling rates used are far higher than those encountered in the laboratory, and as we have seen in Chapter 2, the value of T_g observed is a sensitive function of the cooling rate. Molecular dynamics simulations of the structure of glasses, furthermore, suffer from the limitation that only central forces acting between atoms are employed, restricting the scope of such simulations strictly to rare-gas or ionic solids. If covalent, non-central interactions could be introduced into the interatomic potential, the method is a potentially powerful modelling technique.

3.3.6 Summary

Structural modelling can be a valuable aid to determining the structure of amorphous solids. A model built using knowledge about the chemistry (i.e. local order) of the system, and whose computed scattering properties (e.g. structure factor, RDF etc.) fit well those derived from experiment, can give considerable additional insight into the structure, which otherwise cannot be obtained. An example might be the relationship for a CRN between the ring-size and the dihedral-angle distributions and features in the RDF, such as the peaks corresponding to the third and fourth coordination shells.

Many structural models have been *physical* models, either of the 'ball-and-stick' type (for CRNs and other models of covalent materials) or random packings of spheres (for metallic solids), because visualization of the model is most important. Nevertheless, for convenience many models have been constructed using a computer, particularly DRP models, often adatom-by-adatom to simulate real growth. Computer methods play an important role in the modification (e.g. strain-energy relaxation) or transformation of such models, as well as in Monte Carlo simulations in which models are constructed by the random positioning of atoms starting from either a crystalline or another amorphous structure. Computer techniques also figure prominently in *molecular dynamics* simulations, in which the dynamics of a fluid subject to central (i.e. ionic) interatomic potentials are followed as a function of time. By lowering the 'temperature' during a simulation run, the liquid can be vitrified. The considerable disadvantages of this technique for the study of the structure of glasses (as opposed to that of liquids) are that the simulated rates of cooling are much higher than those encountered in the laboratory, and as yet, non-central (i.e. covalent) interatomic potentials have not been used.

It should be stressed finally that a structural model, of whatever type, is always an idealization of the true structure of an amorphous solid. Unlike a crystal, there is no unique structure of an amorphous material, and furthermore, most of the structural models built so far have not taken account of the growth processes and the resulting morphologies which occur in real amorphous systems.

3.4 Medium-range structure

So far, we have considered those experimental techniques which probe primarily the short-range structure of amorphous solids, i.e. atomic correlations occurring in the first and second coordination shells in the range up to 5 Å, and have discussed the detailed interpretation of the data produced by such experiments. We must now address the considerably more difficult problem of longer range correlations, the so-called medium-range order. Of course, most of the experimental techniques discussed in sections 3.2.1–3.2.5 (except for EXAFS) are sensitive in principle to medium- as well as to short-range order. The important question is thus whether there are in fact any medium-range correlations in the range 5–20 Å in amorphous materials. The answer is emphatically yes, but we must be careful to distinguish between correlations arising from a variety of causes. The fact that peaks at large distances are observed in the RDFs of amorphous solids obtained by scattering experiments (see Figs. 3.22 and 3.23) is itself indicative that complete randomness has not set in beyond the first coordination shell. This is particularly true for covalent materials, where it is an indication that the dihedral angle distribution is not random (i.e. flat) but that certain values are preferred over others. Even the random packed glassy metals exhibit features in the RDF behond the first coordination shell. However, these peaks arise from *average* correlations existing within the amorphous structure at certain distances, and this case should be distinguished from another manifestation of medium-range order, the existence of well-defined clusters of atoms, perhaps made up from many basic structural units. Another aspect allied to these features is the existence of wrong (like) bonds and the chemical ordering, or otherwise, of atoms comprising an alloy.

We have touched on the dihedral angle distribution and its effect on the RDF previously, together with the use that may be made of modelling studies to account for features in the RDF at distances beyond the first coordination sphere. We will therefore concentrate for the rest of this section on the other aspects, cluster formation and chemical ordering.

3.4.1 Cluster formation

This subject is one of the more controversial aspects of the structure of amorphous solids at present, and as such we can do no more than to present the experimental evidence and the arguments for the presence of such large structural units.

One of the strongest pointers to the existence of medium-range correlations is the existence of the so-called 'pre-peaks' in the scattering intensity of a variety of amorphous solids (i.e. prior to the first peak in $F(k)$ characteristic of the nearest-neighbour separation). These peaks invariably are sharp and occur at about 1 Å^{-1}; an example is shown in Fig. 3.28 for amorphous $GeSe_2$. The position of these peaks

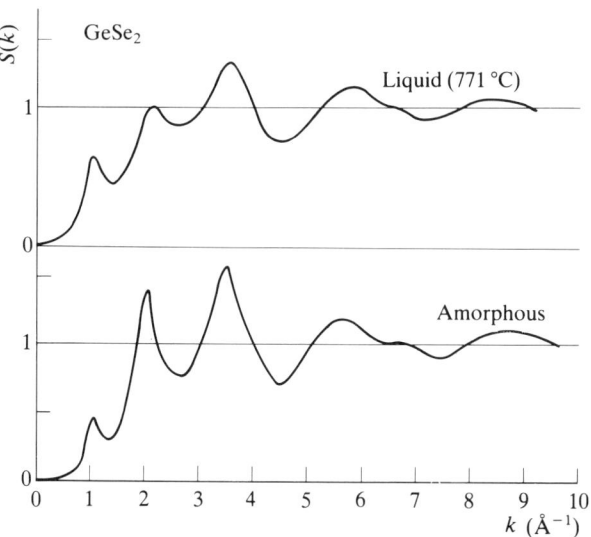

Fig. 3.28 Structure factor for amorphous and liquid $GeSe_2$ (Uemura *et al.* 1978).

in reciprocal space implies that Fourier components of period $\simeq 2\pi/k \sim 6$ Å are involved. Such peaks have been observed in the arsenic chalcogenides (As–S and As–Se systems) and in the germanium chalogenides (Ge–S and Ge–Se systems) as well as elemental B, P, As and Sb, but *not* in Ge, Si or SiO_2. The fact that many of the materials have crystalline polymorphs which have a layer structure with an interlayer separation of ~ 4–6 Å led early workers to suppose that the glasses were composed of microcrystallites and diffraction from the layers produced the pre-peaks. In view of the evidence that the peaks are observed even in the liquid state (cf. Fig. 3.28) this point of view is untenable. So too is the notion that a common structural feature could be responsible for all materials since they have completely differing structures, although interestingly all are low-dimensional solids having small values of the average coordination number. The correlation length, D, over which the separations reponsible for the peak at $k \sim 1$ Å^{-1} are maintained can be gauged from the width of the peak by assuming the Scherrer equation, linking crystallite size to diffraction line profile (see e.g. Warren 1969), is valid in this case:

$$D = \frac{K\lambda}{\beta \cos \theta} \quad [3.63]$$

where K is a constant (~ 0.9), λ is the wavelength of the scattering radiation, β is the half-width of the diffraction line (in radians) and θ is the Bragg angle. Typically, values for $D \sim 15$ Å are inferred from the widths of the pre-peaks for the glasses listed above.

In general, it can be fairly safely stated that the correlations responsible for the pre-peaks are not understood. There is one exception, however, and that is the case of evaporated amorphous As_xS_{1-x} films (Daniel *et al*. 1979). As mentioned previously (sect. 3.3.2), these are believed to be amorphous molecular solids. The scattering intensity of the as-deposited films can be well fitted by a model of a random packing of approximately spherical As_4S_4 molecules; in particular, since the effective molecular diameter of As_4S_4 is ~ 5.5 Å, this probably accounts for the position of the pre-peak. The pre-peak shifts to a position characteristic of the (melt-quenched) bulk glass either on thermal annealing (Daniel *et al*. 1979) or on illumination with band-gap light (de Neufville *et al*. 1973/4) – this is termed the *irreversible* photo-structural effect. Both these processes are believed to involve polymerization and cross-linking of the molecular units to form a fully three-dimensionally bonded solid. The correlations giving rise to the pre-peak in melt-quenched glasses cannot therefore be due to intermolecular separations, and a residual layer-like structure has been suggested as being responsible, either a vestige of crystal-like layers or a stacking of layered clusters or 'rafts' (see later). However, it appears that cation–cation distances are predominantly involved; anomalous scattering results for glassy $GeSe_2$ show that Ge–Ge correlations are principally responsible (Fuoss 1980), and a similar conclusion can be inferrred from a comparison of neutron and X-ray scattering data for As_2S_3 (since $f_{As} > f_S$ but $b_{As} \simeq b_S$) where the pre-peak in the X-ray data appears to be stronger and better defined than in the neutron data (Daniel *et al*. 1979). Any plausible model for the pre-peak must therefore be able to account for this feature. A further interesting facet is that well-annealed films or bulk glasses of Ge and As chalcogenides also undergo a photo-structural change, evinced by a change in position and width of the pre-peak, and this effect is *reversible* since by annealing to T_g the material is restored to its unilluminated state (see e.g. Tanaka 1980). The mechanism for this phenomenon is not understood.

An experimental technique, in principle sensitive to medium-range structural organizations and which has not been mentioned so far is high-resolution electron microscopy (HREM). Two configurations employed in imaging are shown in Fig. 3.29: these are the so-called 'bright-field' mode in which the undiffracted beam and

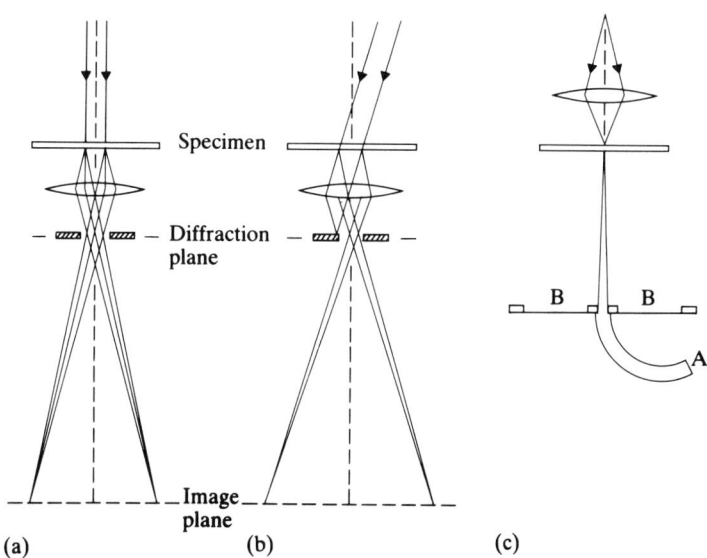

Fig. 3.29 Electron microscope imaging modes: (a) and (b) are for conventional microscope in bright-field and dark-field modes respectively – the diffraction and image planes are indicated; (c) scanning transmission electron microscopy (STEM) geometry. Bright-field images (with energy analysis) can be obtained from the axial detector A, dark-field images from the annular detector B. Microdiffraction patterns can be obtained either by post-specimen scanning or by photographing a fluorescent screen at the annular detector position.

the first diffraction ring are accepted by the aperture and are both used to construct a real-space image (Fig. 3.29(a)), and the 'dark-field' mode in which the undiffracted beam is excluded and only the first diffraction halo is used. Real-space images can be recorded by photographing the image formed in the image plane, and the diffraction pattern formed in the diffraction plane can also be photographed if required. The technique is potentially powerful because only very small areas of the specimen ($\sim 30 \times 30$ Å^2) are probed since the focused electron beam spot is so small, and thus it is potentially sensitive to medium-range structural details, unlike the other diffraction techniques in which gross averaging (over areas of mm^2 or cm^2) necessarily occurs.

Interest was quickened in this technique as a result of the observation of small regions of parallel fringes in the real-space images of several amorphous solids, C, Ge, Si, As and SiO$_2$ and certain metal glasses (see Fig. 3.30). The presence of such

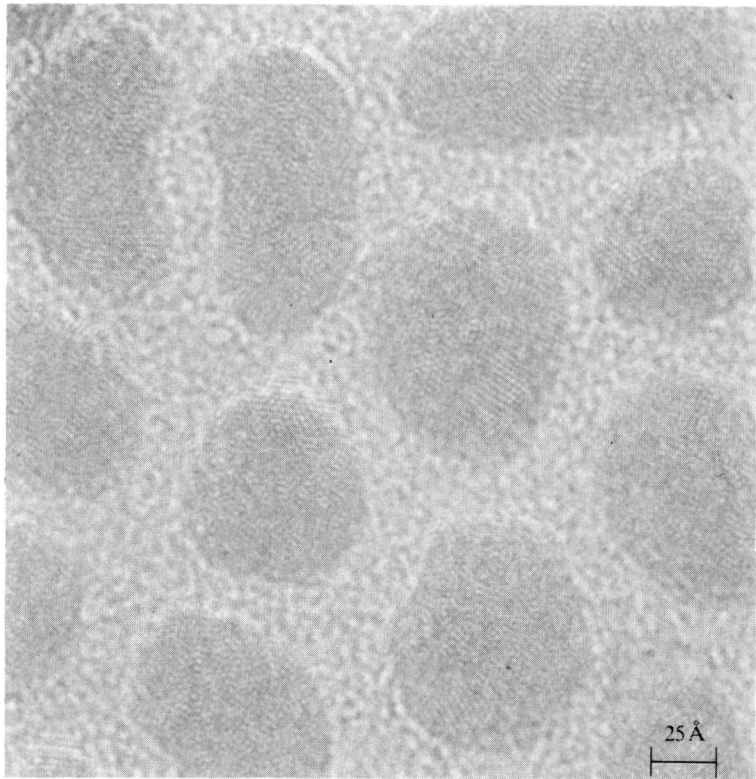

Fig. 3.30 High resolution electron micrograph of particles of an evaporated amorphous palladium-silicon alloy recorded at an accelerating voltage of 500 kV. Small fringe packets with spacings around 2.3 Å are visible in most particles (P. H. Gaskell and D. J. Smith (1980) *J. Microscopy* **119**, 63).

fringes can be interpreted as signalling the presence of regions of coherently diffracting material having linear dimensions up to ~15 Å. These were originally taken as evidence for the presence of microcrystallites, a conclusion which was considerably weakened by the finding that under certain conditions the fringes can be an experimental artefact. However, it appears now that some fringes are genuine, but are only observed in very thin samples of thickness less than ~50 Å where diminution of contrast due to randomly oriented coherent regions (overlap) is minimized. The question remains whether random structures, such as CRNs, can account for the fringy images; Alben *et al.* (1976) have demonstrated that certain tetrahedral CRNs do have planar correlations occurring accidentally in certain directions, but it is not certain whether the predicted contrast is sufficiently high to account for the experimental images. It is interesting to note that several of the amorphous materials which have been reported as exhibiting fringes in HREM images (e.g. C, As) have layered crystalline polymorphs, and the fringes may represent a manifestation of residual layer-like, medium-range order in the amorphous phase.

Strong evidence for the existence of large structural units has come from vibrational spectroscopy. As an example, we show in Fig. 3.31 Raman spectra for the

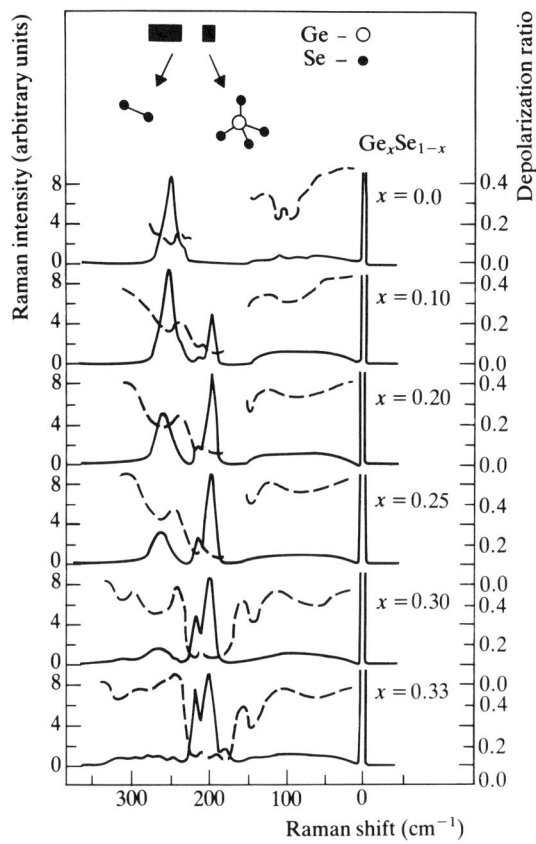

Fig. 3.31 Raman (solid line) and depolarization (dashed line) spectra of Ge_xSe_{1-x} excited with a Kr^+ laser (7993 Å line). Note the appearance of the 'companion' line at $\simeq 220\,cm^{-1}$ with increasing Ge content. Also shown schematically are the relevant local atomic clusters and the frequency range of their bond-stretching modes (Nemanich *et al.* 1977).

a-Ge_xSe_{1-x} system ($0 < x < 0.33$), which can be regarded compositionally as the progressive dilution of —$GeSe_4$ units by Se. A single Raman-active mode is expected for a decoupled tetrahedral unit on a simple bond-stretching and bond-bending model (the symmetric breathing mode with Ge atoms fixed), the intensity of which would be expected to scale linearly with x as the tetrahedra are broken up by the Se excess. Indeed a peak is observed at $\sim 200\,cm^{-1}$ (in good agreement with calculation) and which scales in the manner expected. However, another peak is also observed at $\sim 220\,cm^{-1}$ which has a much stronger dependence of intensity on composition, and is therefore anomalous in this respect. This feature has been ascribed to the presence of large, preferred clusters in the stoichiometric glass which are more rapidly broken up than the basic —$GeSe_4$ tetrahedra on addition of excess Se. Two models have been suggested (Fig. 3.32); a symmetric ring containing 6 Ge and 6 Se atoms (Nemanich *et al.* 1977) and a larger 'raft' containing 12 Ge atoms and 30 Se atoms (similar to the local structure of the β-crystalline form) and which is bordered by Se—Se 'wrong' bonds (see e.g. Phillips 1981). The anomalous peak is proposed to arise, respectively, from symmetric displacements of the ring, at

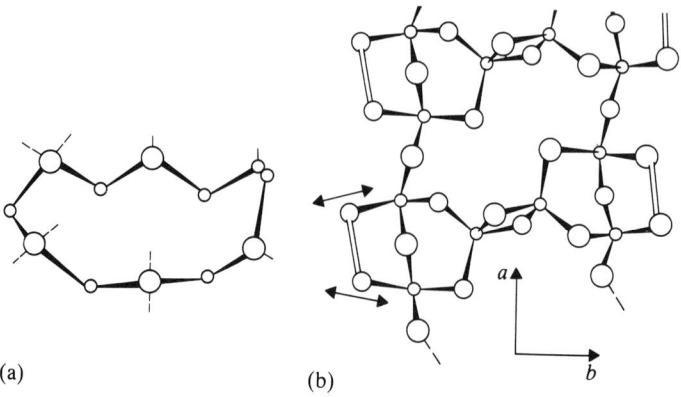

Fig. 3.32 Proposed medium-range structure in a-GeSe$_2$:
(a) six-membered ring (Nemanich *et al.* 1977);
(b) 'raft' bordered by Se—Se wrong bonds (Phillips 1981).

frequencies calculated to lie in the range 217–224 cm^{-1}, or from local motions of the wrong-bonded Se atoms of the raft (although no calculations for the frequency of this mode have been made). Phillips has also suggested, by analogy, similar large clusters in arsenic chalcogenide glasses.

The observation of a very sharp polarized Raman-active mode at 808 cm^{-1} in vitreous B$_2$O$_3$ has led to the suggestion that planar 'boroxol' rings (three —BO$_3$ triangles linked at the apices) are present, the Raman mode arising from symmetric breathing motions of the three O atoms in the ring (Krogh-Moe 1969). However, the presence of a large concentration of boroxol rings in the structure seems to be ruled out on density considerations (Williams and Elliott 1982), although the existence of such a sharp Raman peak is unusual, if not unique, among oxide glasses.

3.4.2 Chemical ordering

Thus far, we have tended to concentrate on alloy systems having stoichiometric compositions. However, an interesting problem arises for those compositions off stoichiometry and which perforce must contain 'wrong' bonds, as to what proportion of the various bond types occur. While this can be viewed as a problem in short-range structure, it has considerable bearing on the maintenance of any medium-range order.

The simplest case to consider is a covalent binary alloy system $A_{1-x}B_x$ which, if the atoms A and B are in columns a and b of the periodic table, will have ideal coordinations of $8-a$ and $8-b$, respectively, satisfying their normal valence and the '$8-N$' rule (strictly valid only for groups IV–VII). We neglect for the time being any coordination defects, such as dangling bonds. In general, there can be A—A, A—B and B—B bonds in an alloy of arbitrary concentration, and two models exist which describe the distribution of such bond types (Lucovsky *et al.* 1977). The first is the *random covalent network* (RCN) model which, as its name implies, treats the distribution of bond types as being purely statistical, determined only by the local coordinations $y_a = (8-a)$ and $y_b = (8-b)$ and the concentration x. Thus, any effects

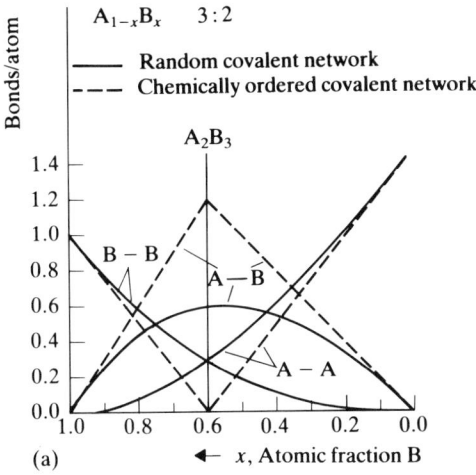

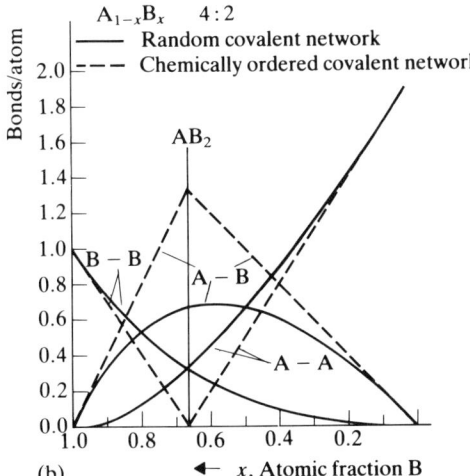

Fig. 3.33 Bond-counting statistics (after Lucovsky and Hayes 1979) for (a) 3:2 networks (e.g. As$_2$Se$_3$) and (b) 4:2 networks (e.g. SiO$_2$).

leading to preferential ordering (e.g. relative bond energies) are neglected. The RCN model admits A—A, A—B and B—B bonds at all compositions except at $x = 0$ and 1. The other approach is the *chemically ordered network* (CON) model which assumes that heteropolar A—B bonds are favoured; a completely chemically ordered phase thus occurs at a composition $x_c = y_a/(y_a + y_b)$ (e.g. As$_2$Se$_3$, GeS$_2$). In contrast to the RCN model, only A—A and A—B bonds are allowed for A-rich compositions ($0 < x < x_c$), and conversely B—B and A—B bonds only for $1 < x < x_c$. The bond statistics for the two models are shown in Fig. 3.33 for (3:2) and (4:2) alloys. There is considerable evidence from vibrational spectra that the CON model is obeyed for most chalcogenide systems.

The RCN and CON approaches are obviously appropriate only for covalently bonded systems and not for random close-packed structures such as metallic or ionic glasses. Nor are they appropriate for covalent network-forming materials which are alloyed with network modifiers (sect. 2.5) for which two cases may be distinguished: 'network rupture' in which the number of fully coordinated network-forming anions (e.g. O) is progressively *decreased* and the connectivity of the network is broken on addition of modifier (e.g. the $Na_2O:SiO_2$ system; Fig. 2.11), and 'network transformation' in which the coordination of the network-forming cation is progressively *increased* (e.g. the $Na_2O:B_2O_3$ system in which the coordination of B, $\langle B \rangle$, increased from 3 to 4, and $Na_2O:GeO_2$ in which $\langle Ge \rangle$ increases from 4 to 6).

II MACROSCOPIC STRUCTURE

In our discussion of the structure of amorphous solids so far, we have implicitly assumed the structure to be homogeneous on a macroscopic scale and concerned ourselves solely with a consideration of the fine details of short- and medium-range order. However, many amorphous solids exhibit gross inhomogeneities on the macroscopic scale, such as voids in vapour-deposited material or phase separation for certain compositions of multicomponent melt-quenched glasses. A discussion of the experimental techniques sensitive to such coarse features is therefore desirable, and we follow this with a consideration of the factors that can give rise to macroscopic structure in amorphous materials.

3.5 Experimental techniques

3.5.1 Microscopy

The most direct evidence for gross structural inhomogeneities comes from microscopy. If the different regions are sufficiently large (say greater than 10 μm), optical microscopy can be used to study the macroscopic structure. In most cases, however, macroscopic features occur on considerably smaller length scales and the greater resolving power of electron microscopy must be used. In certain cases where there are *density* fluctuations (as, for example, in the void network of evaporated a-Ge), these may be observed directly by electron microscopy (see Fig. 3.24(a)). In the other major type of macroscopic structure, that involving *compositional* fluctuations, the phase-separated regions may be revealed by etching, thus transforming the compositional variation into density variations which are observable by electron microscopy. For example, glasses in the $Na_2O-B_2O_3-SiO_2$ system containing more than 20 wt.% of B_2O_3 separate into two interconnecting phases, one being silica rich; the other is preferentially attacked by hydrochloric acid, leaving behind the siliceous skeleton which is very apparent in electron microscopic images. A similar system is $PbO-B_2O_3-Al_2O_3$, and a micrograph of the phase-separated structure is shown in Fig. 3.34.

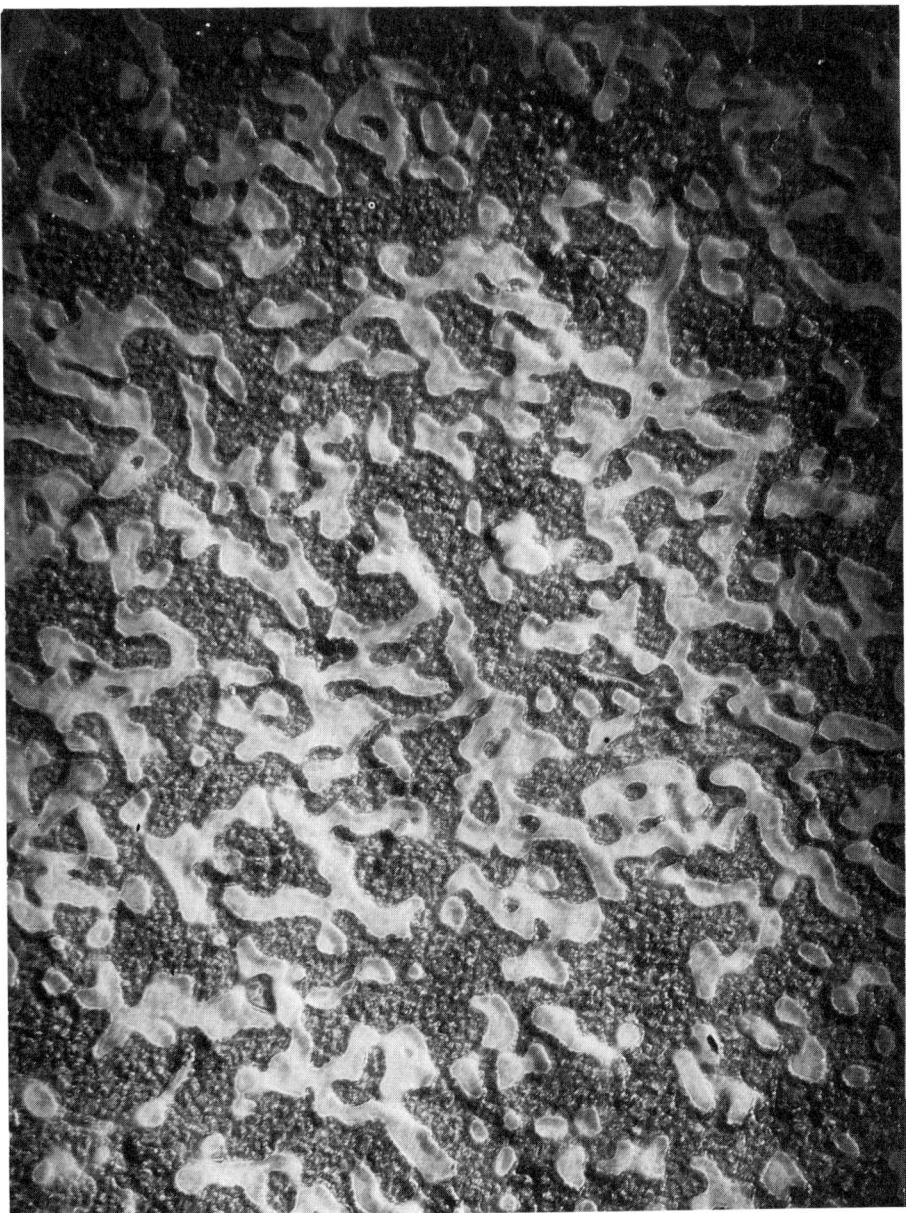

Fig. 3.34 Electron micrograph of phase separation in 77% B_2O_3–18% PbO–5% Al_2O_3. The micrograph is of a replica of an etched fracture surface (magnification × 50 000). (Micrograph by courtesy of Professor J. Zarzycki.)

3.5.2 Small-angle scattering

In addition to the scattering of radiation by atoms as discussed in section 3.2.1, larger scale compositional or density fluctuations can also coherently scatter radiation but in a different region of k-space to that for conventional (sometimes called 'wide-angle') scattering. By the reciprocity relation between real and reciprocal space, the scattering from large-scale features will be confined to small values of k, i.e. 'small-angle scattering'. For example, the scattering from a collection of spheres of diameter D will be confined between the angles $\theta = 0$, where all the centres scatter in phase, and $\theta \sim \lambda/2D$ where destructive interference occurs. Thus, the presence of macroscopic inhomogeneities is signalled by a large increase in the scattered intensity as k tends to zero, and furthermore the functional form of the intensity can yield information on the size, shape and number of such inhomogeneities.

The small-angle scattering intensity is calculated in a similar manner to that for wide-angle scattering (sect. 3.2.1), except that (e.g. for X-ray scattering) an assumed constant density deficit, $\Delta\rho$, is introduced which is the difference in electron density between the two phases of material (whether between void and matrix, or two compositionally different phases). The small-angle X-ray scattering (SAXS) is calculated for an inhomogeneous system modelled by a collection of particles of one type of material embedded in a homogeneous matrix of the other. Since the structure factor $S_h(k)$, and hence the X-ray scattering intensity, for a *homogeneous* medium tends to zero as $k \to 0$, the SAXS intensity (in electron units) can be taken to a very good approximation to be due solely to the dispersed particles

$$I_{SA}(k) = \Delta f^2 |S_p(k)|^2 \qquad [3.64]$$

where Δf is the difference in scattering amplitude for the two phases, and $S_p(k)$ is the structure factor for the particles only. This can be written in the form

$$I_{SA}(k) = (\Delta\rho)^2 \left| \int \rho_c(r) \, e^{ik \cdot r} \, d^3r \right|^2 \qquad [3.65]$$

where $\rho_c(r)$ is defined such that for r inside the particle, $\rho_c = 1$ and for r outside the particle, $\rho_c = 0$. The final expression for the SAXS intensity for an assembly of equally but arbitrarily shaped particles of maximum dimensions, a, is (Letcher and Schmidt 1966):

$$I_{SA}(k) = (\Delta\rho)^2 V(a) N(a) \int_0^\infty 4\pi r^2 \gamma_0(r) \frac{\sin kr}{kr} \, dr \qquad [3.66]$$

where $V(a)$ is the volume of a particle, $N(a)$ is the concentration of the particles and $\gamma_0(r)$ is defined by:

$$\gamma_0(r) = \int \rho_c(R) \rho_c(R+r) \, d^3R \qquad [3.67]$$

which is the probability that for an arbitrary point R within a particle, a vector of unit length drawn in an arbitrary direction also lies within the particle. Therefore, $\gamma_0(r=0) = 1$ and $\gamma_0(r \geq a) = 0$. For a set of particles which do not have the same dimensions (i.e. 'polydisperse'), the factor $\gamma_0(r)$ in [3.66] is replaced by:

$$\gamma(r) = \int_0^\infty N(a) V(a) \gamma_0(r) \, da \qquad [3.68]$$

Comparison of [3.66] for SAXS with the corresponding expression for conventional (wide-angle) scattering ([3.14]) shows that in the latter case, it is the local variation in (atomic) density which determines the scattering, whereas for SAXS it is variations in *shape* which governs the scattering intensity. In a similar fashion to the Fourier inversion of the wide-angle scattering data to extract the reduced RDF ([3.17]), the SAXS data may also be inverted to isolate the shape factor $\gamma(r)$:

$$\gamma(r) = \frac{1}{2\pi^2(\Delta\rho)^2 r} \int_0^\infty k \sin kr \, I_{SA}(k) \, dk \quad [3.69]$$

While this gives some information on the dimensions of the dispersed particles (from the value of r at which $\gamma(r) \to 0$) a more useful quantity is $N(r)$, which is much more difficult to obtain in the general case. The expression for the SAXS intensity of a dilute dispersion of particles of volume v may be approximated as (Guinier and Fournet 1955):

$$I_{SA}(k) = Nv^2(\Delta\rho)^2 \exp\left(-\frac{R^2 k^2}{3}\right) \quad [3.70]$$

where R is the 'electronic radius of gyration' given by:

$$R^2 = \frac{\int r^2 \rho(r) \, dv_r}{\int \rho(r) \, dv_r} = \frac{\int r^2 \, dv_r}{\int dv_r} \quad [3.71]$$

for a homogeneous particle of density $\rho(r) = \rho_0$. For example, the radius of gyration of a spherical particle of diameter L is given by $R = (3/5)^{1/2} L/2$. Thus, examination of [3.70] shows that a plot of $\ln I_{SA}(k)$ v.s. k^2 should be linear, of slope $-R^2/3$; this is the so-called 'Guinier plot'. In practice, the Guinier plot often deviates from linearity which is an indication of a polydispersed system of particles of varying shapes and sizes. The slope as $k \to 0$ gives a value for the weighted radius of gyration R_G defined by:

$$R_G^2 = \frac{\sum_i R_i^2 v_i^2 P_i}{\sum_i v_i^2 P_i} \quad [3.72]$$

where R_i, v_i and P_i are the gyration radius, volume and fraction of particles of species i. The determination of particle shapes from Guinier plots is not straightforward, and the interested reader is referred to the book by Guinier and Fournet (1955). An expression involving the concentration and volume of particles can be obtained from the zero intercept of the Guinier plot:

$$I_{SA}(0) = (\Delta\rho)^2 Nv^2 \quad [3.73]$$

or from the integrated intensity given by:

$$Z = \frac{1}{2\pi^2} \int_0^\infty k^2 I_{SA}(k) \, dk \quad [3.74]$$

since it can be shown that:

$$Z = (\Delta\rho)^2 Nv \quad [3.75]$$

It should be stressed that the Guinier approximation only holds for an inhomogeneous system composed of a matrix containing *widely* separated particles which each scatter independently. This is not so for the situation where one phase is not dilute relative to the other.

Other forms of radiation can also be used in small-angle scattering experiments, and in recent years with the advent of nuclear reactor sources of neutrons, small-angle neutron scattering (SANS) has been much employed. This technique has the advantage that neutrons of long wavelength (e.g. ~ 10 Å) can be used, thereby still satisfying the condition for small k-values, but considerably easing the problems of collimation encountered for small angles.

Note that a finite amount of small-angle scattering is to be expected in certain cases even if no obvious gross macroscopic inhomogeneities, such as voids or phase separation, are present. Variations in density can remain in melt-quenched glasses as configurational (but not vibrational) fluctuations frozen-in on cooling the melt through the glass-transition temperature, T_g. Thermodynamic fluctuation theory predicts that the density fluctuations in a single component liquid are related to the isothermal compressibility κ_T by:

$$\frac{V\langle(\Delta\rho)^2\rangle}{\rho^2} = k_B T \kappa_T \qquad [3.76]$$

where V is the volume. However, the density fluctuations in a glass are the same as those existing in the liquid at T_g and so the structure factor becomes (see problem 3.3):

$$S(0) = \rho k_B T_g \kappa_T(T_g) \qquad [3.77]$$

Thus, the small-angle scattering for a glass does not in fact tend to zero as k tends to zero, but instead approaches the 'compressibility limit' given by [3.77]. The SAXS from glassy SiO_2 and GeO_2 is shown in Fig. 3.35, whence it can be seen that the intensity decreases monotonically from the first peak to a small but finite intercept. Values for the density fluctuation $\langle(\Delta\rho)^2\rangle$ obtained from experiment and calculated using [3.77] agree to within a factor of 1.5 (Pierre *et al.* 1972).

3.6 Examples of macroscopic structure

3.6.1 Defects in growth morphology

Amorphous materials, particularly prepared by vapour deposition on cold substrates, are often not macroscopically homogeneous because of growth morphology defects. We have in fact alluded to several instances of this previously: e.g. the network of voids in vapour-deposited a-Ge and a-Si (Fig. 3.24(a)); the columnar growth morphology of obliquely deposited films of a-Ge and a-Ge chalcogenides (sect. 1.3.1 and Fig. 1.5) and also glow-discharge deposited a-Si:H (Fig. 3.24(b)).

The existence of small voids in evaporated a-Ge, for example, is strikingly demonstrated by SAXS, as shown in Fig. 3.36. The small-angle scattering increases dramatically as k tends to zero, and can be ascribed to voids with radii between 5 and 40 Å which occupy about 5% of the total volume; the density deficit relative to the crystal obtained by weighing is in reasonable agreement at $\sim 11\%$. Samples

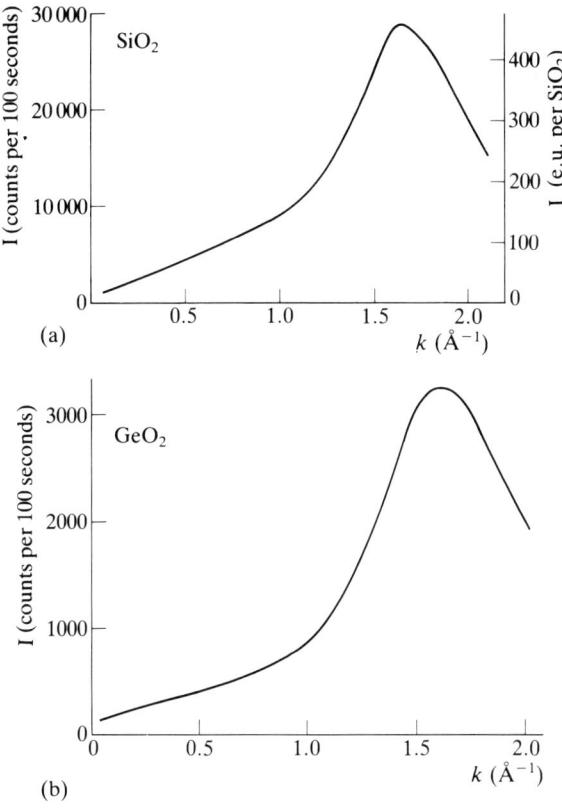

Fig. 3.35 Small-angle X-ray scattering for two oxide glasses showing the 'compressibility' limit at $k=0$ (Pierre et al. 1972).

obtained by sputtering or electrolytic deposition exhibit a progressive decrease in the magnitude of the SAXS (although the density deficit measured by weighing is comparable), suggesting that the majority of the void space in these two cases lie in voids less than ~ 6 Å in diameter, for which the SAXS extends under the first peak in $I(k)$ and hence is obscured. Notice particularly from Fig. 3.36 that the SAXS from chalcogenide glasses (and vapour-deposited thin films) shows no evidence for the presence of voids, but instead tends to the compressibility limit, as for SiO_2 and GeO_2 (Fig. 3.35). An increase in the SAXS is however observed in vitreous SiO_2 after irradiation by high-energy neutrons. The increase in the SAXS is uniform with k and the very sharp rise at small angles as observed in evaporated a-Ge is not seen. Interestingly, the density of the glass increases by $\sim 3\%$ after irradiation; the density of the crystalline form (quartz), on the other hand, decreases by $\sim 14\%$ to the *same* density on amorphization for high doses. The SAXS and the increase in density of the neutron-irradiated glass both disappear on annealing to near T_g. The origin of this effect is believed to be the 'thermal spikes' induced by the neutron bombardment which cause local volumes of damage. The paramagnetic nature of the defects induced by neutron bombardment will be discussed in Chapter 6.

Columnar growth has been observed in many amorphous and crystalline

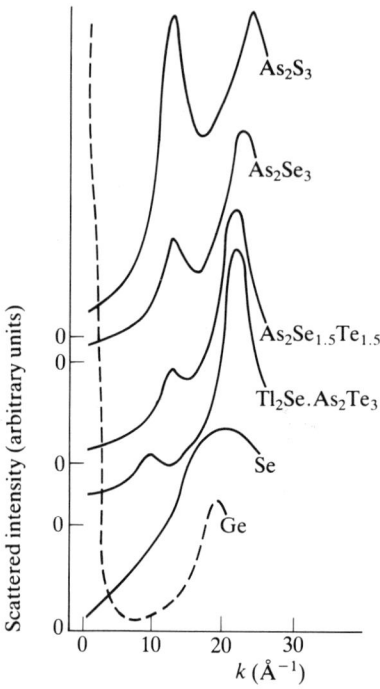

Fig. 3.36 Small-angle X-ray scattering for chalcogenide glasses and amorphous germanium (Bishop and Shevchik 1974). Note the absence of SAXS for the chalcogenide glasses, but the large intensity at small k for a-Ge arising from the presence of small voids.

vapour-deposited thin films (Leamy et al. 1980). Small-angle scattering again is a sensitive and quantitative probe for such macroscopic structural features. The orientation of such columns with respect to the substrate (and consequently with respect to the original vapour beam direction) can be ascertained by tilting the film relative to the X-ray beam, or the electron beam in an electron microscope. The small-angle scattering intensity approximately follows the shape transform of the assembly of particles ([3.64]), which is a torus for the case of parallel, uniformly separated columns. The column orientation may thus be ascertained by the position at which the diffraction plane of the tilted plane bisects the toroid. The orientation and dimensions of the columns may be measured in this single experiment, since the separation between columns is proportional to the reciprocal of the radius of the torus. Intriguingly, the column orientation is *not* found to be parallel to the angle of incidence of the vapour, but rather the following relation is generally obeyed:

$$2 \tan \beta = \tan \alpha \qquad [3.78]$$

where α and β are the angles between the vapour beam and the substrate normal, and the column orientation and the substrate normal, respectively (see Fig. 1.4). Leamy et al. (1980) discuss this phenomenon in more detail.

3.6.2 Phase separation

An obvious example of a macroscopic inhomogeneity produced by phase separation is the nucleation and growth of crystallites within an amorphous solid. The presence of these can certainly be detected by the techniques mentioned in section 3.5, particularly since the crystallites would produce sharp Bragg peaks in any scattering experiment if they were of sufficient size. However, we will not consider this situation further, but restrict our attention to the compositional (i.e. glassy → glassy) phase segregations that can occur because of immiscibility of the melt in certain systems.

To discuss this problem in detail, we need to consider phase separation (or immiscibility) of liquids. The simplest example concerns a binary system, AB, which can exist in the liquid phase either as a single phase or as a two-component mixture, depending on the temperature. Phase separation will occur when the free energy of the phase-separated mixture is less than that of the single homogeneous phase. The Helmholtz free energy, F, is related to the internal energy, U, and the entropy, S, at any temperature, T, by:

$$F = U - TS \qquad [3.79]$$

The entropy term can be regarded as arising from the entropy of mixing at any composition c:

$$S_{mix} = -Nk_B[c \ln c + (1-c) \ln (1-c)] \qquad [3.80]$$

If U for the single phase is greater than for a mixture of the components, the free energy curve will exhibit minima at *two* compositions in general, since the $-TS$ curve is a symmetric function of composition having a minimum at $c = \frac{1}{2}$, whereas U generally is an asymmetric function of composition having a maximum, but the endpoints for the pure components are not equal. Thus the free energy–composition curve in this case is as shown in Fig. 3.37(b). Between the compositions marked a and b, the free energy of the phase-separated mixture at c is lower than that of the single homogeneous phase (given by the solid curve), and hence phase separation takes place in the melt. The composition range a–b at that temperature marks the 'miscibility gap'. The phase diagram for the system is thus as shown in Fig. 3.37(a). At a certain critical temperature, T_{CO}, the two minima in the F v.s. c curve coalesce and a single homogeneous phase is now stable for all $T \geqslant T_{CO}$; this temperature is known as the 'upper consolute' temperature.

Complications can arise if crystalline phases are formed below a certain temperature. The phase diagram for such a case is shown in Fig. 3.38(a). At temperature T_1 the liquid has the lower F for all c, and since it exhibits a single minimum, a homogeneous liquid phase is formed. At T_2 two minima have developed and so between compositions a and b two liquid immiscible phases form, but for compositions very close to pure B the solid phase has the lower energy. At T_3 the solid phase and a single component phase are stable for compositions along the tangent line. For a supercooled liquid, such as might form a glass, the immiscibility limits extend below the liquidus and are simply continuously extrapolated. The composition q is called the 'immiscibility limit' and is important for the consideration of glass formation in such a system, since for compositions to the right of q

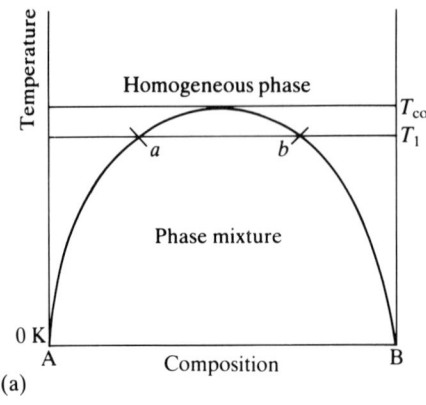

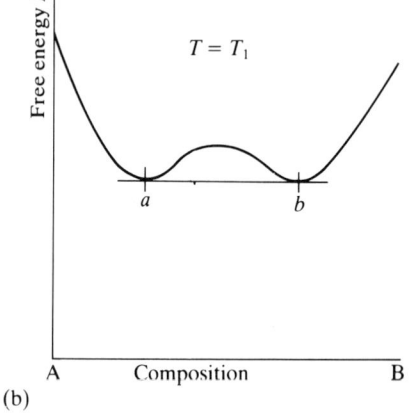

Fig. 3.37 Phase diagram (a) and composition dependence of the free energy (b) for a simple binary AB which can exist either as a single phase or a two-component mixture in the liquid phase depending on the temperature. The temperature marked T_{co} in (a) is the upper consolute temperature, and the composition range a–b in (b) marks the miscibility gap (after Rawson 1967).

a two-phase melt is formed which will produce a two-phase glass on cooling. For both this and the former case, if a glass is formed by quenching the immiscible melt, a very coarse-grained phase-separated glass ensues, which is noticeably inhomogeneous to the naked eye (i.e. opalescent).

The final case we will consider which can lead to liquid immiscibility is indicated in Fig. 3.39, in which the two-liquid immiscibility region (i.e. the upper consolute temperature) is only encountered for temperatures *below* the liquidus. For temperatures below T_2, a solid crystalline phase is coexistent with a single liquid phase for the phase diagram given, but if the liquid is supercooled until the immiscibility region is encountered (e.g. at T_3), the liquid (or the glass) will show a propensity to separate into two phases. This 'metastable immiscibility' gives rise to phase separation on a very small scale and such glasses appear homogeneous (transparent) to the naked eye, but not, of course, in electron microscopy or small-angle scattering. Its likelihood is signalled by the liquidus having an approximately

Examples of macroscopic structure

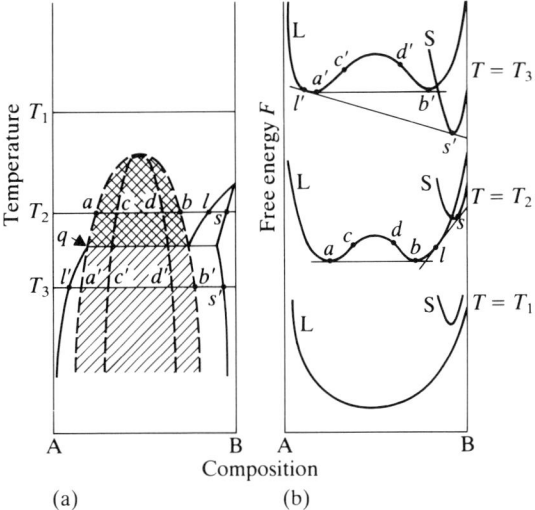

Fig. 3.38 (a) Schematic phase diagram showing the phase boundary and spinodal (dotted line) of a two-liquid immiscibility region (cross-hatched) with a sub-liquidus extension (hatched), together with the solid solubility behaviour of an unrelated third crystalline phase. The composition q is termed the immiscibility limit.
(b) Free energy–composition curves of the liquid (L) and solid (S) phases for the temperatures shown in (a) (after Rawson 1967).

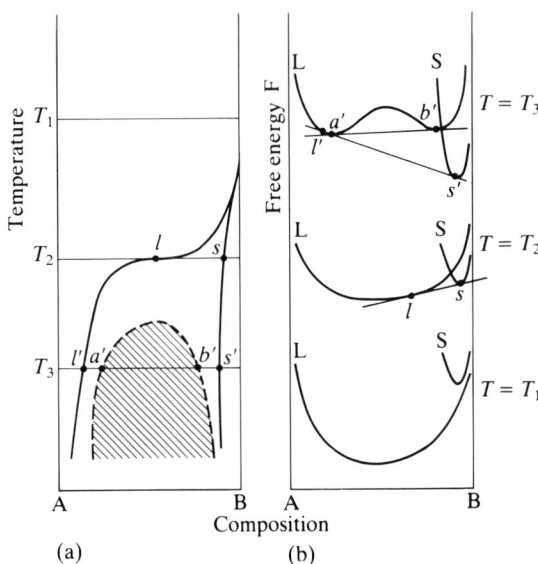

Fig. 3.39 (a) Schematic phase diagram showing an entirely sub-liquidus two-liquid immiscibility region (hatched) and an unrelated third crystalline phase.
(b) Free energy–composition curves of the liquid (L) and solid (S) phases for the temperatures shown in (a) (after Rawson 1967).

127

flat section or a point of inflection. It is this 'microphase separation' extending over some hundreds of ångströms, and the forms of separation, with which we concern ourselves for the remaining part of this chapter. It is of some considerable commercial and technological importance, since the Vycor process for the manufacture of non-crystalline silica utilizes the phase separation into almost pure silica and sodium borate phases which occurs in Na_2O–B_2O_3–SiO_2 glasses; the borate regions are selectively removed by acid leaching and the resulting silica framework compacted by subsequent heat treatment. In addition, glass–ceramic materials are made by controlled crystallization of one portion of a microphase-separated glass.

The texture of a microphase-separated glass is dependent to a certain degree, particularly during the initial stages of segregation, on the mechanism and dynamics of decomposition. Two distinct mechanisms may be distinguished which differ in their dependence of free energy on composition. Gibbs originally demonstrated that a phase is stable (or metastable) to an infinitesimal fluctuation in composition if the chemical potential of each component increases with increasing density of that component, or equivalently $(\partial^2 F/\partial c^2) > 0$. In this region, only fluctuations greater than a certain size are stable and hence can lead to phase separation; all others spontaneously disappear. Thus if $(\partial^2 F/\partial c^2) > 0$ phase separation occurs by a process of 'nucleation and growth'. Conversely, for a system for which $(\partial^2 F/\partial c^2) < 0$, the single phase is unstable to *all* fluctuations, of any size, and hence phase separation does not involve a nucleation step, being instead diffusion limited (so-called 'spinodal decomposition'). The boundary in a binary phase diagram separating the unstable from the stable region is given by the locus of $(\partial^2 F/\partial c^2) = 0$, and is known as the 'spinodal' (see Figs 3.38 and 3.39).

Cahn (1965) has given a theory for the initial stages of the spinodal decomposition in terms of the growth of certain Fourier components of composition fluctuations. The preferred growth of only a few of these spatial components results in the formation of a quasi-periodic texture of the phase-separated glass of wavelength corresponding to the wavevector of the favoured component. This can be seen as follows. The free energy of an inhomogeneous solution containing small concentration gradients can be written as:

$$F = \int [f(c) + \kappa(\nabla c)^2] \, dV \qquad [3.81]$$

where $f(c)$ is the free energy density for a homogeneous medium of concentration c, and $\kappa(\nabla c)^2$ represents the increase in free energy arising from a non-uniform concentration which, like surface tension, minimizes the extent of the interfacial region. The free energy density may be expanded about the average composition, c_o, by means of a Taylor series:

$$f(c) = f(c_o) + (c - c_o)\left(\frac{\partial f}{\partial c}\right) + \frac{(c - c_o)^2}{2}\left(\frac{\partial^2 f}{\partial c^2}\right) + \cdots \qquad [3.82]$$

Since by definition $\int (c - c_o) \, dV = 0$, the free energy difference between the homogeneous and inhomogeneous phases is:

$$\Delta F = \int \left[\frac{1}{2}\left(\frac{\partial^2 f}{\partial c^2}\right)(c-c_o)^2 + \kappa(\nabla c)^2 \right] dV \qquad [3.83]$$

Consider the composition fluctuation to be made up from a series of Fourier components such as

$$(c-c_o) = A \cos \beta x \qquad [3.84]$$

The contribution of this component to the free energy difference is obtained by substitution in [3.83] and integrating

$$\Delta F_\beta = \tfrac{1}{4} V A^2 \left[\left(\frac{\partial^2 f}{\partial c^2}\right) + 2\kappa\beta^2 \right] \qquad [3.85]$$

The critical wavenumber β_c below which spinodal growth occurs is given by:

$$\beta_c^2 = -\frac{1}{2\kappa}\left(\frac{\partial^2 f}{\partial c^2}\right) \qquad [3.86]$$

The kinetics of the initial stages of spinodal decomposition can be obtained by solving the diffusion equation in a linearized form:

$$\frac{\partial c}{\partial t} = M\left(\frac{\partial^2 f}{\partial c^2}\right)\nabla^2 c - 2M\kappa\nabla^4 c \qquad [3.87]$$

where M is an atomic mobility ($M>0$). The solution to this equation is:

$$(c-c_o) = \exp\left[R(\beta)t\right]\cos(\boldsymbol{\beta}\cdot\boldsymbol{r}) \qquad [3.88]$$

where

$$R(\beta) = -M\left(\frac{\partial^2 f}{\partial c^2}\right)\beta^2 - 2M\kappa\beta^4 \qquad [3.89]$$

The amplification factor, $R(\beta)$, is positive in the spinodal region and since $M = D/(\partial^2 f/\partial c^2)$, it can be seen from [3.89] that in this region the diffusion constant, D, is *negative* (i.e. 'uphill' diffusion). This is because the attraction between like species is so great that the flux of atoms opposes the concentration gradient. Substitution of $M = D/(\partial^2 f/\partial c^2)$ into [3.89] yields:

$$R(\beta) = -D\beta^2(1-\beta^2/\beta_c^2) \qquad [3.90]$$

which is sharply peaked at the 'spinodal wavenumber' of the system

$$\beta_m = \beta_c/\sqrt{2} \qquad [3.91]$$

The spatial components of concentration fluctuations of this wavenumber receive maximum amplification in the early stages of the spinodal growth and hence give rise to a phase-separated glass having a texture of characteristic wavelength $\Lambda_m = 2\pi/\beta_m$. The type of texture expected is shown schematically in Fig. 3.40, which is a section through a computed two-phase structure (the minor phase occupying 24 vol. %) calculated by the addition of a number of random sine waves all having the same wavelength Λ (Cahn and Charles 1965).

The attraction of Cahn's theory is that it is formulated in reciprocal space, and can therefore be tested directly by small-angle scattering experiments. The scattered

Structure

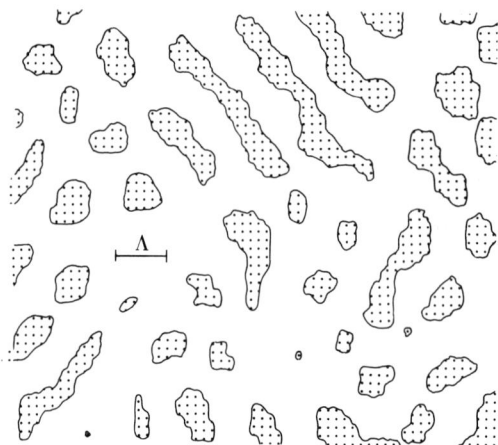

Fig. 3.40 Computed two-phase structure characteristic of spinodal decomposition (Cahn and Charles 1965).

intensity at wavevector k is proportional to $|A(\beta, t)|^2$ for the concentration fluctuation, and furthermore, since $k = \beta$

$$\frac{4\pi \sin \theta}{\lambda} = \frac{2\pi}{\Lambda} \qquad [3.92]$$

where 2θ and λ are the scattering angle and wavelength of the probing radiation, respectively. Thus, a glass which is metastably homogeneous originally and which, after heat treatment at a given temperature, for a given time, decomposes spinodally should exhibit a diffraction intensity which from [3.88] should obey the equation:

$$I(k, t) = I(k, 0) \exp[2R(k)t] \qquad [3.93]$$

Furthermore, since $R(\beta)$ is sharply peaked at β_m, an intensity maximum is expected in the small-angle scattering spectrum at $k = \beta_m$, and this peak should not shift in position with increasing time of heat treatment. Figure 3.41 shows the SAXS spectra for a B_2O_3–PbO–Al_2O_3 glass quenched from 1150°C and heat treated at 450°C for varying lengths of time (Zarzycki and Naudin 1969). The spinodal mechanism is clearly indicated for the initial heat treatments. A plot of $R(k)/k^2$ v.s. k^2 should be linear ([3.90]); from such a plot, the value for the spinodal wavelength $\Lambda_m = 130$ Å was deduced. At longer times, the peak in the SAXS shifts to smaller values of k, indicative of a coarsening process where Cahn's theory is no longer appropriate. Such a process might be the 'Ostwald ripening mechanism' where differences in solute concentration due to various particle sizes (or local curvature of the interface for a fully connected biphasic system) set up concentration gradients which cause the growth of large particles at the expense of smaller ones.

3.6.3 Summary

Amorphous materials in general are *not* as homogeneous as structural models would indicate, and such gross structural imperfections may be detected using optical or electron microscopy or small-angle (X-ray or neutron) scattering.

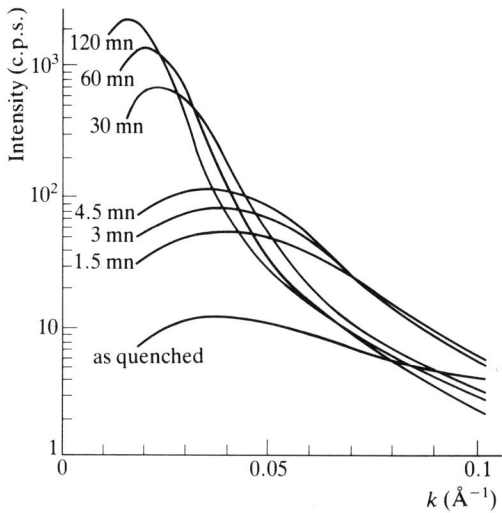

Fig. 3.41 Small-angle X-ray scattering from a 76 $B_2O_3 \cdot 19PbO \cdot 5Al_2O_3$ glass, quenched from 1150°C and heat-treated at 450°C for varying lengths of time (Zarzycki and Naudin 1969).

Amorphous thin films prepared by vapour deposition are almost invariably structurally inhomogeneous to some degree. Voids are often present, and in certain extreme cases can coalesce, resulting in a columnar growth morphology for the film. Glasses prepared by melt-quenching are not macroscopically homogeneous either, but at the very least possess density fluctuations, characteristic of a liquid, and frozen-in at T_g. These density fluctuations are responsible for the non-zero intercept of the structure factor at $k=0$ (the 'compressibility limit').

In certain cases, melt-quenched glasses can undergo varying degrees of phase separation, depending on the miscibility of the components in the melt. If the melt is immiscible (two-phase), the resulting phase-separated glass will be very coarse-grained. If the immiscibility region occurs instead at temperatures below the liquidus, supercooling the melt below this temperature to form the glass will result in a very fine-grained 'microphase' separation. In this case, phase separation can occur either by nucleation and growth, or by a diffusion-limited process called spinodal decomposition.

Problems

3.1 A useful reformulation of the expression for the scattering intensity from a *binary* alloy has been given by Bhatia and Thornton (1970). In this, the individual partial structure factors are replaced by *number–concentration* correlation functions, $S_{NN}(k)$, $S_{NC}(k)$ and $S_{CC}(k)$, where

$$S_{NN}(k) = x_1^2 I_{11}(k) + x_2^2 I_{22}(k) + 2x_1 x_2 I_{12}(k) \qquad [3.1P]$$

$$S_{NC}(k) = x_1 x_2 \{[x_1 I_{11}(k) + x_2 I_{12}(k)] - [x_1 I_{21}(k) + x_2 I_{22}(k)]\} \qquad [3.2P]$$

$$S_{CC} = x_1 x_2 \{1 + x_1 x_2 [I_{11}(k) + I_{22}(k) - 2I_{12}(k)]\} \qquad [3.3P]$$

Structure

(i) Show that the binary form of [3.30], viz.

$$I_{eu}/N - \langle f^2 \rangle = x_1^2 f_1^2[I_{11}(k)-1] + x_2^2 f_2^2[I_{22}(k)-1] + 2x_1 x_2 f_1 f_2[I_{12}(k)-1]$$

can be written in the Bhatia–Thornton form as:

$$I_{eu}/N = \langle f \rangle^2 S_{NN}(k) + 2\langle f \rangle \Delta f S_{NC}(k) + \frac{[\langle f^2 \rangle - \langle f \rangle^2]}{x_1 x_2} S_{CC}(k) \quad [3.4P]$$

where

$$\Delta f = f_1 - f_2 \quad [3.5P]$$

The number–concentration functions have the following physical significance. $S_{NN}(k)$ represents the *topological* short-range order (the analogue of crystalline Bragg peaks due to the average lattice occupied by A and B atoms in a binary alloy). It oscillates around unity and tends to that value as $k \to \infty$. $S_{CC}(k)$ describes the chemical short-range order and for a random distribution of A and B atoms, $S_{CC}(k) = x_1 x_2$; otherwise it oscillates about this value and tends to it as $k \to \infty$. *Positive* values of $S_{CC}(0) - x_1 x_2$ indicates a preference for *like* nearest neighbours (or clustering), whereas *negative* values indicate a preference for *unlike* neighbours (or chemical short-range order). Finally, $S_{NC}(k)$ is zero for random mixtures; otherwise it oscillates about this value and vanishes as $k \to \infty$.

(ii) Show that by suitable isotopic substitution (e.g. in the Ni–Ti system – see sect. 7.2.1 and Table 7.2) the following correlation functions are obtainable *directly* from neutron scattering experiments (in which $f \equiv b$):

(a) $b_1 = b_2$ gives $S_{NN}(k)$
(b) $x_1 b_1 + x_2 b_2 = 0$ ('zero alloy') gives $S_{CC}(k)$
(c) $b_1 = 0$ or $b_2 = 0$ ('null elements') give $I_{22}(k)$ and $I_{11}(k)$, respectively.

3.2 Obtain the equations describing the bond statistics for AX_2- and A_2X_3-type glasses for both the random covalent network and chemically ordered network models (see Fig. 3.33).

3.3 Derive the 'compressibility limit' of the scattering intensity ([3.77]) by considering the density fluctuations in a liquid.

(a) Consider a small volume V_o in a liquid containing N_o molecules. Show that N_o can be written as:

$$N_o = \sum_{i=1}^{N} P(r_i) \quad P(r) = 1 \quad r \text{ in } V_o \quad [3.6P]$$
$$= 0 \quad r \text{ not in } V_o$$

and hence that

$$\langle N_o \rangle = n V_o \quad [3.7P]$$

where n is the number density and

$$\langle N_o^2 \rangle = \int_{V_0} \rho_1(\mathbf{r}_1)\, d\mathbf{r}_1 + \iint_{V_0} \rho_2(r_1, r_2)\, d\mathbf{r}_1\, d\mathbf{r}_2 \qquad [3.8P]$$

where ρ_1 and ρ_2 are the single particle and pair distribution functions respectively. Thus show that the fluctuations in the number of molecules is:

$$\frac{\langle N_o^2 \rangle - \langle N_o \rangle^2}{\langle N_o \rangle} = 1 + \frac{4\pi}{n} \int_{V_0} [\rho_2(r) - n^2] r^2\, dr \qquad [3.9P]$$

Deduce the relation of the right-hand side of [3.9P] with the X-ray scattering intensity ([3.14]).

(b) The grand partition function, for a system containing N particles is defined by

$$\Xi = \sum_N \lambda^N Z_n \qquad [3.10P]$$

where $\lambda = e^{\beta \mu}$ (μ is the chemical potential and $\beta = 1/k_B T$) and Z_n is the partition function $Z_n = \sum_k \exp(-\beta E_{N,k})$. It is a standard result that the average number of particles in the system is

$$\langle N \rangle = \frac{\partial \log \Xi}{\partial (\beta \mu)} \qquad [3.11P]$$

Hence show that $\partial \langle N \rangle / \partial (\beta \mu) = \langle N^2 \rangle - \langle N \rangle^2$ and using the expression for the chemical potential

$$N\, d\mu = -S\, dT + V\, dp \qquad [3.12P]$$

obtain finally [3.77], $S(0) = \rho k_B T \kappa_T$.

Bibliography

P. H. Gaskell (ed.) (1977), *Structure of Non-Crystalline Materials*, vol. I (Taylor and Francis).

P. H. Gaskell, J. M. Parker and E. A. Davis (eds) (1983), *Structure of Non-Crystalline Materials*, vol. II (Taylor and Francis).

Y. Waseda (1980), *Structure of Non-Crystalline Materials: Liquids and Amorphous Solids* (McGraw Hill).

J. Wong and C. A. Angell (1976), *Glass: Structure by Spectroscopy* (Dekker).

A. C. Wright (1974), *Adv. Struct. Res. Diffr. Meth.* **5**, 1.

A. C. Wright and A. J. Leadbetter (1976), *Phys. Chem. Glasses* **17**, 122.

4 Vibrations

4.1	**Introduction**
4.2	**Vibrational excitations**
4.3	**Computational methods**
4.3.1	Introduction
4.3.2	Negative eigenvalue method
4.3.3	Equation-of-motion method
4.3.4	Recursion method
4.3.5	Summary
4.4	**Experimental probes**
4.4.1	Introduction
4.4.2	Infra-red spectroscopy
4.4.3	Raman spectroscopy
4.4.4	Inelastic neutron scattering
4.5	**Low-temperature properties**
4.5.1	Introduction
4.5.2	Specific heat
4.5.3	Thermal conductivity
4.5.4	Theoretical models
	Problems
	Bibliography

4.1 Introduction

In the last chapter we discussed at length the static arrangement of atoms in an amorphous solid, i.e. its structure. In this chapter, we turn our attention to the dynamical behaviour of such atoms and discuss the nature of the vibrational excitations of the atoms from their equilibrium positions.

Atomic vibrations in a solid, whether amorphous or crystalline, are manifested in many ways. In non-metallic systems, heat is transported via vibrational excitations and the specific heat capacity is determined by the spectrum of allowed vibrational modes. Under certain circumstances, electromagnetic radiation can interact with vibrational excitations, thereby affecting the transmission or scattering of light, and making possible the use of infra-red and Raman spectroscopies as vibrational probes. Furthermore, neutrons can be inelastically scattered from condensed matter, either losing or gaining energy to and from atomic vibrations during the scattering event, and thus are a direct probe of the density of vibrational states.

In many respects the vibrational behaviour of amorphous solids is very similar to that of the corresponding crystalline forms, with the exception that selection rules for transitions are relaxed and sharp features in the density of vibrational modes are smeared out due to the lack of periodicity. The next three sections of this chapter are therefore devoted to a discussion of the nature of vibrational excitations in amorphous materials and how the vibrational properties may be calculated, and of the experimental probes that may be employed to explore the vibrational characteristics of amorphous solids. Where amorphous materials differ markedly from their crystalline counterparts is in the realm of vibrational properties at very low temperatures, the heat capacity, thermal conductivity and other vibrationally dependent properties being singularly anomalous for the amorphous state. These low-temperature anomalies appear to be present in most amorphous solids and are unique to the non-crystalline phase; they form the basis of the last section of this chapter.

The reader may prefer to concentrate at first on the section describing vibrational excitations in amorphous (and crystalline) solids (sect. 4.2), followed by a perusal of the sections describing vibrational calculations (sects. 4.3.1 and 4.3.5) and experimental probes of vibrational excitations (sects. 4.4.1–4.4.3). Finally on a first reading, section 4.5 dealing with low-temperature thermal properties should be studied, since the so-called thermal anomalies (in comparison with crystals) appear to be a universal feature of the amorphous state.

4.2 Vibrational excitations

At the outset one must ask the question whether, in the absence of long-range order, propagating vibrational excitations exist. The answer of course is in the affirmative, since it is common experience that window glass transmits both sound and heat. The difference between amorphous and crystalline materials essentially lies in the fact that periodicity cannot be used to simplify the dynamical equations by the introduction of Bloch states.

The calculation of dynamical properties of three-dimensional solids is not a trivial problem (see e.g. Born and Huang 1954 and Maradudin et al. 1971). In general two approximations are widely made, the 'adiabatic' or Born–Oppenheimer approximation, and the 'harmonic' approximation. The former has the potential energy written only in terms of nuclear (not electronic) coordinates, since electronic motion is so much more rapid than nuclear motion (because the ratio of electronic to nuclear mass is of the order of $\sim 10^{-3}$), or equivalently the energies of electronic excitations (in non-metals) are typically larger ($\sim 1\,\text{eV}$) than those characteristic of vibrational excitations ($\sim 10^{-3}\,\text{eV}$). The harmonic approximation is that in which the potential is expanded in powers of the displacement from equilibrium only up to the first non-vanishing, i.e. quadratic, term. The use of the harmonic approximation ensures, as we shall see later, that the elementary vibrational excitations are non-interacting and independent; if terms in the lattice potential higher than quadratic are included, interaction between excitations is introduced, leading to a shift in the frequencies and finite lifetimes of the excitations.

The Hamiltonian for lattice vibrations may be written generally as (see e.g. Madelung 1978):

$$\mathcal{H} = \sum_{n\alpha i} \frac{M_\alpha}{2} \dot{s}_{n\alpha i}^2 + \tfrac{1}{2} \sum_{\substack{n\alpha i \\ n'\alpha' i'}} \Phi_{n\alpha i}^{n'\alpha' i'} s_{n\alpha i} s_{n'\alpha' i'} \qquad [4.1]$$

where n runs over N, the number of Wigner–Seitz cells in the volume of the material, α runs over r, the number of atoms in the basis, i is one of the three Cartesian coordinates, M_α is the mass of the α-th atom, and $s_{n\alpha}$ is the instantaneous displacement from equilibrium. The force constant matrix $\Phi_{n\alpha i}^{n'\alpha' i'}$ in [4.1] has $3Nr$ rows and columns and represents the force in the ith direction acting on the αth atom in the nth cell when the α'th atom in the n'th cell is displaced by unit distance in the i'th direction. Certain general 'symmetry relations' link various components of the force constants even in the absence of periodicity, such as the symmetric nature of the Φ's with respect to interchange of indices; furthermore the potential energy is invariant to infinitesimally small translations or rotations of the solid.

The equation of motion can be written as:

$$M_\alpha \ddot{s}_{n\alpha i} = -\sum_{n'\alpha' i'} \Phi_{n\alpha i}^{n'\alpha' i'} s_{n'\alpha' i'} \qquad [4.2]$$

and if a solution to this is sought which is periodic in time, i.e.

$$s_{n\alpha i}(t) = M_\alpha^{-1/2} u_{n\alpha i} \exp(-i\omega t) \qquad [4.3]$$

where $u_{n\alpha i}$ are time independent, the equation of motion may be written as:

$$\omega^2 u_{n\alpha i} = \sum_{n'\alpha' i'} D_{n\alpha i}^{n'\alpha' i'} u_{n'\alpha' i'} \qquad [4.4]$$

where the real symmetric matrix $\mathbf{D}$ $(=\Phi/\sqrt{(M_\alpha M_{\alpha'})})$ has $3Nr$ real *eigenvalues* ω_j^2. The *eigenvectors*, u, in [4.4] are characterized by the index j, and for each eigenvalue ω_j there are $3Nr u_{n\alpha i}^{(j)}$, which are termed the 'normal modes'. The introduction of 'normal coordinates', q_j, which are related to the displacements $s_{n\alpha i}(t)$, allows the reduction of the Hamiltonian into *independent* normal modes, and is only possible in the harmonic approximation in which the matrix Φ (or $\mathbf{D}$) is able to be diagonalized (see e.g. Madelung 1978). Thus, the coupled individual oscillations of the atoms are replaced in a formal sense by *decoupled* collective excitations which are known as 'phonons' in quantum-mechanical language.

It is to be noted that nothing we have said so far about the vibrations of a solid has depended on the assumption of a periodic lattice, and hence the discussion is completely general, and can apply to amorphous solids as well as to crystalline materials. The vibrational problem thus reduces formally to the diagonalization of the matrix $\mathbf{D}$ to find the $3Nr$ eigenvalues.

The presence of lattice periodicity has two effects: the number of equations to be solved is reduced, and the introduction of a wavevector q, related to the size of the unit cell and characterizing the eigenvalues, is made possible. Thus, $D_{n\alpha i}^{n'\alpha' i'}$ must be the same for all cells and is only a function of the difference in cell indices, $D_{\alpha i}^{\alpha' i'}(n'-n)$ for a periodic lattice. The eigenvectors may be written in the Bloch form as:

$$u_{n\alpha i} = c_{\alpha i} \exp(i\mathbf{q} \cdot \mathbf{R}_n) \quad [4.5]$$

where $\mathbf{R}_n$ is a point in the Wigner–Seitz cell. These two consequences of periodicity cause [4.4] to become

$$\omega^2 c_{\alpha i} = \sum_{\alpha' i'} D_{\alpha i}^{\alpha' i'}(\mathbf{q}) c_{\alpha' i'} \quad [4.6]$$

where the matrix is defined by

$$D_{\alpha i}^{\alpha' i'}(\mathbf{q}) = \sum_{n'} \frac{1}{\sqrt{(M_\alpha M_{\alpha'})}} \Phi_{\alpha i}^{\alpha' i'}(n'-n) \exp[i\mathbf{q} \cdot (\mathbf{R}_n - \mathbf{R}_{n'})] \quad [4.7]$$

Hence, periodicity has reduced the set of equations from $3Nr$ to $3r$ in number, and consequently there are $3r$ eigenvalues, $\omega_j(\mathbf{q})$, at each value of $\mathbf{q}$. Since there are exactly N values of $\mathbf{q}$ in the Brillouin zone because of the imposition of the cyclic boundary conditions imposed on the system, there are $3Nr$ different $\omega_j(\mathbf{q})$ corresponding to the number of internal degrees of freedom of the crystal. The number of atoms in the basis, r, determines the number of 'branches' in the Brillouin zone, which thus number $3r$.

The type of vibrations in a given branch of the set of phonon dispersion curves can be distinguished by the behaviour of $\omega_j(\mathbf{q})$ as $\mathbf{q}$ tends to zero (i.e. in the infinite wavelength limit). In general, there are three branches whose frequencies go to zero as $\mathbf{q}$ tends to zero; these are known as 'acoustic' branches, and at $\mathbf{q}=0$ the vibrations of the basis atoms in each cell are obviously in the same direction. The remaining $3(r-1)$ branches are referred to as 'optic' branches, and the frequencies tend to a finite value as $\mathbf{q}$ tends to zero; at $\mathbf{q}=0$, the basis atoms vibrate against each other. In addition, at certain symmetry points, or along high symmetry directions in the Brillouin zone, the modes can be further subdivided into 'longitudinal' or 'transverse', depending on the direction of the eigenvector with respect to $\mathbf{q}$. At a general value of $\mathbf{q}$ in the Brillouin zone, however, the lattice vibrations are not strictly longitudinal or transverse, nor are acoustic branch vibrations completely in phase or optic branch modes completely out of phase.

How is this situation altered for amorphous solids, for which the simplifying consequences of periodicity are no longer appropriate? In the low frequency, long wavelength limit, the precise arrangement of atoms is of little consequence and the amorphous solid appears isotropic and acts as an elastic medium, and the vibrational excitations can properly be described as acoustic phonons with the exception that the waves are more heavily damped (or equivalently the phonon lifetime is shorter). At higher frequencies and lower wavelengths, the modes in general will no longer be plane waves but instead they will be much more localized than the corresponding crystalline modes. Furthermore, and more importantly, the wavevector $\mathbf{q}$ is no longer a good quantum number, and is not useful in classifying vibrational modes for amorphous solids. This can be seen in a simple manner. For a single crystal, the size of the first zone in the Brillouin scheme is given by $2\pi/a$, where a is the size of the unit cell. An amorphous solid can be viewed as having a structure equivalent to a unit cell of infinite size, and hence the Brillouin zone reduces to the point $q=0$ and so q becomes redundant as a label of the modes.

Thus, in amorphous solids the concepts of Brillouin zones and phonon dispersion curves lose their meaning (as is the case also for electron states, as will be seen in the next chapter). The dependence of the vibrational frequency ω on the scattering vector q *can* be ascertained from inelastic neutron scattering experiments (sect. 4.4.4), but the dispersion does not follow a single curve for a given branch as in a crystal. Instead, taking an acoustic branch as an example, the locus of points may fall close to a single curve at very low values of q, but at higher values the curve becomes smeared out and ill-defined (see Fig. 4.1). This is a consequence of structural

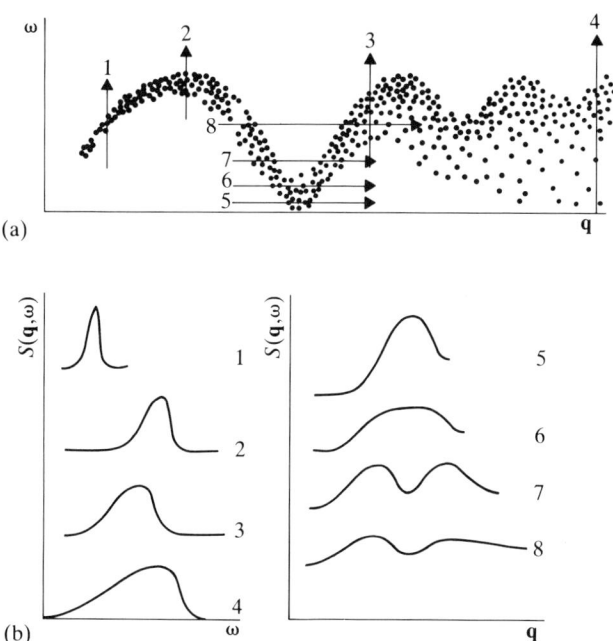

Fig. 4.1 (a) Schematic illustration of a dispersion relation $\omega(q)$ for an acoustic branch in a glass. The form of the scattering law, i.e. the dynamic structure factor $S(q, \omega)$ is shown in (b) for the various experimental loci through the curve in (a) (Leadbetter 1973).

disorder; peaks in the structure factor $S(q)$ act in the scattering process like smeared-out reciprocal lattice vectors. The loci through the dispersion relation in Fig. 4.1 correspond to different forms of the scattering law $S(q, \omega)$, and these are shown also in the figure.

We see therefore, that ω–q relations are not meaningful in describing vibrational excitations in amorphous solids. Is there a quantity which is used to describe phonon modes in crystalline materials which can also be sensibly used for amorphous solids? The answer is yes, and the quantity is the 'density of states' $\rho(\omega)$, where $\rho(\omega)\,d\omega$ is the number of states with frequencies between ω and $\omega+d\omega$. This

Vibrations

can be defined for a crystal as:

$$\rho(\omega) = \frac{V_g}{(2\pi)^3} \int_{S_\omega} \frac{dS_\omega}{|\nabla_k \omega|} \quad [4.8]$$

where N/Vg is the reciprocal volume of a Wigner–Seitz cell (equal to the volume of a Brillouin zone divided by $(2\pi)^3$, and S_ω is a surface of constant ω. The density of states for a crystal exhibits 'Van Hove singularities', which are discontinuities in the slope of $\rho(\omega)$ in three dimensions whenever $\nabla \omega_k = 0$, i.e. where the dispersion curve is flat in ω–q space. The calculated density of states for diamond cubic Si (or Ge) is shown in Fig. 4.2, where the Van Hove singularities are clearly seen. Also shown are

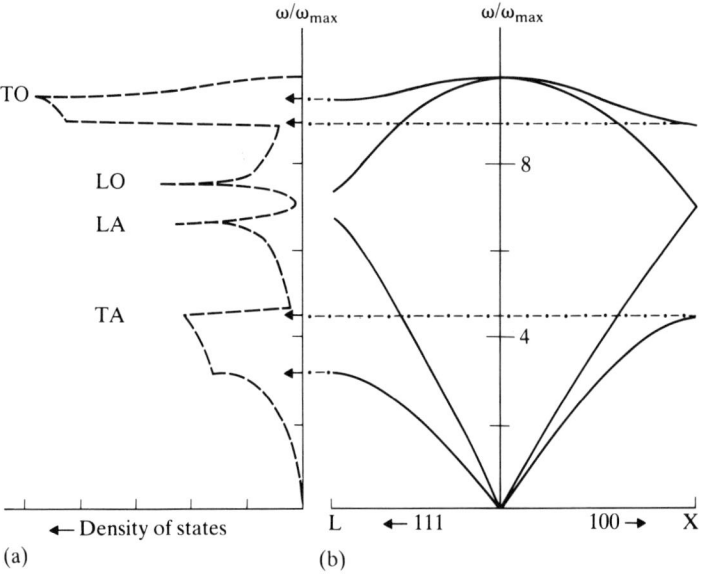

Fig. 4.2 (a) Density of states and (b) phonon dispersion curves for diamond cubic Si (or Ge) calculated using the Born potential. The dotted lines indicate the origin of some Van Hove singularities in k-space. The maximum frequency is given by $\omega_{max} = \sqrt{(8\alpha/M)}$ (Thorpe 1976).

the phonon dispersion curves computed using a Born potential (Thorpe 1976):

$$V_{ij} = \tfrac{3}{4}\beta_B \sum_{i,j} [(\boldsymbol{u}_i - \boldsymbol{u}_j) \cdot \boldsymbol{r}_{ij}]^2 + \tfrac{1}{4}(\alpha_B - \beta_B) \sum_{i,j} (\boldsymbol{u}_i - \boldsymbol{u}_j)^2 \quad [4.9]$$

where $\boldsymbol{u}_i$ are displacement vectors, $\boldsymbol{r}_{ij}$ is a unit vector joining sites i and j; the first term represents bond stretching and the second represents bond bending. The origins of some singularities in $\rho(\omega)$ with respect to features in the dispersion curves are indicated by dotted lines. An alternative way of defining the density of states is in terms of a Green function formalism (Thorpe 1976):

$$\rho(\omega) = \sum_k \delta(\omega - \omega_k) = -\frac{1}{\pi} \operatorname{Im} \sum_k \frac{1}{(\omega - \omega_k)} \quad [4.10]$$

For crystals, the sum over k is over all branches and covers the first Brillouin zone: [4.10] is also valid for an amorphous solid, but now k becomes merely a label for the vibrational eigenstates and is no longer a vector and has no other physical significance.

As an example of a vibrational density of states of an amorphous solid, we show in Fig. 4.3 $\rho(\omega)$ for amorphous Ge or Si, calculated for the 201-atom CRN model of Steinhardt *et al.* (1974), and compared with the density of states calculated for the crystalline diamond cubic polymorph (see also Fig. 4.2). As will be seen in

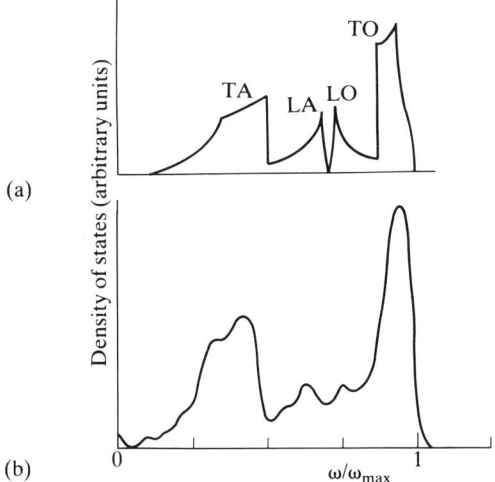

Fig. 4.3 Comparison of density of states calculated for (a) diamond cubic Si (or Ge) (Kelly 1980) and (b) for a CRN (the 201-atom model of Steinhardt *et al.* 1974) (Beeman and Alben 1977).

section 4.4, extraction of $\rho(\omega)$ from experimental data is often not straightforward, and so for simplicity we show in Fig. 4.3 calculated curves obtained using the methods detailed in section 4.3. It can be seen immediately that the forms of the two curves are remarkably similar, and this is a general feature of amorphous solids. The features in the crystal density of states resulting from the periodicity, namely the Van Hove singularities, are smeared out in the amorphous density of states, but otherwise the correspondence between peaks in $\rho(\omega)$ is very close, although the peaks are broadened for the amorphous case. This immediately suggests that it is *short-range order* which essentially determines the vibrational properties, and thus if this is the same for amorphous and crystalline polymorphs, as is the case for Ge or Si, one expects $\rho(\omega)$ to be similar. Indeed we shall see in the next section that the vibrational characteristics of a *single* tetrahedral unit can account for many of the features in $\rho(\omega)$ for Ge or Si.

4.3 Computational methods

4.3.1 Introduction

We have seen that the quantity describing vibrational excitations which is most suitable for an amorphous solid and which can be compared more or less directly with experiment is the density of states. Thus, it is this, rather than say the eigenvectors, to which calculations are directed. The problems in calculating vibrational (or electronic) properties of amorphous materials are principally twofold, both of which are due essentially to the lack of periodicity. The first problem is what to take as the structure; the equilibrium positions of all atoms are required in order to compute the force-constant matrix $\Phi_{n\alpha i}^{n'\alpha' i'}$ on which all calculations are based. The difficulties in ascertaining the structure of amorphous solids were explored in Chapter 3 and the most common starting point is to use the coordinates of a structural model (e.g. a CRN) whose scattering properties match experiment well, assuming such a model exists. The second problem lies in the diagonalization of the force-constant matrix. We have seen that the presence of periodicity reduces the number of eigenvalues from $3Nr$ to $3r$ through the introduction of the Bloch formalism for the eigenvectors. Such a simplification is not available for calculations for amorphous solids. Furthermore, the number of atoms, r, in the 'unit cell' (i.e. the CRN model), of an amorphous solid is very large (~ 500) and so matrices containing several thousand elements are involved, leading to considerable computational difficulties. However, this problem may be eased to a certain degree by the use of only a small cluster (of several tens of atoms) in the interior of the model, since it is the local structure, rather than medium-range structure which determines many of the features in $\rho(\omega)$; sophisticated matrix manipulation techniques also serve to speed diagonalization.

Certain limits to the spectral range of $\rho(\omega)$ can be obtained from mathematical theorems relating to certain forms of the interatomic potentials. For instance, for a perfectly tetrahedral structural unit, and for a bond-stretching potential of the Keating form ([4.11]) which is positive semidefinite, the upper bound to $\rho(\omega)$ is given by $\omega_{max} = \sqrt{8(\alpha_K/M)}$, where α_K is the Keating central force constant, and the lower bound is at $\omega = 0$ (Weaire and Alben 1972). These limits are thus determined by the local geometry, and are the same for crystalline and amorphous solids alike. The small deviations from local tetrahedral symmetry (viz. bond-angle distortions) are expected to alter this limit by only a few per cent. These limits to $\rho(\omega)$ correspond to the transverse optic (TO) and transverse acoustic (TA) branches of the crystal. The features between these limits (corresponding to the longitudinal optic, LO, and acoustic, LA, branches of the crystal) are dependent on the detailed topology of the network, e.g. the distribution of ring sizes. For instance, the dip between LA and LO bands in $\rho(\omega)$ for diamond cubic Si or Ge (Fig. 4.2) is a consequence of the presence of even-membered (sixfold) rings. The preservation, or otherwise, of such features in the density of states for amorphous counterparts has been used to infer details concerning the ring statistics of such materials.

As might be expected, most of the theoretical techniques that have been used to explore vibrational properties of amorphous solids have been numerical in character. No general analytical methods exist which can handle topological disorder in three dimensions, although certain forms of disorder, e.g. 1-D disordered chains, or substitutional disorder, are amenable to such analysis.

All numerical approaches are formulated to solve the eigenvalue problem given by [4.4]. However, a variety of interatomic potentials have been used. All have tended, for simplicity, to be restricted to nearest-neighbour interactions only, and some have been further restricted to account only for stretching forces. Generally, however, a non-central component is included for covalent materials to represent the bond-bending forces needed to stabilize the structure against shear. Several types of potential have been used. The first is the Born model ([4.9]) which we have already mentioned, and which is axially symmetric about the bond connecting two nearest-neighbour atoms because of the threefold symmetry of the other bonds in the perfectly tetrahedral case. The second type of potential, widely used, is due to Keating (1966):

$$V_{ij} = \tfrac{3}{4}\alpha_K \sum_{ij} [(\boldsymbol{u}_i - \boldsymbol{u}_j) \cdot \boldsymbol{r}_{ij}]^2 + \tfrac{3}{16}\beta_K \sum_{i(j,k)} [(\boldsymbol{u}_i - \boldsymbol{u}_j) \cdot \boldsymbol{r}_{ik} + (\boldsymbol{u}_i - \boldsymbol{u}_k) \cdot \boldsymbol{r}_{ij}]^2 \qquad [4.11]$$

where the sums run over all atoms i and their distinct neighbours j and k and where, as in [4.9] $\boldsymbol{u}_i$ are displacement vectors and $\boldsymbol{r}_{ij}$ are unit vectors joining sites i and j. The bond-stretching and bond-bending force constants are written as α_K and β_K respectively, to differentiate them from the corresponding Born terms. This potential has the advantage that it is *rotationally* invariant because it only depends on the scalar product of nearest-neighbour vectors. Also rotationally invariant is the 'valence-force-field' model, in which the scalar product is replaced by the corresponding bond angle:

$$V_{ij} = \tfrac{1}{2}\alpha_{VFF} \sum_{i,j} \Delta r_{ij}^2 + \tfrac{1}{2}\beta_{VFF} \sum_{i(jk)} r_{io}^2 \cos\theta_{jk} \Delta\theta_{jk} \qquad [4.12]$$

where r_{io} is the equilibrium bond length and θ_{jk} is the bond angle subtended at atom i by atoms j and k. Note that all these forms of interatomic potential only take into account topological disorder through the atom position coordinates. Quantitative disorder, manifested by differences in force constants between atoms due to local structural variations, is almost always neglected.

A further problem with vibrational computations for amorphous solids is the effect on the calculated modes of the surface of the model cluster used. Of course, there is no difficulty for perfectly crystalline materials for which periodic boundary conditions may be invoked. However, for models simulating the structure of amorphous solids containing typically 500 atoms a significant number of atoms lie on or near the surface, and spurious surface modes can easily arise if calculations are performed on the entire cluster without proper regard for boundary conditions. Such problems can be reduced if only interior atoms are used in the calculation, if the density of states is projected out locally as in the 'recursion' method (sect. 4.3.4), or if a 'Bethe lattice' is grafted on to the surface unsatisfied bonds of a CRN. A Bethe lattice (also called a Cayley tree) has the same coordination as the network itself, except that there are *no* closed rings of bonds (Fig. 4.4). The use of such a mathematical device eliminates localized surface modes and does not introduce additional structure into the density of states since $\rho(\omega)$ for a Bethe lattice is smooth and featureless. An additional limitation imposed by the use of finite size models is

Vibrations

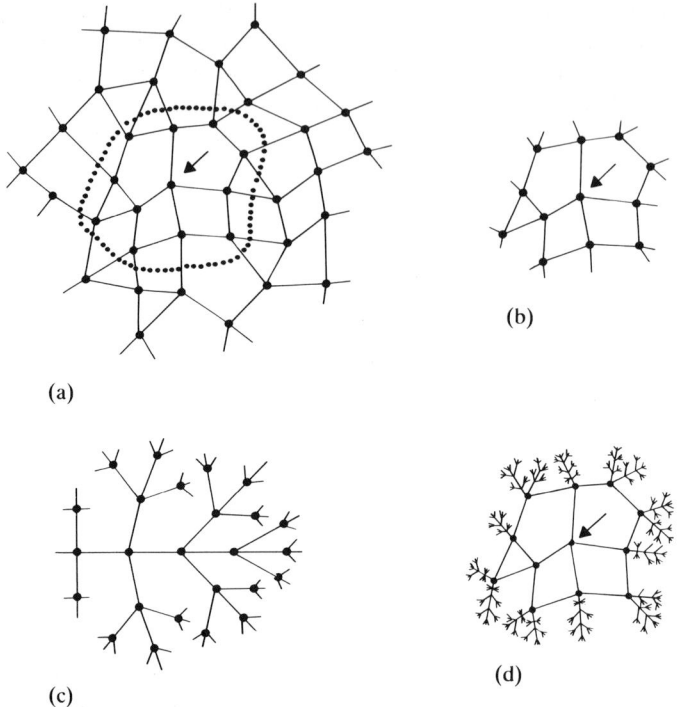

Fig. 4.4 The cluster-Bethe-lattice method (Joannopoulos 1979) for calculating vibrational (or electronic) densities of states of amorphous structures. A given atom is chosen as a reference point (a) and a cluster removed from the network (b). A Cayley tree appropriate for the coordination of the network (c) is attached to each dangling bond of the cluster (d).

that the low frequency acoustic-type modes below, say, 50 cm^{-1} are not amenable to calculation for several reasons; principally this is because such modes have very long wavelengths which are comparable in extent to the size of the model, but also because the local potentials ([4.9], [4.11] and [4.12]) are insufficient to describe the long-range interactions acting in such modes.

As mentioned previously, the theoretical investigation of vibrational modes and calculation of the density of states reduces to the solution of the set of eigenvalue equations given by [4.4]. A variety of numerical techniques have been developed for this problem, and we briefly mention them in the following, starting with the oldest method.

4.3.2 Negative eigenvalue method

This method utilizes a theorem introduced by Dean and Martin (1960) which offers an efficient method of computing the distribution of eigenvalues of a real symmetric matrix; extensive reviews of the method are given by Dean (1972) and Bell (1972).

The equation of motion ([4.4]) can be written in matrix terms as:

$$(\mathbf{D} - \mathbf{I}\omega^2)\mathbf{U} = 0 \qquad [4.13]$$

where **D** as before is the real symmetric force constant matrix, **U** is the unitary matrix, whose elements are the eigenvectors $u_{n\alpha i}$, which consequently obeys the relation $\mathbf{U}^T\mathbf{U} = \mathbf{I}$ where **I** is the unit matrix. If we consider the matrix $\mathbf{A} = \mathbf{D} - \omega^2 \mathbf{I}$, then the negative eigenvalue theorem states that the number of negative eigenvalues (i.e. physical solutions) of the $n \times n$ matrix **A**, denoted by $\eta(\mathbf{A})$ is given by

$$\eta(\mathbf{A}) = \sum_{i=1}^{m} \eta(\mathbf{X}_i) \qquad [4.14]$$

if **A** can be written in the partitioned form:

$$\mathbf{A} = \begin{bmatrix} \mathbf{A}_1 & \mathbf{B}_2 & & 0 \\ \mathbf{B}_2^T & \mathbf{A}_2 & \mathbf{B}_3 & \\ 0 & & \mathbf{B}_m^T & \mathbf{A}_m \end{bmatrix} \qquad [4.15]$$

where

$$\mathbf{X}_1 = \mathbf{A}_1 - \omega^2 \mathbf{I}_1 \qquad [4.16]$$

and

$$\mathbf{X}_i = \mathbf{A}_i - \omega^2 \mathbf{I}_i - \mathbf{B}_i^T \mathbf{X}_{i-1}^{-1} \mathbf{B}_i \quad (i = 2, 3 \ldots)$$

and $\mathbf{A}_i$ is a symmetric square matrix of order 1_i, $\mathbf{I}_i$ is a unit matrix of order 1_i, and $\sum_{i=1}^{m} 1_i = n$. The iteration process implicit in [4.16] involving sequential partitioning of matrices as in [4.15] is simplified computationally for the case of a sparse matrix like **A**, since only the number of columns and rows equal to one less than the number of non-zero elements of **A** from the main diagonal need be considered in the transformation; all others are zeroes. If the partitioning is chosen to separate out just the first row and column, the $\mathbf{X}_i$ are scalars, and $\eta(\mathbf{A})$ is simply the number of *negative* X_i.

The importance of this theorem is that $\eta(\mathbf{A})$ is just the number of squared frequencies of the vibrational system less than ω^2, i.e. the integrated density of states for squared frequencies, and is therefore related directly to the (integrated) density of states. Thus, $\rho(\omega)$ can be obtained in a considerably simpler computational manner than is necessary for diagonalization of the (large) matrix **D**. The negative eigenvalue method suffers from the disadvantage that only the frequency spectra are obtained directly, rather than individual eigenvalues or eigenvectors. However, in practice this is of little consequence since $\rho(\omega)$ is generally the desired quantity; however if a knowledge of the character of the vibration corresponding to a certain feature in the density of states is required, the eigenvector may be calculated using the inverse of [4.13], having obtained a good approximation to the eigenvalue by using the iteration procedure for small intervals around the chosen frequency.

The method outlined above has been applied to the (4:2) glass-forming systems SiO_2, GeO_2 and BeF_2, which are expected to have the same local geometry, using as a basis the CRN models built to simulate the structure of a-SiO_2 (Bell 1972, Dean 1972). Figure 4.5(a) shows typical results of density of states calculations for SiO_2 using fixed atom boundary conditions for the surface atoms. Three well-defined bands are observed, at about 400, 700 and 1100 cm^{-1} which agree in position fairly well with the features at 465, 800 and 1100 cm^{-1} observed in the IR

Vibrations

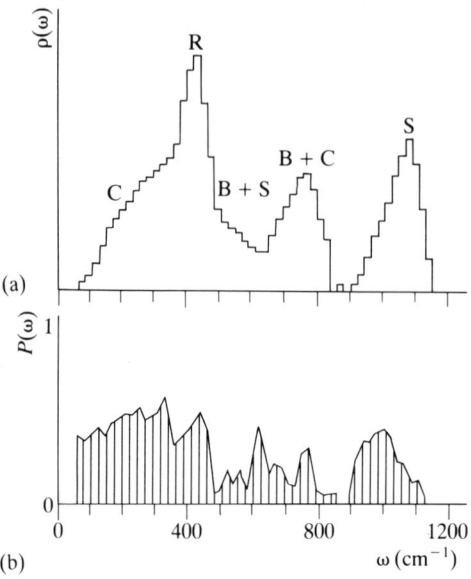

Fig. 4.5 Density of states (a) and participation ratio (b) of vibrational modes calculated for the Bell and Dean CRN of a-SiO$_2$ with fixed boundary conditions (Bell 1972).

spectrum. The various features in $\rho(\omega)$ have been identified with certain predominant types of atomic motion; S refers to stretching, B to bending, R to rocking of the oxygen atoms respectively, and C refers to those modes which involve a large amount of cation motion.

The spatial extent of modes, which in an amorphous solid can span practically the entire range observed in crystals from infinitely extended plane-wave-like excitations to highly localized vibrations associated with point defects, can be quantified to some extent by means of the 'participation ratio' $P(\omega)$. This describes the proportion of the total number N of atoms which contribute effectively to a given mode:

$$P(\omega) = \frac{M_1^2}{M_0 M_2} \quad [4.17]$$

where for a mode at frequency ω_j, M_r is the rth moment of the kinetic energy of the mode

$$M_r(\omega_j) = \sum_{i\alpha} |u_{i\alpha}^{(j)}|^{2r} \quad [4.18]$$

A mode involving only one atom (i.e. localized) has a value $P = 1/N$, whereas an extended mode in which all atoms participate equally gives $P = 1$; the sinusoidal waves characteristic of crystals give $P \sim 0.5$. The participation ratios for the various features in $\rho(\omega)$ for SiO$_2$ are shown in Fig. 4.5(b), whence it can be seen that modes in the low frequency acoustic region ($\omega < 500$ cm^{-1}) are reasonably extended, with $P \sim 0.5$. Higher frequency (optic) modes are much less extended, and there are strong indications of distinct localization ($P \rightarrow 0$) at the band edges. This feature of localization at band edges will re-emerge when we discuss the nature of *electron* states in amorphous materials (Ch. 5). Calculations by Bell and Dean were also made for cases in which the boundary conditions were free rather than fixed. It was found that additional modes appeared in $\rho(\omega)$ in the spectral gap at ~ 900 cm^{-1} observed for the fully connected case (see Fig. 4.5(a)), which were extremely localized in character and were caused by the motion of non-bridging oxygens; the occurrence of highly localized states in the gap in the density of states for defect configurations is also a feature of electronic states.

4.3.3 Equation-of-motion method

This method, as its name implies, concentrates on the time evolution of dynamical quantities; a review of the technique has been given by Beeman and Alben (1977). The equation of motion ([4.2]) is solved by integration forward in time to obtain the displacements up to a time interval T, given values for the initial displacements $s_{i\alpha}^o$ and taking the velocities equal to zero at $t=0$. It is assumed that all modes contribute equally to the ensuing motion from the random starting positions. The function $R(\omega)$ is then computed (i.e. the time dependence is Fourier transformed):

$$R(\omega) = \frac{2}{\pi} \int_o^T \sum_{i\alpha} A_{i\alpha} s_{i\alpha}(t)(\cos \omega t) \exp(-\lambda t^2) \, dt \qquad [4.19]$$

where $A_{i\alpha}$ and $s_{i\alpha}^o$ are chosen such that $R(\omega)$ represents the density of states, IR spectrum or Raman spectrum, as appropriate. For $R(\omega)$ to be the density of states, the following choice is made (Beeman and Alben 1977):

$$s_{i\alpha}^o = \sqrt{2} \cos \theta_{i\alpha} \qquad [4.20(a)]$$

$$A_{i\alpha} = \sqrt{2} \cos \theta_{i\alpha} \qquad [4.20(b)]$$

where $\theta_{i\alpha}$ are random angles distributed uniformly between 0 and 2π. The choices for $s_{i\alpha}^o$ and $A_{i\alpha}$ that cause $R(\omega)$ to represent IR or Raman spectra will be given in the next section. The upper limit T of the integral in [4.19] is taken to be about 30 periods of the fastest oscillation, and the damping factor λ is taken to equal $3/T^2$. This has the result that the computed spectra are broadened by the use of a finite time interval; the value of T is chosen to give a broadening of about 4% of the full spectral width. An average over several computed $R(\omega)$ for several sets of $\theta_{i\alpha}$ is necessary to achieve a reasonable spectral function.

The equation-of-motion method possesses several advantages, the most notable being that smoothed spectral functions ($\rho(\omega)$, etc.) are obtained directly from the coordinates of a model without the need for diagonalization of large matrices. It is also economical in computer time and store. The method gives results which are similar to those produced by the 'recursion' method to be discussed next; a typical result is shown in Fig. 4.6 where $\rho(\omega)$ computed for the Greaves–Davis (1974) threefold coordinated CRN is compared with the density of states for a-As obtained from inelastic neutron scattering by Leadbetter et al. (1976). General features are reproduced reasonably well, but the sub-structure in the lower acoustic peak does not appear because of the simple (Born) force law used.

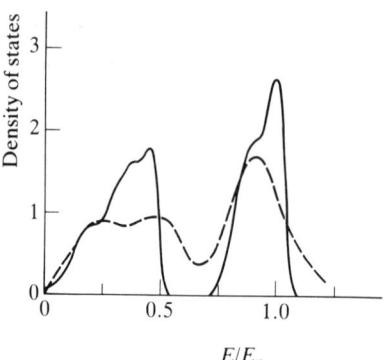

Fig. 4.6 Comparison of the density of states for a-As calculated using the equation-of-motion method applied to the Greaves–Davis CRN model (Beeman and Alben 1977) (solid line) with the density of states obtained experimentally by inelastic neutron scattering (Leadbetter et al. 1976) (dashed line).

4.3.4 Recursion method

We have already seen that it is generally the *local* structure which determines the gross features of the vibrational density of states; longer range structure only serves to introduce fine detail (cf. periodicity giving rise to Van Hove singularities). One computational technique that capitalizes on this feature for amorphous materials is the recursion method. This was originally developed for calculations of the *electronic* structure of disordered materials (see the review by Heine et al. 1980), but was extended to the calculation of vibrational excitations by Meek (1976). The *local* density of modes, i.e. the density of states weighted by the amplitude squared of each mode at the atom considered, is calculated and an average taken over 10 or so central atoms in a model cluster to yield a representative spectrum.

The mathematical details of the method go beyond the level of this book, and so the interested reader is referred to the extensive discussion by Heine and co-workers (1980). In brief, though, it offers an attractive computational technique for obtaining the required density of states for, say, the dynamical matrix **D** ([4.4]), without recourse to direct calculation of the eigenvectors or eigenvalues. The Hamiltonian for vibrational (or electronic) excitations of an amorphous cluster yields a discrete spectrum of eigenvalues, rather than a band, and so the density of states is given by a sum over such states, as in [4.10]. It can be shown that in terms of the Green function (or resolvent operator):

$$\mathbf{G}(\omega^2) = (\omega^2 \mathbf{I} - \mathbf{D})^{-1} \qquad [4.21]$$

an equivalent way of writing the local density of modes is:

$$\rho_{ox}(\omega^2) = \frac{-1}{\pi} \lim_{\varepsilon \to 0} \mathrm{Im}\,[G_{oo}(\omega^2 + i\varepsilon)] \qquad [4.22]$$

where the density of the modes is 'projected' on to the squared *x*-displacement of atom O (denoted by $|0\rangle$) and G_{oo} is the diagonal matrix element of **G**, i.e. $\langle 0|G|0\rangle$. The local density of states can now be calculated directly for **G** by choosing a new set of orthonormal basis vectors $|u_n\rangle$, of the Hermitian matrix **D**, such that the first is chosen to be the particular displacement for which is wanted $\rho(\omega)$, i.e. $|u_o\rangle = |0\rangle$. The new basis of vectors $|u_n\rangle$ are defined by the following recursion relation (hence the name given to the method):

$$b_{n+1}|u_{n+1}\rangle = (\mathbf{D} - a_n)|u_n\rangle - b_n|u_{n-1}\rangle \qquad [4.23]$$

Expressing $\mathbf{D}$ in this basis set, it is then possible to show that $[\omega^2\mathbf{I} - \mathbf{D}]$ may be inverted *analytically* in the form of a continued function:

$$G_{oo}(\omega^2) = \cfrac{1}{\omega^2 - a_0 - \cfrac{b_1^2}{\omega^2 - a_1 - \cfrac{b_2^2}{\cdots}}} \qquad [4.24]$$

where the a_n, b_n are defined according to the recursion relation. The computation is very efficient since only three vectors are needed at any one time ([4.23]), and the continued fraction is stopped at some finite value of $n = N$ (say 15–25). Although the accuracy of the final density of states does increase with increasing N, the higher $|u_n\rangle$ represent displacements of atoms far removed from the origin atom, and therefore have little effect on the local density of states. It has been found from experience that the recursion method produces results very similar to those given by the equation-of-motion method, and there is therefore little to choose between them.

4.3.5 Summary

The presence of disorder in a solid affects the vibrational excitations in a variety of ways. The most important consequence of the lack of periodicity is that a reciprocal lattice can no longer be defined, and therefore any excitations of the solid (whether vibrational or electronic) cannot be described simply with respect to the underlying lattice by means of the Bloch formalism. This has the result that the phonon (or equivalently electron) wavevector is no longer a good quantum number, and therefore the selection rules obeyed by the wavevector in any interaction of the phonon (or electron) in a perfect crystal are relaxed in an amorphous solid. Furthermore, the description of vibrational excitations in terms of dispersion curves, ω v.s. q, (or equivalently band structures, E v.s. k, for electrons), valid for single crystals, becomes inappropriate for non-crystalline solids. Two additional consequences of the presence of disorder for excitations (both vibrational and electronic) are that disorder-scattering reduces the lifetime (or equivalently the mean free path), and if the amount of disorder is very great, the excitations may become spatially localized rather than extended; an example of this is an impurity vibrational mode.

Although excitations of an amorphous solid cannot be described in terms of a dispersion curve (or band structure), the *density of states* remains a valid description, albeit more broadened than in the crystalline case. Thus, calculations of vibrational properties are generally directed towards obtaining this quantity. The 'brute-force' method is simply to diagonalize the dynamical matrix for all atoms in a model cluster, and the considerable computational difficulties which this involves can be simplified somewhat by using the negative eigenvalue method (section 4.3.2). A more elegant solution is to recognize that it is the *short-range* order which essentially determines the major features in the density of states (again true for both vibrations and electrons). This is the basis of the recursion method (sect. 4.3.4) (and the cluster-Bethe-lattice method – sects. 4.3.1 and 5.2.1), in which the *local* density of states is projected out onto the squared displacement of a given origin atom. Contributions from atoms far from the origin atom perturb only slightly the resulting density of states.

4.4 Experimental probes

4.4.1 Introduction

A variety of experimental techniques may be employed to probe the vibrational structure of solids. These range from various light absorption and scattering experiments (IR, Raman and Brillouin) which are available in well-equipped laboratories, to inelastic neutron scattering experiments which require a central facility (e.g. a nuclear reactor) as a source of neutrons. These techniques are equally applicable to crystalline or amorphous solids, but the lack of periodicity characteristic of non-crystalline materials changes somewhat the way photons or neutrons interact with the phonons, as will be discussed in the following.

4.4.2 Infra-red spectroscopy

Infra-red spectroscopy measurements usually employ either a grating (or prism) spectrometer together with a source (e.g. a tungsten lamp) and detector (e.g. a thermocouple or photo-cell), or an interferometric spectrometer which has a Michelson interferometer replacing the monochromator and more sophisticated data processing facilities to transform the data into a frequency spectrum. The measured transmission T is related to the IR absorption coefficient α and the reflectivity R by [3.59] for the case $\alpha d \gg 1$ where d is the sample thickness. Therefore two independent measurements of T and R are required to extract α uniquely.

As already mentioned in section 3.2.5, IR absorption takes place when there is a change in dipole moment associated with the vibrational mode excited. For non-polar, covalently bonded materials, there is no static dipole moment and so any induced dipole moment must arise from a *dynamic* effect involving atomic displacements and the resulting bond compressions and extensions. The dipole moment vector M is proportional to the displacement u_i (to first order) and according to one model can be written as (Alben *et al.* 1975):

$$M = \sum_{i(j,k)} (r_{ik} - r_{ij})[(u_i - u_j) \cdot r_{ij} - (u_i - u_k) \cdot r_{ik}] \qquad [4.25]$$

or in terms of the bond compression $C_{ij} = (u_i - u_j) \cdot r_{ij}$ as:

$$M = 2 \sum_i \left(\sum_j C_{ij} \right) \left(\sum_k r_{ik} \right) \qquad [4.26]$$

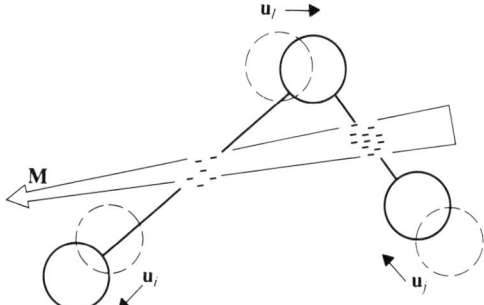

Fig. 4.7 Dynamic bond-charge model for the dipole moment which can give rise to IR activity (Alben *et al.* 1975). Bond charge moves from extended to compressed bonds, resulting in a local electric-dipole moment **M**. The local moments cancel for modes of the diamond cubic structure, but do not cancel for amorphous structures.

where u_i and r_{ij}, etc., have the same meaning as in [4.9]. The origin of the dynamic dipole moment can be seen with reference to Fig. 4.7; charge flows into a bond under compression and out under tension. Note from [4.25] or [4.26] that the induced dipole moment is zero for perfect tetrahedral symmetry since $\sum_{ij} r_{ij} = 0$, and hence crystalline Ge and Si should not be, and indeed are not IR-active; in fact *all* elemental crystals with two atoms per unit cell exhibit a vanishing first-order dipole moment and are IR-inactive. Thus, we can see that one consequence of disorder is that modes become IR-active that otherwise would be inactive.

Another way of stating this consequence of the lack of long-range order is that the selection rules for the wavevector **k** that are obeyed in crystals are relaxed for amorphous solids. In crystals, as a consequence of periodicity, we have seen that **k** is a good quantum number. For this case, we must have conservation of momentum, or equivalently wavevector, i.e. $\mathbf{k}+\mathbf{q}=\mathbf{k}'$, where **q** corresponds to the photon momentum and **k** and **k**' are the incident and scattered phonon wavevector respectively. However, since $|\mathbf{q}| \ll |\mathbf{k}|$, only phonons very near $\mathbf{k}=0$ (i.e. at the centre of the Brillouin zone) are excited in crystals, giving sharp lines in the IR spectrum. For amorphous materials, in which **k** is *not* a good quantum number, the **k**-selection rule breaks down, and *all* modes can contribute (not necessarily equally) to the IR (or Raman) activity. Thus, broad bands in the IR spectrum are expected for amorphous materials, qualitatively similar to the phonon density of states but weighted by the matrix element for the transition (proportional to the square of the dipole moment). The broad features in the spectrum of amorphous germanium and the degree of correspondence between experimental spectra and a calculated density of states for the Steinhardt 201-atom CRN, are shown in Fig. 4.8.

Attempts have been made to calculate the IR spectra expected for various amorphous materials. The fine details in spectra produced by these calculations depend somewhat crucially on two aspects: the quality of the structural model; and the frequency dependence of the matrix elements (or dipole moment) that are used. Beeman and Alben (1977) have given values for $s_{i\alpha}^{\circ}$ and $A_{i\alpha}$ which, when used with [4.19], yield $R(\omega)$ proportional to the IR absorption $\alpha(\omega)$:

$$s_{i\alpha}^{\circ} = \sum_{\gamma} \varepsilon^{\gamma} e_{i\alpha}^{\gamma}/m_i \qquad [4.27(\text{a})]$$

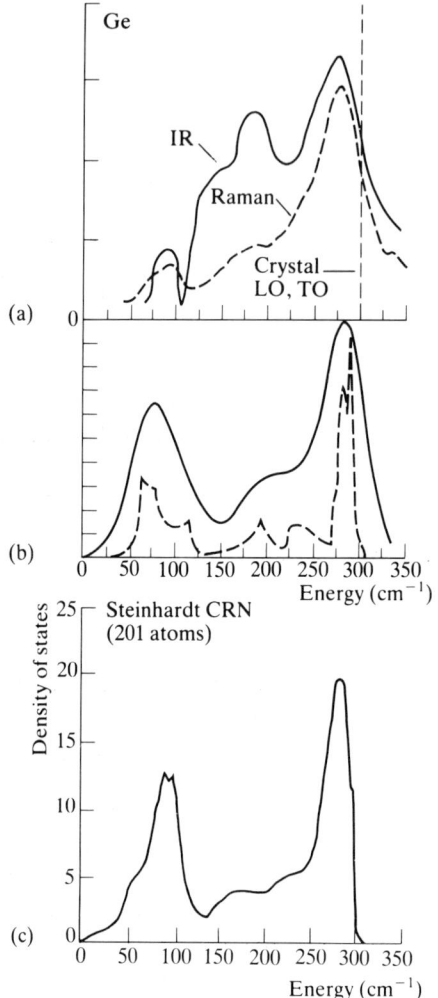

Fig. 4.8 Comparison of (a) IR and Raman spectra of a-Ge with (b) the crystalline density of states obtained from inelastic neutron scattering measurements (dotted line) and a broadened density of states (solid line) (Alben et al. 1975), and with (c) the density of states calculated for the Steinhardt CRN by Meek using the recursion method (Heine 1980).

$$A_{i\alpha} = m_i s_{i\alpha}^o \qquad [4.27(b)]$$

where ε^γ is the γ component of a unit vector in the direction of polarization of the incident light, and $e_{i\alpha}^\gamma$ is the 'dynamic charge' which gives the dipole moment in the γ-direction associated with a displacement of atom i in direction α; the model used for $e_{i\alpha}^\gamma$ was that given by [4.25]. Thus, the IR absorption is given by:

$$\alpha(\omega) = \frac{2\pi^2}{c} \frac{R(\omega)}{V} \qquad [4.28]$$

where c is the speed of light and V the volume of the sample. An example of such a calculation for a-Ge is shown in Fig. 4.9, with the measured IR spectrum for

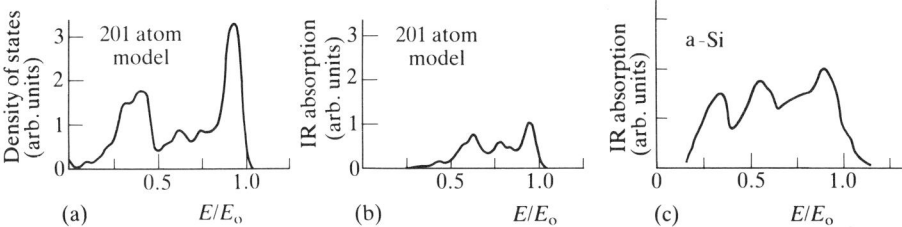

Fig. 4.9 Calculated IR spectrum for a-Si (b) compared with (c) the experimental IR spectrum (for which $E_0 = 521$ cm^{-1}). The calculated density of vibrational states (a) calculated for the Steinhardt 201 atom CRN model is also shown (Beeman and Alben 1977).

comparison. Agreement is reasonable but not quantitative, and it must be admitted finally that the decomposition of experimental IR spectra to give the density of states (or alternatively the calculation of $\alpha(\omega)$ from first principles) depends crucially on the model for the dynamic charge that is used, and at present this is too crude to give anything other than qualitative results.

4.4.3 Raman spectroscopy

Raman and IR spectroscopies are usually discussed together, although the microscopic processes that are responsible are very different, making the two techniques (at least for crystals) complementary. This is because vibrational modes are Raman-active only if they are associated with a change in polarizability (rather than the change in dipole moment for IR transitions), and so modes in (centrosymmetric) crystals which are IR-inactive are Raman-active. This complementary relation is generally lost in amorphous materials since, as we have seen, *all* modes contribute (but not equally) to the IR activity and hence also to the Raman activity.

Raman spectroscopy has been revolutionized by the advent of lasers producing monochromatic, high-intensity light with which the weak Raman scattering may easily be detected. A typical Raman set-up consists of a laser, double grating monochromator and a detector (e.g. a photomultiplier), together with data processing facilities such as photon counting electronics. The double grating monochromator is required for spectral purity since the light is down-scattered to $\omega_L - \omega$, the 'Stokes' line, or up-scattered to $\omega_L + \omega$, the 'anti-Stokes' line, by some small frequency ω corresponding to a typical phonon frequency (of the order of a few hundred cm^{-1}) from the central laser line at ω_L. To further decrease interference from the laser line, the experiment is often carried out in the geometry shown in Fig. 4.10 for a scattering angle of 90°. The polarization of both the incident and scattered radiation affects the intensity, and this dependence is made plain by labelling the scattered intensity as being vertically polarized (V) or horizontally polarized (H) with respect to the scattering plane; thus I_{VH} refers to the scattered intensity measured in the H direction for an incident beam polarized is the V direction. The polarization properties of the Raman spectrum are of utmost importance in the study of single crystals since it is then possible to determine the structure of the Raman tensor of each mode and hence its symmetry character. This is not possible for anything other than single crystals; for amorphous materials, the 'depolarization

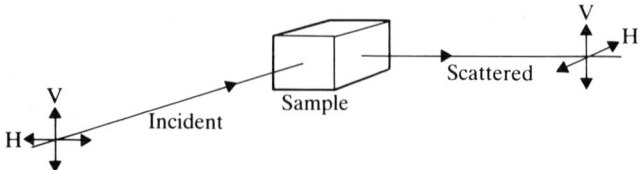

Fig. 4.10 Geometry employed in Raman scattering experiments. The polarization directions are labelled V (vertical) and H (horizontal).

ratio', $p(\omega) = I_{HV}/I_{VV}$, is often used and can take values between 0 and 3/4. Raman lines for which $p = 3/4$ are referred to as depolarized, whereas those for which $p < 3/4$ are referred to as polarized; the dependence of the depolarization ratio on Raman shift can give information about the nature of the active vibrational modes. For instance, if the Raman spectrum is plotted as $(I_{VV} - 4I_{HV}/3)$ only the pure polarized component will contribute, since all depolarized modes give an identically zero contribution in such a plot.

Raman scattering is a complicated phenomenon and for more detail the interested reader is referred to Brodsky (1975). The induced polarizability is a second rank tensor (component $E_{i\alpha}^{\gamma\beta}$), which is a linear function of the atomic displacements $\boldsymbol{u}_i$. Beeman and Alben (1977) have considered a model involving three different mechanisms contributing to the Raman scattering for tetrahedrally bonded materials:

$$\mathbf{E}_1 = \sum_{ij} [\mathbf{r}_{ij}\mathbf{r}_{ij} - \tfrac{1}{3}\mathbf{I}]\boldsymbol{u}_i \cdot \boldsymbol{r}_{ij} \qquad [4.29(a)]$$

$$\mathbf{E}_2 = \sum_{ij} \{\tfrac{1}{2}(\mathbf{r}_{ij}\boldsymbol{u}_i + \boldsymbol{u}_i\mathbf{r}_{ij}) - \tfrac{1}{3}\mathbf{I}\}\boldsymbol{u}_i \cdot \boldsymbol{r}_{ij} \qquad [4.29(b)]$$

$$\mathbf{E}_3 = \sum_{ij} \mathbf{I}\boldsymbol{u}_i \cdot \boldsymbol{r}_{ij} \qquad [4.29(c)]$$

where $\mathbf{I}$ is the unit dyadic. Note that [4.29(a)] and [4.29(c)] depend only on the bond compression C_{ij}, and the last two expressions vanish in the case of perfect tetrahedral symmetry ($\sum_{ij} \boldsymbol{r}_{ij} = 0$), although not necessarily for amorphous structures in which bond-angle distortion exists. The problem is that the relative contribution of each individual mechanism to the Raman activity is not known, although an estimate for the purposes of calculation may be made by noting that only [4.29(c)] gives a purely polarized spectrum. To let $R(\omega)$ represent Raman scattering in the equation-of-motion method, the choice for $s_{i\alpha}^o$ and $A_{i\alpha}$ are:

$$s_{i\alpha}^o = \sum_{\gamma,\beta} \varepsilon_I^\beta \varepsilon_s^\gamma E_{i\alpha}^{\gamma\beta}/m_i \qquad [4.30(a)]$$

$$A_{i\alpha} = m_i s_{i\alpha}^o \qquad [4.30(b)]$$

where ε_I^β and ε_s^γ are components of the unit polarization vectors of the incident and scattered light respectively. The differential cross-section for Stokes-shifted Raman scattering of incident light of frequency ω_L into scattered light of frequency $\omega_s = \omega_L - \omega$ is:

$$\frac{d^2\sigma(\omega)}{d\Omega\, d\omega} = \frac{\omega_L \omega_s^3 \hbar}{c^4} \frac{[n(\omega)+1]}{2\omega} R(\omega) \qquad [4.31]$$

where $d\Omega$ is an element of solid angle, and $n(\omega)$ is the Bose occupation number:

$$n(\omega) = [\exp(\hbar\omega/k_B T) - 1]^{-1} \qquad [4.32]$$

The use of [4.30]–[4.32], together with [4.19] enables the Raman scattering to be calculated for a given model and set of force constants.

In the reduction of *experimental* data, the scheme proposed by Shuker and Gammon (1970) is often used. They showed by consideration of the Fourier transform of the space–time autocorrelation function (related to fluctuations in the dielectric tensor) that the Stokes component of the *reduced* Raman intensity could be written as:

$$I_R(\omega) = \frac{\omega \sum_b c_b \rho_b(\omega)}{(1 + n(\omega))(\omega_L - \omega)^4} \qquad [4.33]$$

(The corresponding equation for the anti-Stokes component has $(1 + n(\omega))$ replaced by $n(\omega)$ and $(\omega_L - \omega)$ replaced by $(\omega_L + \omega)$.) The assumption made is that the matrix elements are *constant* within a given band of modes, b, and therefore the reduced Raman intensity should directly mirror the density of states $\rho(\omega)$.

The expression of the Raman intensity in reduced form is important since it thereby completely removes a spurious peak below 100 cm^{-1} (e.g. in vitreous SiO$_2$) originally interpreted as a vibrational mode. However, the assumption of frequency-independent matrix elements, and the division of the spectrum into separate bands due to different types of vibrational mode, is very poor for many materials. Different frequency dependences of the coupling constants $c_b(\omega)$ in [4.33] have been proposed to try and improve the fit to experimental data compared with that given by the Shuker–Gammon formula (see Lannin 1977).

A connection with the structural use of Raman scattering (sect. 3.2.5) may be made for studies on glassy As$_2$S$_3$. We saw in the last chapter that chalcogenide glasses, for which the chalcogen bond angle approaches 90°, exhibit vibrational modes characteristic of isolated molecular units. For these, there will be two bond-stretching modes, one antisymmetric and therefore IR-active, and the other symmetric and therefore Raman-active; these are shown in Fig. 4.11(a). For comparison, plotted in Fig. 4.11(b) is the frequency dependence of the depolarization ratio I_{HV}/I_{VV}; the higher frequency, more symmetric vibrations give more nearly polarized scattering as expected.

4.4.4 Inelastic neutron scattering

Thermal neutrons typically have energies of the order 25 meV (~ 200 cm^{-1}) and can thus readily lose energy to or gain energy from phonons having frequencies in the range ~ 50–1500 cm^{-1}. In contrast, X-rays have energies much greater than typical phonon energies, and inelastic events are therefore not important. An extensive discussion of the use of neutron scattering as a vibrational probe for glasses is given by Leadbetter (1973).

As we saw in the last chapter (sect. 3.2.1), the occurrence of inelastic scattering naturally means that the structure factor becomes frequency dependent, $S(\mathbf{Q}, \omega)$, and consequently the pair correlation function takes on a dynamic form, $G(\mathbf{r}, t)$

Vibrations

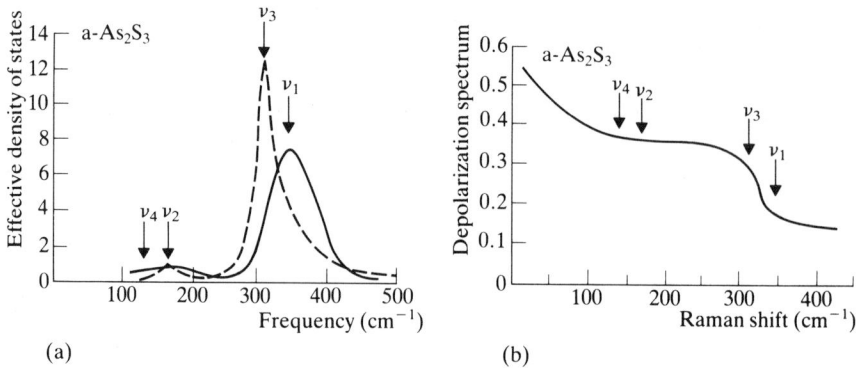

Fig. 4.11 (a) Comparison of the reduced Raman spectrum (solid line) of a-As_2S_3 with the IR absorption spectrum $\omega^2\varepsilon_2$ (dashed line) which also gives an effective density of states. The arrows mark the normal mode frequencies of an AsS_3 molecule (Lucovsky 1972).
(b) Depolarization spectrum $p(\omega) = I_{HV}/I_{VV}$ of a-As_2S_3 (Kobliska and Solin 1973).

([3.21] and [3.22]). The total correlation function can be divided into two parts, corresponding to self and distinct correlations, $G(r,t) = G_s(r,t) + G_D(r,t)$. For experiments at high Q, the important contribution to $G_D(r,t)$ is at small r, and here it tends to zero as only the origin atom is found at later times close to its position at $t=0$, since all atoms vibrate about their quasi-equilibrium sites and do not approach closer than the interatomic distance r_o. The neglect of the distinct term in the expression for the neutron scattering cross-section (describing the interference of neutrons scattered by *different* atoms) is termed the 'incoherent approximation', and obtains when $Qr_o \gg 1$. The simplification that this approximation brings about is that the scattering becomes proportional simply to the phonon density of states; this is not true for coherent scattering in which only certain discrete energy (as well as momentum) transfers are allowed (analogous to Bragg scattering).

The scattering cross-section in the incoherent approximation may be written for neutron energy loss (Leadbetter 1973) as:

$$\frac{d^2\sigma_{inc}}{d\Omega\, dE'} = \frac{k'}{k}\sum_j \delta(\omega - \omega_j)\frac{(n(\omega)+1)}{2\omega_j}\left[\frac{\langle b^2\rangle}{M}|\mathbf{Q}\cdot\mathbf{u}_j(k)|^2 e^{-2W}\right] \qquad [4.34]$$

where primed and unprimed terms (E and k) refer to scattered and incident quantities, respectively, $n(\omega)$ is the Bose occupation number ([4.32]), M is the atomic mass, and W is the Debye–Waller factor ([3.28]). The summation is over individual modes j, and [4.34] has been written for one atom type only. For multicomponent systems, all quantities within square brackets become atom-type dependent and a further sum must be taken over all atoms within a unit of composition (e.g. SiO_2). Equation 4.34 may be simplified further by assuming that chemically equivalent atoms are dynamically equivalent. Averaging over all orientations of $\mathbf{Q}$ with respect to $\mathbf{u}$ gives $\langle|\mathbf{Q}\cdot\mathbf{u}_j(k)|^2\rangle = \frac{1}{3}Q\langle u^2\rangle$, where the mean-square vibrational amplitude $\langle u^2\rangle$ is in fact frequency and atom-type dependent. The summation of the delta functions in [4.34] yields the one-phonon density of states,

$$\sum_j \frac{(\omega-\omega_j)}{\omega_j} = \frac{3N}{\omega}\rho(\omega)$$

where $3N$ is the total number of phonon states. Thus, [4.34] becomes:

$$\frac{d^2\sigma_{inc}}{d\Omega\,dE'} = \frac{k'}{k}\frac{Q^2(n(\omega)+1)N\rho(\omega)}{2\omega}\left[\frac{\langle b^2\rangle}{M}\langle u^2\rangle e^{-2W}\right] \quad [4.35]$$

A function $g(Q,\omega)$ may be defined in terms of experimentally known quantities:

$$g(Q,\omega) = \frac{d^2\sigma}{d\Omega\,dE'}\frac{k}{k'}\frac{2\omega}{N(n(\omega)+1)Q^2} \quad [4.36]$$

Coherence effects are manifested by a Q-dependence of $g(Q,\omega)$, but if the incoherent approximation is valid, $g(Q,\omega)$ is equal to the density of states but weighted by the frequency dependent term $[(\langle b^2\rangle/M)(\langle u^2(\omega)\rangle)e^{-2W}]$ (for each atom in the composition unit, if appropriate).

Obviously, interpretation is simplest for monatomic systems, and an example of some inelastic neutron results is shown in Fig. 4.12 for the case of amorphous arsenic and two of the crystalline modifications, orthorhombic and rhombohedral (Leadbetter *et al.* 1976). It can be seen that all forms of As have well-separated acoustic and optic bands, and in addition the amorphous and orthorhombic forms exhibit a distinct dip in the lower acoustic band, a feature not shared by the rhombohedral form; on the basis of recursion method calculations, Davis *et al.* (1979) ascribe the filling in of this dip to a predominance of bonds in the staggered configuration ($\phi = 180°$) in the rhombohedral polymorph.

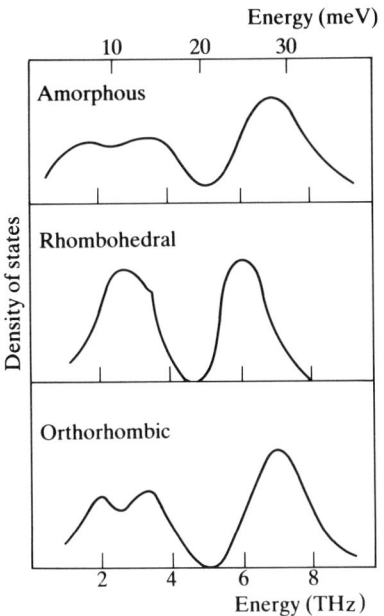

Fig. 4.12 Densities of states obtained from inelastic neutron scattering data for various structural modifications of arsenic (Leadbetter *et al.* 1976).

4.5 Low-temperature properties

4.5.1 Introduction

It is a striking feature of glasses that, at low temperatures, they exhibit markedly different behaviour from their crystalline counterparts in phonon-related properties, such as specific heat capacity, thermal conductivity, acoustic and dielectric absorption. The purpose of this section is to describe these most interesting and somewhat disparate phenomena under the one umbrella, and to show how far these low-temperature properties may be understood in terms of a simple phenomenological model.

The thermal properties of *crystalline* solids at low temperatures are well understood in terms of Debye's theory in which, assuming the distribution of phonons is cut off at some frequency with which a characteristic temperature θ_D may be associated, the famous T^3 temperature dependence of the heat capacity, as T tends to zero, results:

$$C_v = \frac{234 N k_B T^3}{\rho \theta_D^3} = \frac{2\pi^2}{5} \frac{k_B^4 T^3}{\hbar^3 \rho v_D^3} \qquad [4.37]$$

where ρ is the mass density, N is the atomic number density and v_D is given by a weighted average of longitudinal and transverse sound velocities:

$$\frac{3}{v_D^3} = \frac{1}{v_L^3} + \frac{2}{v_T^3} \qquad [4.38]$$

One would expect that at low temperatures, as the phonon mean free path and wavelength increased, the structural disorder characteristic of a glass would become unimportant, and in this continuum approximation the thermal properties should be similar to those of crystals, i.e. Debye-like. As Zeller and Pohl (1971) first found, this is far from the truth. Figures 4.13(a) and (b) show the specific heat and thermal conductivity of vitreous silica as a function of temperature compared with data for crystalline silica (α-quartz). The specific heat of the glass decreases much more slowly with temperature than the Debye T^3 prediction at low temperatures (for the temperature range $0.1\,K < T < 1\,K$) and can be expressed instead in the form:

$$C_v = aT + bT^3 \qquad [4.39]$$

where b exceeds the Debye prediction ([4.37]). The thermal conductivity is also markedly different (Fig. 4.13(b)) being a monotonically decreasing function of temperature for the glass, in contrast to the peaked behaviour characteristic of crystals.

The remaining sections of this chapter discuss the low-temperature behaviour of individual phonon-related properties in detail, finishing with a discussion of a model which goes a long way to explaining these anomalies for glasses. An extensive review of these phenomena is given in the book edited by W. A. Phillips (1981).

4.5.2 Specific heat

The behaviour of the specific heat, in common with other thermal properties, can conveniently be separated into two temperature regimes, above and below 2 K,

Low temperature properties

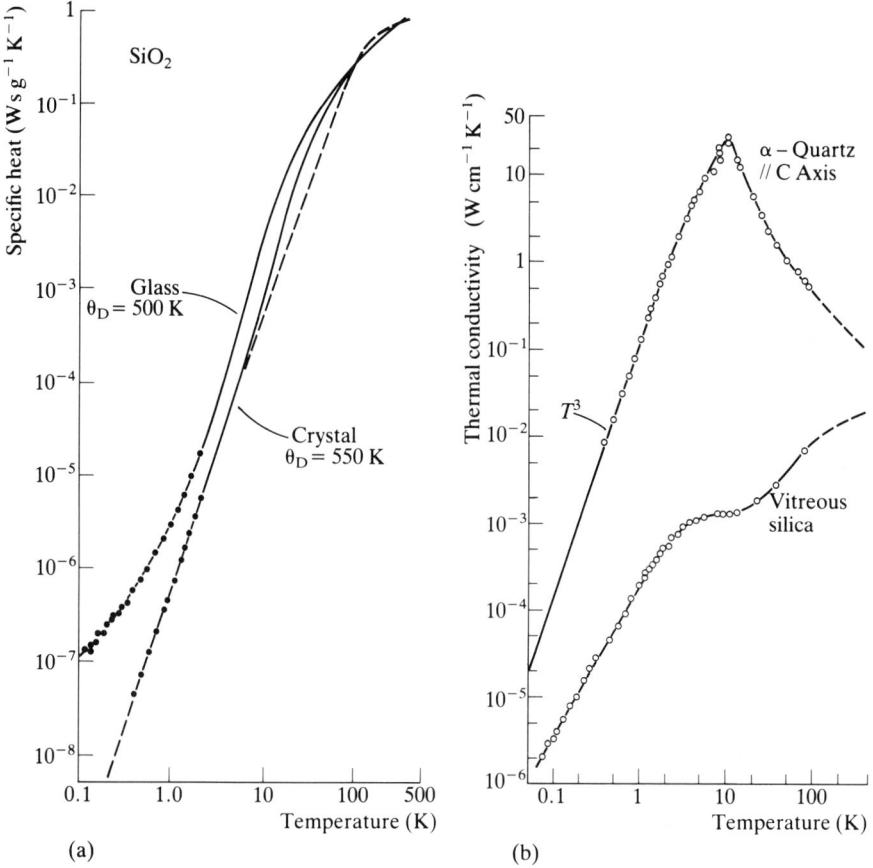

Fig. 4.13 (a) Specific heat of vitreous SiO_2 compared with that of quartz. Note the 'anomalous' specific heat at low temperatures varies almost linearly with T (R. O. Pohl in Phillips (1981)).
(b) Thermal conductivity of vitreous silica compared with that of quartz (Zeller and Pohl 1971).

since theories for the former are often extensions of existing (crystalline) models whereas those describing the latter case are distinct and novel (Phillips 1978).

Dealing first with the case of $T > 2$ K, Fig. 4.14 shows typical behaviour for three materials of very different structure, namely SiO_2, Se and polyethylene. Both crystalline and amorphous forms exhibit a peak in the heat capacity plotted as C/T^3 (or equivalently C/C_{Deb}) v.s. T. In the crystalline case, this 'excess' heat capacity is explained by a higher density of states of the lowest TA branch near the edge of the first Brillouin zone. The hump is also observed for the amorphous materials, but shifted to lower values of reduced temperature T/θ_D. An excess heat capacity has been observed for all (non-metallic) amorphous materials so far studied, and the peak generally occurs at a different value of T/θ_D to that of the corresponding crystal. The cause of the shift is not known but must be a very subtle effect. Two crystalline forms of silica, quartz and cristobalite, exhibit peaks at different temperatures, and for these the bonding topology is the same out to the third nearest neighbour, and this is true also, on average, for the glass. Vibrational calculations are at present too crude for a theoretical understanding to be gained for the

159

Vibrations

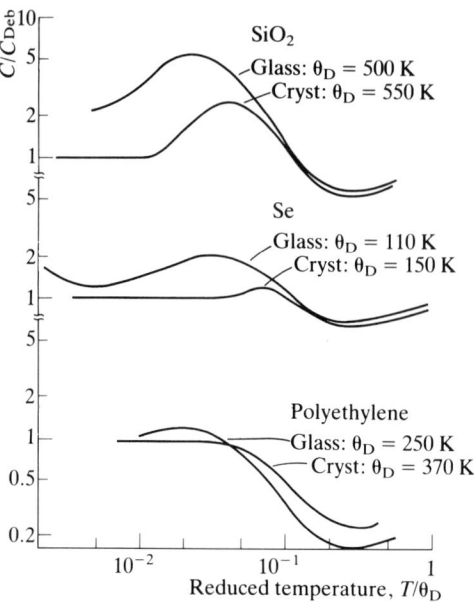

Fig. 4.14 Specific heat of glasses above 2 K. The normalized specific heat C/C_{Deb} is plotted versus reduced temperature T/θ_D for SiO_2, Se and polyethylene in both glassy and crystalline forms. Note the 'excess' heat capacity manifested as a hump in these curves (R. O. Pohl in Phillips 1981).

behaviour of the low-frequency vibrational density of states, and so the origin of these anomalies remains obscure.

Another feature of note in Fig. 4.14 is the observation that the curves of C/C_{Deb} v.s. T do not tend to unity, the Debye limit, for the amorphous materials for $T < 2$ K. Instead, as was seen in Fig. 4.13(a), the heat capacity decreases more slowly than the Debye T^3 law, and can be approximated well by [4.39]; a plot of C/T v.s. T^2 should plot linearly according to this equation, giving an intercept C_1 at $T=0$. The 'anomalous linear specific heat' has been observed in a wide variety of amorphous materials, silica and silicates, chalcogenide glasses, various organic polymers and even epoxy resins. An example of a non-crystalline insulator in which the linear term is very small is bulk amorphous arsenic (Jones *et al.* 1978). This and amorphous Ge films which behave similarly are interesting since they do not possess any flexible structural units, such as the bridging oxygen in SiO_2. The linear term in the specific heat appears to be predominant in disordered materials containing low coordinated atoms. Note that we have said *disordered* rather than amorphous; disordered (e.g. neutron irradiated) *crystals* have also been found to have an anomalous specific heat, although perfect crystals do not. We might add at this juncture that [4.39] is often only approximately obeyed; in practice, deviations from strict linearity of the anomalous term are often found (Fig. 4.15).

So far we have addressed the case of amorphous insulators – what of glassy metals? The problem here is that the lattice contribution to the specific heat is masked by the electronic contribution which also behaves linearly in T. The results of an elegant experiment to eliminate the electronic effect in order to reveal any possible anomalous specific heat are shown in Fig. 4.16. Glassy $Zr_{70}Pd_{30}$, prepared

Low temperature properties

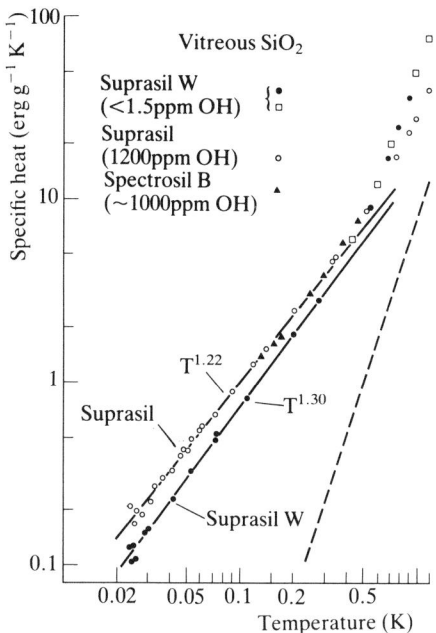

Fig. 4.15 Specific heat of a-SiO$_2$ above 25 mK. Suprasil and Spectrosil B have large OH concentrations but small metal concentrations. Suprasil W has small OH and metal concentrations, but 230 p.p.m. Cl and 290 p.p.m. F. The dashed line is the Debye specific heat, $C_{\text{Deb}} = 8T^3$ erg g^{-1} K^{-4} (R. O. Pohl in Phillips 1981).

in ribbon form by melt extraction, is a superconductor below $T_c = 2.53$ K. Thus, below T_c no unpaired normal electrons exist, and hence the electronic contribution to the specific heat is removed. It is found that below T_c this material *does* appear to exhibit an anomalous linear specific heat as given by [4.39], with the value of the constant 'a' very similar to those found in insulating glasses.

In summary, it appears that an anomalous linear term in the specific heat is a characteristic feature of most glass-forming solids, with the constant 'a' being in the range 1–5 J g^{-1} K^{-1}. However, at very low temperatures, strict linearity is often not observed and the magnitude of the specific heat is often sample dependent (see Fig. 4.15).

4.5.3 Thermal conductivity

Thermal transport in crystalline non-metallic materials is by means of propagating phonons; the thermal conductivity may be written as:

$$\kappa = \tfrac{1}{3} \sum_i \int_0^{\omega_D} C_i(\omega) v_i(\omega) \ell_i(\omega) \, d\omega \qquad [4.40]$$

where the sum is over the three phonon modes, v and ℓ are the phonon velocity and mean free path, $C(\omega)$ is the contribution to the lattice specific heat by phonons of frequency ω, and integration is from $\omega = 0$ to ω_D. The temperature dependence of κ

161

Vibrations

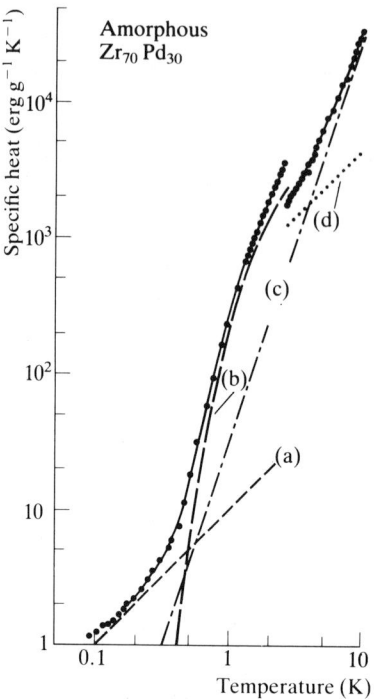

Fig. 4.16 Specific heat of the superconducting amorphous metallic alloy $Zr_{70}Pd_{30}$ as a function of temperature (J. E. Graebner, B. Golding, R. J. Schutz, F. S. L. Hsu and H. S. Chen (1977) *Phys. Rev. Lett.* **39**, 1480). The superconducting transition at $T_c = 2.53$ K is clearly evident. Curve (a) is the linear term in the specific heat; curve (b) is the exponential decrease of the specific heat due to the normal electrons below T_c; curve (c) is a T^3 extrapolation of the specific heat near 10 K; curve (d) represents the electronic contribution to the specific heat above T_c.

for a crystal is typically peaked (Fig. 4.13(b)). Above 10 K, phonons are scattered by intrinsic (phonon–phonon umklapp) processes, and hence ℓ increases as T decreases since the number of phonons decreases. Eventually, the mean free path ℓ becomes comparable to the sample dimensions and is hence constant, and for lower temperatures κ follows the specific heat and therefore behaves as T^3.

In contrast, the behaviour of amorphous solids is completely different (Fig. 4.13(b)). No peak in κ is observed, but instead a plateau region occurs at ~10 K, and the low-temperature region behaves approximately as T^2, rather than T^3. The cause of the plateau region B in Fig. 4.17(a) is not well understood; however a rapid decrease in the phonon mean free path with frequency, as shown schematically in Fig. 4.17(b), does account for this feature although its cause is not understood. Region A of κ v.s. T is accounted for by a constant ℓ at high frequencies. The T^2 temperature dependence of κ is explained if ℓ has a T^{-1} temperature dependence (cf. [4.40]); Zeller and Pohl (1971) first suggested that this behaviour might be associated with the anomalous linear specific heat, and this is discussed in more detail in the next section.

We note finally that values of the temperature exponent of κ in region C often differ from two, values in the range 1.8–2 commonly being found. Furthermore, the thermal conductivities of amorphous materials in the T^2 region are often very

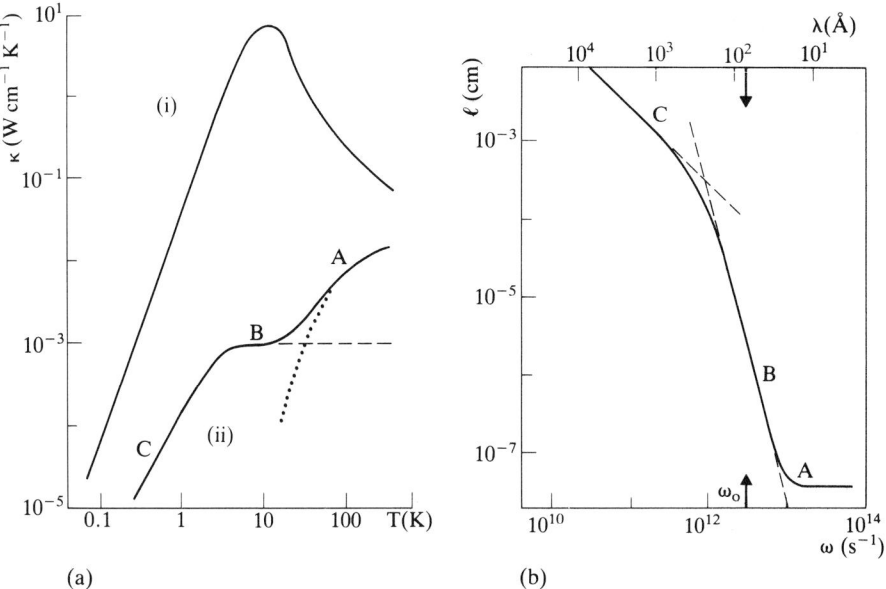

Fig. 4.17 (a) Schematic illustration of the temperature dependence of the thermal conductivity of (i) crystalline quartz and (ii) vitreous silica. Note the plateau region B in the curve for the glass. (b) Frequency dependence of the intrinsic phonon mean free path proposed to account for the κ v.s. T, curve (ii), in (a). Region C is drawn with a slope -1, and region B at ω_0 is drawn with a slope -4; the whole curve is quantitatively appropriate for a borosilicate glass (A. C. Anderson in Phillips 1981).

similar (in contrast to the linear specific heat), as can be seen from Fig. 4.18, and may therefore be more of an 'intrinsic' property and less dependent on the impurity content.

4.5.4 Theoretical models

The low-temperature thermal properties mentioned above that are observed in glasses are so different from those observed in crystals that considerable theoretical effort has been expended in attempting to understand them. As a consequence, a variety of theories abound, proposing that, for example, the effects are electronic (due to localized electron states), or due to damped phonons, or phonon scattering from structural inhomogeneities or dislocations. While in most cases these are capable of accounting for a single thermal property, none can explain all the observed low-temperature properties self-consistently. The model which is most successful in accounting for much of the observed behaviour is the 'tunnelling state model' (or the 'two-level system – TLS') proposed independently by Anderson *et al.* (1972) and Phillips (1972).

The tunnelling state model takes as its central hypothesis the assumption that, in a glassy system, a certain number of atoms (or groups of atoms) can occupy one of two local equilibrium positions; the atoms can tunnel between these positions and therefore move within a double-well potential such as that shown schematically in Fig. 4.19. In general, such a potential will be asymmetric, characterized by the

Vibrations

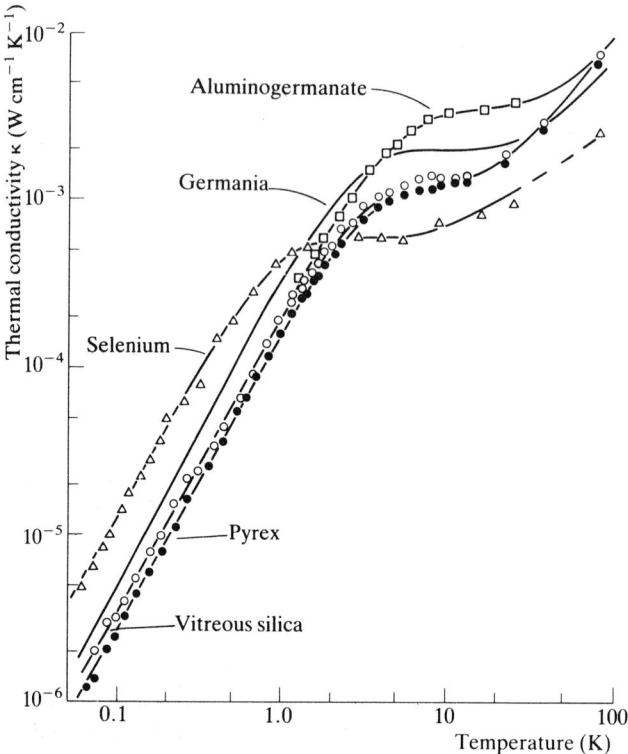

Fig. 4.18 Thermal conductivity as a function of temperature for a variety of glasses (Zeller and Pohl 1971).

'asymmetry energy' Δ and the barrier height V_o, both of which are expected to be random quantities in a glass; the positional coordinate in Fig. 4.19 is termed a configurational coordinate to disguise the fact that the microscopic nature of the state is in fact not known. The energies of the two lowest states of the asymmetric double well (i.e. the energy splitting of the ground state introduced by the tunnelling) have energies $\pm E/2$ where:

$$E^2 = (\Delta^2 + \Delta_o^2) \quad [4.41]$$

where Δ_o, the coupling energy, is the energy difference between the two lowest states in the symmetric case. Δ_o is related to the parameters of the well by

$$\Delta_o = \hbar\omega_o e^{-\lambda} \quad [4.42]$$

where

$$\lambda = (2mV_o/\hbar^2)^{1/2} d \quad [4.43]$$

ω_o is the frequency of oscillation in an individual well and the exponential factor in [4.42] represents the overlap between the wavefunctions for the two wells. This model can account for the low-temperature thermal anomalies since the atomic tunnelling offers a mechanism by which low excitation energies can be generated (10^{-5} eV $< E < 10^{-4}$ eV, corresponding to temperatures between 0.1 and 1 K) which

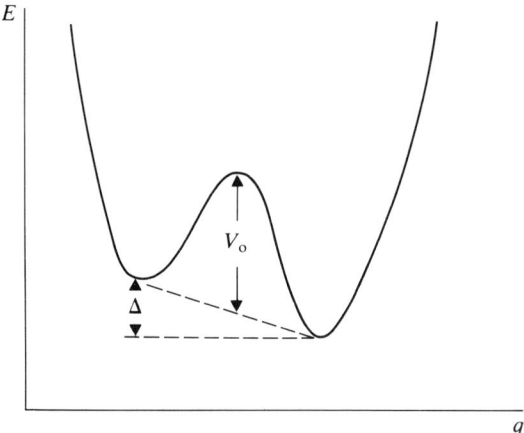

Fig. 4.19 Schematic illustration of the double-well potential characterizing a two-level system plotted as a function of configuration coordinate q. The parameters Δ (the asymmetry energy) and V_0 (the barrier height) are also indicated.

therefore give rise to the anomalous specific heat, and the tunnelling states at the same time act as resonant scattering centres for acoustic phonons, thereby controlling the mean free path and hence the thermal conductivity and, as we will see, the ultrasonic attenuation. The tunnelling state model will now be explored in more detail.

Although the terms 'tunnelling state' and 'two-level system' are often used interchangeably, there is in fact a distinction between the two. The tunnelling state model is the more general and assumes that in an amorphous solid there is a distribution of values of Δ_o (or equivalently V_0) as well as E and averages must be taken over both in any calculation; in practice, the distribution function for Δ_o is not known and if this averaging is neglected, the more simplified two-level model ensues in which E is the only parameter.

The free energy of a single two-level system may be written (Phillips 1972) as:

$$F(E) = -k_B T \ln \left[\cosh (E/2k_B T) \right] \qquad [4.44]$$

and since the specific heat is given by $-T(\partial^2 F/\partial T^2)$, the total specific heat is thus given by an integration over all two-level systems:

$$C = k_B \int_0^\infty n(E) \left(\frac{E}{2k_B T} \right)^2 \operatorname{sech}^2 \left(\frac{E}{2k_B T} \right) dE \qquad [4.45]$$

where $n(E)$ is the density of two-level systems per unit volume and energy interval. A similar equation holds for the case of tunnelling states, but includes also a distribution in Δ_o (or λ). If the density of states is taken to be a constant, n_o, for the range of energies of interest (i.e. 10^{-5} eV $< E < 10^{-4}$ eV), [4.45] can be readily evaluated to give:

$$C = \frac{\pi^2}{6} k_B^2 n_o T \qquad [4.46]$$

Therefore the tunnelling state (or two-level) model accounts naturally for the anomalous linear term in the specific heat with the physically plausible assumption that the density of tunnelling states is independent of energy.

This expression is valid if the specific heat is measured on time-scales which are sufficiently long that even the slowest states can equilibrate. The rate of transitions between the two levels can be written approximately as $\Gamma = \Gamma_o \exp(-2\lambda)$. The minimum value for λ is given by $\lambda_{min}(E) = \ln(\hbar\omega_o/E)$ from [4.41] and [4.42], and the upper limit for a centre which can respond in a time comparable to t is given by $\lambda_{max} = \frac{1}{2} \ln[\Gamma_o t]$. Thus if the experimental time-scale t is short enough that it lies between the minimum and maximum relaxation times of the centres, $\tau_{min} < t < \tau_{max}$, then only a certain proportion of the states will contribute, characterized by a cut-off at λ_{max}, and the specific heat is thus predicted to increase logarithmically with time (Anderson et al. 1972). Experimentally a large increase in temperature (overshoot) is expected in the first few microseconds after a heat pulse is applied (corresponding to a smaller effective heat capacity than that characteristic of millisecond or greater time-scales). The short-time data available are somewhat ambiguous; certainly the changes observed are smaller than expected, but this may be due to experimental difficulties or the fact that τ_{max} is smaller than otherwise expected. Black (1978) has suggested that *two* types of tunnelling states might be present in glasses, termed 'standard' and 'anomalous'; the latter are anomalous in that they contribute only to the heat capacity and no other thermal or acoustic property.

The tunnelling state model seems to account satisfactorily for the linear temperature dependence of the specific heat; data exhibiting a super-linear temperature dependence $\propto T^{1+m}$ (see Fig. 4.15) can be accommodated within the theory by relaxing the condition that $n(E)$ is independent of E, and assuming instead that it is weakly dependent on energy $n(E) \propto E^m$, where $m = 0.1$–0.3.

More seriously, it is not at all clear that the states giving rise to the anomalous specific heat are intrinsic in character; there is considerable evidence from sample purity variations, etc., that perhaps some of the effect may be extrinsic. The density of tunnelling states for SiO_2 determined from specific heat studies using [4.46] is $\sim 10^{21}$ eV^{-1} cm^{-3} ($\sim 10^{46}$ J^{-1} m^{-3}). If the states are assumed to be distributed over a range of energies, ~ 0.1 eV, then this corresponds to the number of states being of the order of 1/250 of the total number of SiO_2 groups, if only intrinsic centres are assumed to be responsible.

The tunnelling state model is also capable of explaining the anomalous T^2 temperature dependence of κ observed in glasses at low temperatures (see region C of Fig. 4.17(a)). The additional scattering centres which progressively limit the mean free path of the phonons at low temperatures can be identified with tunnelling states. Phonons of energy $\hbar\omega = E$, the tunnel splitting, can be scattered by a tunnelling state by a process of excitation from the ground state and subsequent spontaneous decay and emission of an incoherent phonon, i.e. 'resonance scattering'. The mean free path of phonons scattered in such a manner by a two-level system can be shown to be (see W. A. Phillips 1981):

$$\ell_{res} = \frac{\rho v^3}{\pi n_o M^2 \omega} \coth\left(\frac{\hbar\omega}{2k_B T}\right) \qquad [4.47]$$

where v is the average velocity, M is the coupling constant relating E to the strain (~ 3 eV), and in this simplified model the constant density of states n_o is identical with that obtained from the linear specific heat. For thermal phonons, $\hbar\omega \sim k_B T$, and hence $\ell_{res} \sim T^{-1}$ and if this is substituted into [4.40] together with $C \propto T$, the T^2 temperature dependence of κ results. If this scattering mechanism is assumed, there is no plateau in κ as a function of temperature; ℓ_{res} becomes larger without limit with T at fixed ω.

There is an additional phonon scattering mechanism that is exhibited by tunnelling states. In common with any two-level system (e.g. a spin $1/2$ in a magnetic field), the rate at which equilibration of the levels takes place after a change in population is given by T_1^{-1} where (Phillips 1981):

$$T_1^{-1} = \frac{M^2 E^3}{2\pi\rho\hbar^4 v^5} \coth\left(\frac{E}{2k_B T}\right) \qquad [4.48]$$

This process gives a non-resonant scattering mean free path given by (Phillips 1978) $\ell_{nres} \propto T^{-3}$, and the net mean free path is $\ell^{-1} = \ell_{nres}^{-1} + \ell_{res}^{-1}$. The non-resonant term limits the increase in mean free path with temperature resulting from the resonant term, and can thereby give rise to a plateau region as in Fig. 4.17(a), but its magnitude is insufficient to account entirely for the experimentally observed plateau.

As is the case for the specific heat, an energy dependence of the density of tunnelling states can be introduced into the model; if $n(E) \propto E^m$ is assumed as before, $\kappa \propto T^{2-m}$ is obtained, in agreement with many experimental data. As we have seen previously, the thermal conductivity is very similar for a variety of different materials (Fig. 4.18); furthermore it is practically sample independent, unlike the linear specific heat, and is therefore probably more of an intrinsic phenomenon.

Further evidence which confirms the presence of two-level systems in glasses in a striking way comes from acoustic absorption experiments. Data for vitreous silica are shown in Fig. 4.20 (from Phillips 1981) where the inverse mean free path of longitudinal ultrasonic waves (1 GHz) is plotted as a function of temperature. For *low* acoustic intensities, ℓ^{-1} increases as T decreases at very low temperatures. This can be understood in terms of the two phonon scattering processes discussed for thermal conductivity. The resonant inverse mean free path ([4.47]) varies as $\ell_{res}^{-1} \propto \omega^2/T$ for the case of acoustic phonons for which $\hbar\omega \ll k_B T$, and this process gives rise to the rise in attenuation at very low temperatures. The increase in attenuation at higher temperatures is due to the other process for which $\ell_{nres}^{-1} \propto T^3$. The startling observation is that at *high* acoustic intensities, the attenuation decreases continuously with decreasing temperature. This can be explained by the *saturation* of the tunnelling states at high intensities when the two levels become equally populated, and hence the process leading to [4.47], namely spontaneous decay, becomes inoperative. This is incontrovertible evidence that two-level systems, whatever their origin, exist.

Although the model is phenomenologically very successful, there is no agreed explanation concerning the microscopic origin of two-level systems. It is observed

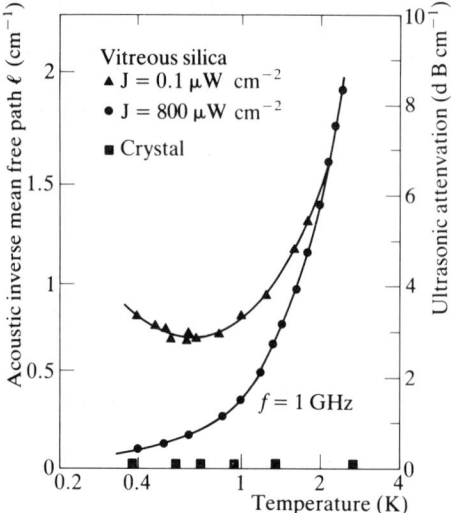

Fig. 4.20 Temperature dependence of the ultrasonic absorption in vitreous silica for longitudinal waves of 1 GHz at low temperatures. Note the decrease in absorption at the lowest temperatures for high acoustic intensities. The absorption of a quartz crystal is also shown for comparison (Hunklinger and Schickfus in Phillips 1981).

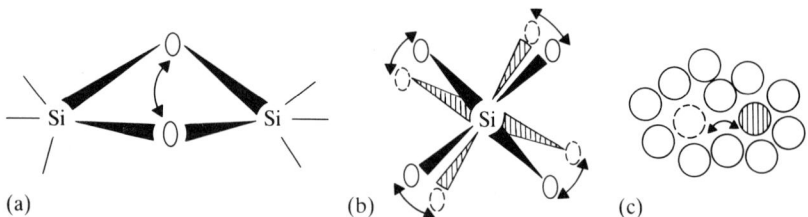

Fig. 4.21 Microscopic model for two-level systems in glasses:
(a) atom flip (e.g. oxygen in oxides or chalcogen in chalcogenides);
(b) structural unit rotation (in e.g. oxides);
(c) atom–vacancy interchange in glassy metals.

that low temperature thermal anomalies are universally found in glasses containing low coordinated atoms. The obvious microscopic cause for a tunnelling system, as in Fig. 4.19, is the flipping of the low coordinated atom (e.g. O in SiO_2) from one metastable position to another (Fig. 4.21(a)). Alternatively, a rigid rotation of, say, an SiO_4 tetrahedron from one position to another (i.e. a cooperative flipping) may be responsible (Fig. 4.21(b)). This cannot be a general explanation, however, since thermal anomalies are also found in metallic glasses which do not have directional bonding. Nevertheless a similar type of atom switching motion may be postulated, the atom moving from one relatively low density region of the close-packed assembly to a similar neighbouring region (Fig. 4.21(c)).

The proposition that impurities are the cause of the tunnelling states in many amorphous systems cannot be dismissed lightly. Indeed, the —OH group, which is a widespread impurity in many silica samples, has a considerable effect on low temperature thermal properties, as can be seen by reference to Fig. 4.15 in which the heat capacity of an a-SiO$_2$ sample containing ~ 1000 p.p.m. OH is some 30% higher than one containing only ~ 1 p.p.m. However, there is also much evidence that a proportion of the effect is intrinsic. What also remains a mystery, besides the microscopic nature of the tunnelling states, is why the phonon coupling constant M should be approximately constant for all materials.

Problems

4.1 Deduce the dynamics of amorphous networks using a Lagrangian formulation assuming only nearest-neighbour central forces to be operative. (See M. F. Thorpe and F. L. Galeener (1980), *Phys. Rev.*, **B22**, 3078 and P. N. Sen and M. F. Thorpe (1977), *Phys. Rev.*, **B15**, 4030 for details.)

Refer atomic displacements to bond directions (described by unit vectors $\hat{q}$) and call the amplitude of the displacement q. Use the labels ℓ, ℓ' to denote *sites* and Δ, Δ' to denote *bonds*, i.e. the unit vectors are $\hat{q}_\Delta(\ell)$ and the displacement amplitudes of the generalized coordinates $q_\Delta(\ell)$. The potential energy, for bond-stretching forces only, can be written in this coordinate system as:

$$V = \frac{\alpha}{2} \sum_{\{\ell, \ell', \Delta\}} [q_\Delta(\ell) + q_\Delta(\ell')]^2 \qquad [4.1P]$$

where each bond is only counted once.

The kinetic energy can be written in terms of conventional, orthogonal coordinates (x, y, z) as

$$T = \frac{M}{2}(\dot{x}^2 + \dot{y}^2 + \dot{z}^2) \qquad [4.2P]$$

(a) For a twofold coordinated atom (e.g. O), construct axes such that $\hat{y}$ bisects the angle, θ, between two bonds $\hat{q}_1$ and $\hat{q}_2$, and $\hat{x}$ is perpendicular to $\hat{y}$ and in the same plane as $\hat{q}_1, \hat{q}_2$. Hence show that

$$T = \frac{M}{2}[a(\dot{q}_1^2 + \dot{q}_2^2) + b(\dot{q}_1 + \dot{q}_2)^2] \qquad [4.3P]$$

where $a = (1 - \cos\theta)^{-1}$, $a + 2b = (1 + \cos\theta)^{-1}$, and the third generalized coordinate $q_3 = z$ has been omitted because for nearest-neighbour central forces, V is independent of q_3.

Similarly show for a tetrahedrally coordinated atom (e.g. Si), that

$$T = \frac{3m}{8}(\dot{q}_1^2 + \dot{q}_2^2 + \dot{q}_3^2 + \dot{q}_4^2) \qquad [4.4P]$$

with the constraint $q_1 + q_2 + q_3 + q_4 = 0$, for a coordinate system such that $\hat{z}$ is antiparallel to q_4 and $\hat{x}$ is the projection of $\hat{q}_1$ on to a plane perpendicular to $\hat{z}$.

The Lagrangian $\mathscr{L}$ is defined by

$$\mathscr{L}([q_i, \dot{q}_i]) = T([\dot{q}_i]) - V([q_i]) \qquad [4.5P]$$

Vibrations

where [ ] denotes the whole set of q_i, $\dot{q}_i$ for all i. The equations of motion are:

$$\frac{d}{dt}\left(\frac{\partial \mathscr{L}}{\partial \dot{q}_i}\right) - \frac{\partial \mathscr{L}}{\partial q_i} = 0 \qquad [4.6P]$$

(b) Deduce the dynamical properties of a tetrahedral network by first showing that the Lagrangian may be written as

$$\mathscr{L} = \frac{3}{8} M \sum_{\ell,\Delta} [\dot{q}_\Delta(\ell)]^2 - \frac{\alpha}{2} \sum_{\{\ell,\ell',\Delta\}} [q_\Delta(\ell) + q_\Delta(\ell')]^2 - \frac{\lambda}{2} \sum_\ell \left(\sum_\Delta q_\Delta(\ell)\right)^2 \qquad [4.7P]$$

where in the last term λ is a Lagrange multiplier, tending to infinity eventually, and which accounts for the constraint at each site ℓ, $\sum_\Delta q_\Delta(\ell) = 0$. Obtain the equation of motion from [4.6P] and look for normal modes of $q_\Delta(\ell)$ and $q_\Delta(\ell')$ having a time dependence $\exp(i\omega t)$ and hence eliminate $q_\Delta(\ell')$. By summing over all ℓ' (neighbours of ℓ), calling $Q(\ell) = \sum_\Delta q_\Delta(\ell)$, and rewriting $\sum_{\ell'} Q(\ell') = \varepsilon Q(\ell)$ show that

$$\varepsilon = \frac{1}{\lambda \alpha}[(\alpha - \tfrac{3}{4} M \omega^2)^2 + 4\lambda(\alpha - \tfrac{3}{4} M \omega^2) - \alpha^2] \qquad [4.8P]$$

Let $\lambda \to \infty$ thereby driving off to infinite frequency the unwanted modes associated with the extra degrees of freedom. Thus finally

$$M\omega^2 = \frac{\alpha}{3}(4 - \varepsilon) \qquad [4.9P]$$

The bonds to ε can be found by noting that $\sum_{\ell'} Q(\ell') = \varepsilon Q(\ell)$ can be written as $\mathbf{M}\mathbf{v} = \varepsilon \mathbf{v}$ where $\mathbf{M}$ is the connectivity matrix ($M_{ij} = 1$, i, j nearest neighbours; $M_{ij} = 0$, otherwise), and is real, symmetric and positive, whereupon the Frobenius theorem can be applied, and it then found that

$$-Z \leqslant \varepsilon \leqslant Z \qquad [4.10P]$$

where Z is the connectivity (4 for Si). Hence, the bounds to the frequencies given by [4.9P] are $0 \leqslant \omega^2 \leqslant 8\alpha/3M$.

(c) By inserting twofold coordinated atoms between the atoms to generate an AX_2 network (e.g. SiO_2) with an A—X—A bond angle θ, use the same Lagrange formalism to deduce that the frequencies of the band edges (i.e. when $\varepsilon = 0, 4$) are given by the equations:

$$\omega_1^2 = \frac{\alpha}{m}(1 + \cos\theta) \qquad \omega_2^2 = \frac{\alpha}{m}(1 - \cos\theta)$$

$$\omega_3^2 = \omega_1^2 + \frac{4\alpha}{3M} \qquad \omega_4^2 = \omega_2^2 + \frac{4\alpha}{3M} \qquad [4.11P]$$

Discuss the behaviour of the modes as $\theta \to 90°$, particularly in relation to an isolated AX_4 molecule, and find the critical bond angle in this model below and above which molecular-like and band-like modes, respectively, exist.

4.2 There is a formal analogy between the two-level system (TLS) used to describe low-temperature excitations in glasses and a spin-$\frac{1}{2}$ particle in a magnetic field, enabling much of the formalism developed for the magnetic case to be used for the two-level system model also.

The Hamiltonian for a TLS in an elastic strain field e and an electric field F can be written as:

$$\mathcal{H} = H_0 + H_1 + H_2 = \frac{1}{2}\begin{pmatrix} E & 0 \\ 0 & -E \end{pmatrix} - \frac{1}{2}\begin{pmatrix} D & 2M \\ 2M & -D \end{pmatrix}e - \frac{1}{2}\begin{pmatrix} \mu & 2\mu' \\ 2\mu' & \mu \end{pmatrix}F \quad [4.12\text{P}]$$

where H_0 represents the unperturbed case (E given by [4.41]), H_1 represents the perturbation due to the strain field and H_2 represents the perturbation due to the electric field. The diagonal and off-diagonal components of the deformation potential (D, M), and the permanent and induced dipole moments (μ, μ') can be written in terms of TLS parameters as:

$$D = \frac{2\gamma\Delta}{E}, \qquad M = \frac{\gamma\Delta_0}{E} \quad [4.13\text{P}]$$

$$\mu = \frac{2p\Delta}{F}, \qquad \mu' = \frac{p\Delta_0}{F} \quad [4.14\text{P}]$$

where $\gamma = \frac{1}{2}(\partial\Delta/\partial e)$ and $p = \frac{1}{2}(\partial\Delta/\partial F)$.

(a) Show that, by analogy with the equivalent equations for magnetic resonance, the dynamics of a TLS subject to an elastic strain field can be written in the form of the Bloch equations:

$$\dot{X} = -X/T_2 - (E + De)Y/\hbar \quad [4.15(a)\text{P}]$$
$$\dot{Y} = (E + De)X/\hbar - Y/T_2 - (2Me/\hbar)Z \quad [4.15(b)\text{P}]$$
$$\dot{Z} = (2Me/\hbar)Y - (Z - Z_e)/T_1 \quad [4.15(c)\text{P}]$$

X and Y represent the induced elastic (or electric) polarization, but Z can only formally be interpreted as a polarization component since the level splitting is not caused by an external field (as in the magnetic case). Z_e is the value of Z in its 'instantaneous equilibrium':

$$Z_e = -\tfrac{1}{2}\tanh\left[(E + De)/2k_B T\right] \quad [4.16\text{P}]$$

(b) Discuss the dynamical solutions of [4.15P] for a TLS subject to a pulsed field of duration τ for various relative magnitudes of T_1, T_2 and τ.

(c) The contribution of N TLS (per unit volume) to the internal energy U can be written as $U = -N(2Me, 0, De) \cdot (X, Y, Z)$. The contribution to the dynamic elastic modulus is then:

$$\partial C = \frac{1}{e}\frac{\partial(\partial u)}{\partial e} = -\frac{N}{e}(2MX + DZ) \quad [4.17\text{P}]$$

Sound dispersion and absorption are then given by:

$$\delta v(\omega) = \left(\frac{1}{2\rho v}\right) \text{Re}\{\partial C(\omega)\} \qquad [4.18(a)P]$$

$$\alpha(\omega) = -\left(\frac{\omega}{\rho v^3}\right) \text{Im}\{\delta C(\omega)\} \qquad [4.18(b)P]$$

where v is the velocity of sound in the absence of TLS. Using the steady state solution of X ([4.15(a)P]) which gives an absorption of resonant character, derive

$$\alpha_{\text{res}}(\omega) = \frac{\pi n_a M^2 \omega}{\rho v^3} \frac{\tanh(\hbar\omega/2k_B T)}{\sqrt{(1+I/I_C)}} \qquad [4.19P]$$

where I is the acoustic intensity, $I_c = \hbar^2 \rho v^3 / 2M^2 T_1 T_2$ and n_a is the constant density of states for the acoustically active centres. Hence account for the saturation behaviour shown in Fig. 4.20.

Bibliography

Vibrational properties

D. Beeman and R. Alben (1977), *Adv. Phys.* **26**, 339.
M. Cardona (ed.) (1975), *Light Scattering in Solids* (*Topics in Applied Physics*, vol. 8) (Springer-Verlag).
S. S. Mitra (ed.) (1976), *Physics of Structurally Disordered Solids* (Plenum).
B. T. M. Willis (ed.) (1973), *Chemical Applications of Thermal Neutron Scattering* (OUP).

Low-temperature properties

W. A. Phillips (ed.) (1981), *Amorphous Solids: Low Temperature Properties* (*Topics in Current Physics*, vol. 24) (Springer-Verlag).

5 Electrons

5.1	**Introduction**
5.2	**Electronic density of states**
5.2.1	Theoretical calculations
5.2.2	Experimental determination
	Photoemission
	Ultra-violet and X-ray absorption
5.3	**Localization**
5.3.1	Introduction
5.3.2	Effects of disorder
5.3.3	One dimension
5.3.4	Two and three dimensions
5.3.5	Minimum metallic conductivity and the mobility edge
5.3.6	Percolation
5.4	**Transport properties**
5.4.1	Direct current electrical conductivity
5.4.2	Thermopower
5.4.3	Hall effect
5.4.4	Ionic conduction
5.5	**Small polarons**
5.6	**Optical properties**
5.6.1	Interband absorption
5.6.2	The absorption edge
5.6.3	Intraionic absorption and luminescence
	Problems
	Bibliography

5.1 Introduction

It has been a great triumph of the application of quantum mechanics to solid-state physics that an understanding of why certain crystals are metals and others are insulators has been achieved. The presence of perfect periodicity greatly simplifies the mathematical treatment of the behaviour of electrons in a solid. The electron states in this case can be written as 'Bloch waves' extending throughout the crystal:

$$\psi(\mathbf{k},\mathbf{r}) = u(\mathbf{k},\mathbf{r}) \exp(i\mathbf{k}\cdot\mathbf{r}) \qquad [5.1]$$

where the function $u(\mathbf{k},\mathbf{r})$ has the periodicity of the crystal lattice (in which a lattice translation vector $\mathbf{R}_1$ connects lattice points):

$$u(\mathbf{k},\mathbf{r}+\mathbf{R}_1) = u(\mathbf{k},\mathbf{r}) \qquad [5.2]$$

and this modulates the term $\exp(i\mathbf{k}\cdot\mathbf{r})$ representing a plane wave. The allowed wavevectors $\mathbf{k}$ of the electrons are intimately related to the symmetry of the underlying crystal lattice since a reciprocal lattice (related to the unit cell parameters) can be established in reciprocal or $\mathbf{k}$-space.

The allowed energies of the electrons can thus be represented by means of a 'band structure' in $\mathbf{k}$-space. A free electron has an energy $E(\mathbf{k}) = \hbar^2 k^2/2m$, but this parabolic dependence is distorted considerably if the electron experiences a scattering potential. Of course for a crystalline solid, these potentials arise from the periodic array of atom centres, and the electron waves can Bragg reflect from the lattice planes. This results in energy gaps opening up at values of k corresponding to certain values of reciprocal lattice vector ($2\pi/a$ for a linear array of atoms of separation a), i.e. at the edge of the 'Brillouin zone'. The occurrence of energy ranges for which there are *no* allowed electron states can be thought of as being due to destructive interference of the Bragg reflected electron waves; in the 'nearly-free electron' model, the size of the energy gap is determined by the Fourier component of the potential corresponding to the Bragg condition (see e.g. Madelung 1978). A simple one-dimensional band structure in the extended zone scheme is shown in Fig. 5.1. The difference between metals and insulators then simply amounts to whether there are sufficient electrons available to fill all the states in a Brillouin zone; if the band is only partly filled the solid is a metal, whereas if the band is completely filled, there is a gap between occupied and unoccupied electron states, and the material is an insulator at $T=0$.

Let us now consider the case of amorphous materials. For these, as we saw in

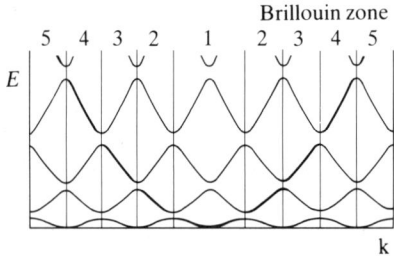

Fig. 5.1 One-dimensional band structure in the extended zone scheme.

Chapter 3, there is no periodicity and hence there can be no reciprocal (or k-) space, and the electrons cannot be represented as Bloch states ([5.1]). Since, as we have seen, the occurrence of band gaps can be viewed as arising from Bragg reflection of electron waves and hence is a direct consequence of periodicity, the question therefore arises: Should band gaps occur in amorphous materials too? The fact that window-glass (silica) is transparent to visible light is direct experimental proof that a band gap $\gtrsim 2$ eV must exist for this material (in fact the gap is nearer 10 eV). Conventional (crystalline) solid-state physics theory is incapable of accounting for this behaviour, and we shall see that concepts more akin to those of chemistry can resolve the dilemma. Before we consider this matter further, it is perhaps pertinent to discuss another consequence of the lack of long-range order on the description of electron states in an amorphous solid.

We have seen that the absence of periodicity in an amorphous solid dictates that there can be no reciprocal space. In this case, electron states cannot be represented by a band structure in the form $E(k)$. This problem is entirely analogous to that encountered for phonon dispersion curves $\omega(q)$ in Chapter 4. The quantity that *is* equally valid as a description of electron states for both crystalline and amorphous solids is, as in the phonon case, the density of states. This can be written in the form:

$$g(E) = \frac{1}{V} \sum_n \delta(E - E_n) \qquad [5.3]$$

where $g(E)$ is the density of states per unit volume per unit energy interval, and V is the volume of the system. The quantity more often used is the integrated density of states:

$$N(E) = \int_{-\infty}^{E} g(E)\, dE \qquad [5.4]$$

We can now address the problem of whether a gap can exist in the density of states of an amorphous covalently bonded material. Weaire and Thorpe (1971) first showed that if short-range interactions between electrons are dominant, then it is the *short-range order* which mainly determines the electronic density of states. In particular they showed using a simple model Hamiltonian for the electron interactions that a gap *is* expected for an ideal tetrahedrally coordinated amorphous solid, providing that the interactions are of a certain magnitude. The amorphous structure is taken to be that in which each atom is in a perfect tetrahedral environment, with presumably a wide distribution of dihedral angles necessary to generate a random network. (Whether in practice a CRN could be constructed without bond-angle distortions remains unclear, but this assumption simplifies the treatment.)

The alternative approach to the nearly-free electron approximation, namely the tight-binding LCAO approach, is adopted, in which the basis functions are localized at each atomic site rather than being extended plane (Bloch) waves. The two interactions considered in the model are an intrasite 'banding' interaction V_1, responsible for the width of the bands, and an intersite 'bonding' interaction V_2, responsible for the separation of bonding and antibonding bands; they are shown

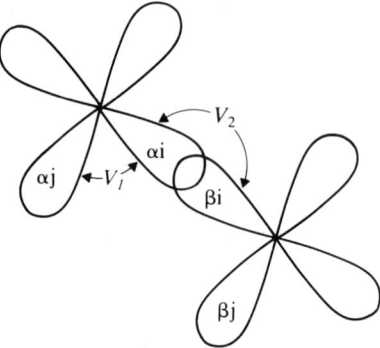

Fig. 5.2 Interactions and basis functions in the Weaire–Thorpe Hamiltonian.

schematically in Fig. 5.2, where the basis functions are taken to be sp³ hybridized orbitals localized at each site. The Hamiltonian is thus written as:

$$\mathcal{H} = \sum_{\alpha i j} V_1 |\alpha i\rangle\langle \alpha j| + \sum_{\alpha \beta i} V_2 |\alpha i\rangle\langle \beta i| \qquad [5.5]$$

All site orbitals are assumed to be orthogonal, and V_1 and V_2 are assumed to be the same for all atoms of the network. By consideration of an isolated atom, it can be shown that V_1 must be negative (so that the s-states are lower in energy than the p-states), and V_2 must also be negative (so that bonding orbitals are lower in energy than antibonding orbitals). Instead of using the states $|\alpha i\rangle$ as a basis, we can equally use bonding (B) and antibonding (A) orbitals associated with pairs of neighbouring atoms:

$$|B, \alpha i\rangle = \frac{1}{\sqrt{2}}(|\alpha i\rangle + |\beta i\rangle)$$

$$|A, \alpha i\rangle = \frac{1}{\sqrt{2}}(|\alpha i\rangle - |\beta i\rangle) \qquad [5.6]$$

If the assumption is made that the valence band is solely constructed from *bonding* orbitals, i.e. $|\psi\rangle = \sum_{\text{bonds}} b_{\alpha i}|B, \alpha i\rangle$, then the expectation value of the energy for this state is:

$$E = \frac{\langle\psi|H|\psi\rangle}{\langle\psi|\psi\rangle} = \frac{\sum b^*_{\alpha' i'} b_{\alpha i} \langle B, \alpha' i'|H|B, \alpha i\rangle}{\langle\psi|\psi\rangle} \qquad [5.7]$$

where the sum is over all pairs of bonds. Contributions of V_2 are obtained from the terms for which $\alpha i = \alpha' i'$, and $\tfrac{1}{2} V_1$ from terms for which $\alpha = \alpha'$, $i \neq i'$, giving:

$$E = V_2 + \tfrac{1}{2} V_1 \left[\sum_\alpha \sum_{i \neq i'} b^*_{\alpha i'} b_{\alpha i} \right] / \tfrac{1}{2} \sum_\alpha \sum_i |b_{\alpha i}|^2 \qquad [5.8]$$

It can be shown (see problem 5.1) that the valence band limits, i.e. E, must lie between $(V_2 - V_1)$ (when each $\sum_i b_{\alpha i} = 0$) and $(V_2 + 3V_1)$ (when each $b_{\alpha i} = b_{\alpha i'}$). Similarly, by assuming the conduction band is constructed solely from *antibonding* orbitals, the band limits are found to be $(-V_2 + 3V_1)$ and $(-V_2 + V_1)$.

Thus, if $|V_2| > 2|V_1|$, there is no overlap between the bands, i.e. a true band gap of magnitude

$$E_g \geq 2|V_2 - 2V_1| \qquad [5.9]$$

must exist. Note that this model is concerned only with the short-range structure and says nothing at all about longer range structure, i.e. the results obtained hold for amorphous *and* crystalline tetrahedral systems alike, irrespective of the presence of periodicity. This is in accord with the chemist's view of covalent bonding; there are four states per atom in the valence (and conduction) band, and hence all bonds are satisfied in the valence band.

Although providing an 'existence theorem' for a gap in the density of states of an amorphous semiconductor, the model is too simple to be expected to give quantitative results. As an example, the band structure calculated for diamond cubic Ge using the Weaire–Thorpe Hamiltonian is compared in Fig. 5.3 with a more sophisticated pseudopotential calculation; also shown is the density of states obtained from such band structures. Agreement is seen to be qualitative for the valence band, but very poor for the conduction band. Note that the Weaire–Thorpe model gives a delta function in the density of states at the top of the valence band (resulting, in the crystalline case, from the flat band in the band structure at the same position), which contains pure p-like bonding states. This region is relatively insensitive to the detailed structure (although the presence of like-atom bonds in an alloy does affect it); in contrast, as we shall see later, the deeper-lying states in the valence band density of states (mainly s-like for the deepest band and mixed s- and p-

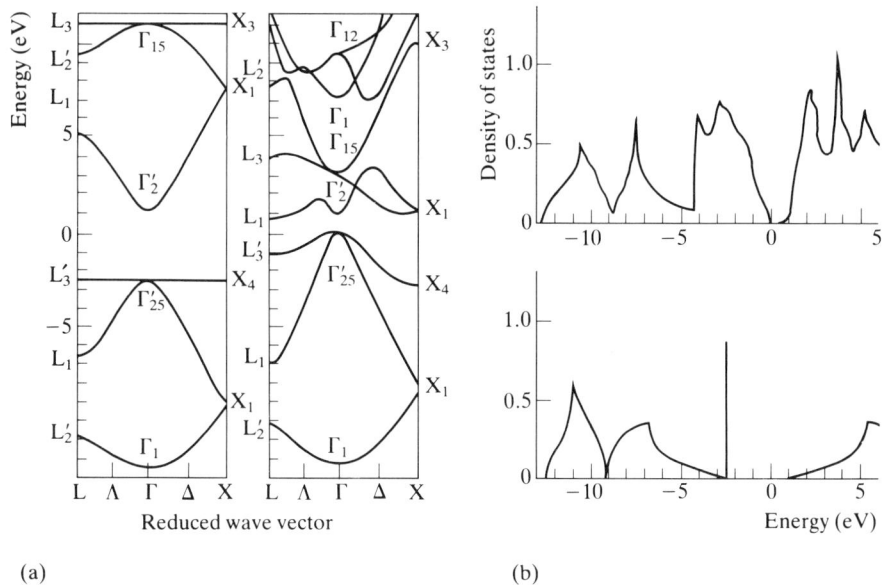

(a) (b)

Fig. 5.3 (a) Band structure of crystalline Ge (diamond cubic structure) calculated using pseudopotential theory.
(b) (top) The density of states for this band structure (the zero of energy marks the Fermi level). The bottom figure in (b) is the density of states calculated using the Weaire–Thorpe Hamiltonian (Weaire *et al.* 1972).

like for the intermediate band) *are* very sensitive to structural variations. In order to improve on these calculations for amorphous solids, a more realistic Hamiltonian is required, which remedies at least two deficiencies, namely the lack of the inclusion of longer range interactions and the absence of variations in the interactions arising from fluctuations in the structure, e.g. bond-angle distortions (see Yonezawa and Cohen 1981).

The object of this chapter is to discuss the electronic properties of amorphous materials, confining the discussion for the most part to amorphous semiconductors, leaving amorphous metals for consideration in Chapter 7. In addition, we will consider almost exclusively 'ideal' amorphous structures, in which there are no structural defects such as dangling bonds or voids, etc., only structural fluctuations such as bond-angle variations or like-atom bonds consistent with a CRN; defect-controlled electronic properties will be dealt with in Chapter 6.

In this chapter, several universal features exhibited by amorphous materials, in particular semiconductors, will be explored. The first is the fact that randomness causes the electron wavevector, k, to be no longer a good quantum number (hence making band structures meaningless); the *density of states*, however, does remain a valid concept and in fact is determined primarily by the local order in an amorphous solid, as we have already seen in this section. The second universal feature is that disorder can, under certain circumstances, cause *localization* of the electronic wavefunction for electron energies through much of the band of states. This is particularly important for amorphous semiconductors where a critical energy may exist in a band, separating localized and extended states at the 'mobility edge', and on a first reading, the reader might prefer to study sections 5.3.2, 5.3.4 and 5.3.5 in this regard: localization effects in amorphous metals are considered later in section 7.3.3. Electrical transport properties in the mobility edge model are discussed in sections 5.4.1 and 5.4.2 (the latter could be omitted on a first reading), and another universal feature of amorphous semiconductors, namely the sign anomaly in the Hall effect, is considered in section 5.4.3. This anomaly is a feature of the localized, rather than extended, nature of the electron wavefunction. Another manifestation of localization, namely self-trapping of the charge carriers to form small polarons, is discussed in section 5.5. Finally, yet another general feature exhibited by amorphous semiconductors, namely the Urbach edge (the exponential spectral dependence of the optical absorption edge), is discussed in section 5.6.2, and the beginning reader may like to read this in conjunction with section 5.6.1.

5.2 Electronic density of states

5.2.1 Theoretical calculations

We have already seen that the allowed electron states for a tetrahedrally bonded amorphous solid can be determined by using the Weaire–Thorpe tight-binding Hamiltonian [5.5]. This is a very simple model, however, and considerable effort has been expended in attempts to improve on this method and to obtain more realistic estimates for the density of states of amorphous solids.

Improvements can be made in two areas:

(a) A more realistic Hamiltonian can be used, involving more interactions than just the two inter- and intrasite terms considered by Weaire and Thorpe (Fig. 5.2), in particular including interactions involving more distant neighbours (see e.g. Bullett and Kelly 1979).

(b) *Topological* disorder can be introduced quantitatively by considering a more realistic structural model than that used by Weaire and Thorpe, which assumed perfect nearest-neighbour tetrahedral order, but did not address the problem of longer range structure (since only *short-range* interactions were included). Although this generality of structure is a feature of the Weaire–Thorpe model, giving rise to an energy gap whatever the structure (if $|V_2| > 2|V_1|$), it is intuitively obvious that details of the structure, such as ring statistics for a covalent solid, would be expected to influence the detailed shape of the density of states. Thus the atomic coordinates from a structural model which fits well experimental scattering data, such as a continuous random network (for a covalent solid) or a dense random packed model, are a better starting point. In this manner, *quantitative* disorder in the form of variations of the value of interactions (e.g. V_1 and V_2) are naturally included as a result of the presence of bond-angle and dihedral-angle variations for a CRN, or packing variations for a DRP model.

The use of large structural models in density of states calculations, however, necessitates the employment of sophisticated numerical techniques in order to cope with the diagonalization of the large matrices involved; a general review of these techniques can be found in Kramer and Weaire (1979). Essentially the same methods can be employed as are used in the *vibrational* density of states calculations that we discussed in section 4.3. The Lanczos method at the heart of the negative eigenvalue method (sect. 4.3.2) and the equation-of-motion method (sect. 4.3.3) for example, have both been used in this regard. More widely used, however, are two equally successful techniques which calculate the *local* density of states of a cluster ([4.22]), rather than effect the diagonalization of large matrices corresponding to the whole cluster. These are the 'recursion' method and the 'cluster-Bethe lattice' method.

The recursion method has been reviewed in section 4.3.4; an extensive discussion of its application to electronic density of states calculations for amorphous solids has been given by Bullett and Kelly (1979) and Kelly (1980). The recursion scheme to calculate the local density of states commences with a choice of $|u_o\rangle$, the orbital of interest (e.g. an sp^3 hybrid for a tetrahedral solid), together with an appropriate Hamiltonian. If the starting vector is constructed from an equal contribution from each orbital in the system, but with a random phase factor $\exp(i\delta)$ for $0 \leq \delta_i \leq 2\pi$, a good approximation to the total density of states is obtained since the starting orbital picks up an equal contribution to the spectrum from each distinct energy. The local density of states is related to the diagonal Green function matrix element $n(E, m) = -1/\pi \, \text{Im} \, \langle \phi_m | (E + i0 - H)^{-1} | \phi_m \rangle$ ([4.22]) where $|\phi_m\rangle = |u_o\rangle$ in the recursion scheme. The average over a unit cell of a *crystalline* local density of states is proportional to the total density of states; for an amorphous solid, on the other hand, an average should be performed over as many atomic sites as possible since each $n(E)$ reflects its own particular environment, although 20 sites

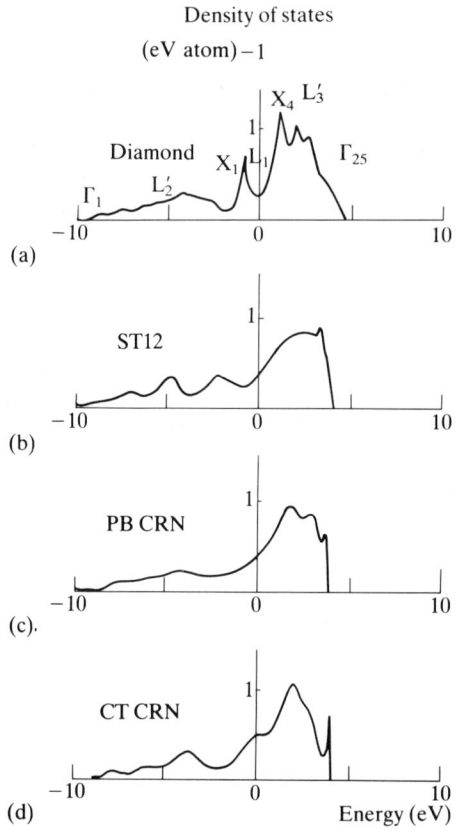

Fig. 5.4 Densities of states calculated using the recursion method for various crystalline modifications and amorphous models of Si (Kelly 1980): (a) diamond cubic; (b) ST12 crystalline modification; (c) Polk–Boudreaux CRN; (d) Connell–Temkin even-ring CRN. The zero of energy is self-energy of an isolated bond.

seem to be sufficient in practice. Boundary effects are minimized by choosing sites near the centre of the cluster since, although interactions with distant sites are included in the continued fraction, their contribution becomes negligibly small if the cluster is large enough (say, a few hundred atoms). Thus, there is no need for structural models with periodic boundary conditions (e.g. the Henderson model) as required by those methods more commonly used for crystalline systems (e.g. using pseudopotentials).

As an example of the recursion method, we show in Fig. 5.4 calculations of the valence-band densities of states for silicon in various structural forms. The two crystalline forms are the diamond lattice and the ST-12 lattice (so named because it has a simple tetragonal unit cell containing 12 atoms), which differ in two respects; the diamond lattice is formed from sixfold rings with each atom having the same tetrahedral bond angle, whereas the higher density ST-12 structure has a variety of ring sizes, even and odd (the smallest being five-membered) with a spectrum of bond angles $\sim \pm 25\%$ about the tetrahedral value, $109°\,28'$. The model amorphous structures employed are the Polk–Boudreaux (even–odd ring) model and the

Connell–Temkin (even ring) model. The first point to note is that the local densities of states of all these structural forms are qualitatively similar, supporting Weaire and Thorpe's contention that it is the short-range order, i.e. the tetrahedral coordination, which determines the gross features. However, there are distinct differences, despite the fact that the same model Hamiltonian was used in each case, which must arise therefore from topological differences between the various structural forms. In particular, the amorphous forms are differentiated from the diamond cubic form by a filling-in of the dip between s-states (at X_1) and a distinct skewing of the p-state distributions at the top of the valence band towards the gap. These trends are also observed experimentally in photoemission studies of Ge and Si (see later) and are shown schematically in Fig. 5.5, lending credence to the validity of the theoretical calculations and suggesting that odd-membered rings must be present in the amorphous phase. Note however from Fig. 5.5(b) that these trends are different for the case of heteropolar systems (e.g. GaAs), principally because of the difference in ionicity of the anion and cation and the consequent avoidance of like-atom bonds and hence odd-membered rings.

The other method commonly used to obtain the local density of states is the 'cluster-Bethe-lattice' method – for a review of this approach see Joannopoulos and Cohen (1976) and Joannopoulos (1979). In this, a small symmetrical cluster (containing a few tens of atoms) has Bethe lattices (or Cayley trees) attached to the surface dangling bonds; the Bethe lattice is characterized by having the same connectivity as the host network, but contains *no* closed rings of atoms. The method is shown schematically in Fig. 4.4, and has the following advantages:

(a) The density of states of a Bethe lattice can be calculated exactly for a variety of model Hamiltonians, and furthermore generally yields a smooth and featureless spectrum.

(b) The local density of states of the atom at the centre of the cluster in a given

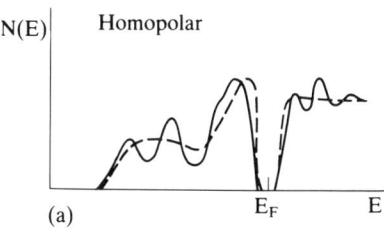
(a)

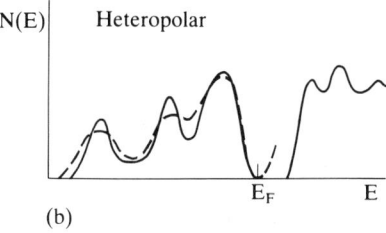
(b)

Fig. 5.5 Schematic density of states for tetrahedrally bonded semiconductors, (a) homopolar and (b) heteropolar, in both the amorphous (dashed line) and crystalline (solid line) phases (after Joannopoulos and Cohen 1976).

environment (say in a fivefold ring) can be calculated analytically, without attendant problems of boundary effects, because of the attached Bethe lattices which simulate the rest of the network.

The local density of states is calculated in a manner similar in spirit to that of the recursion method, namely via a continued fraction, although the details differ. The Green function involving the model Hamiltonian $\mathscr{H}$ may be written as a Dyson equation:

$$\mathbf{G} = (E - \mathscr{H})^{-1} = 1/E + \mathscr{H}\mathbf{G}/E \qquad [5.10]$$

so that:

$$E\langle i|\mathbf{G}|j\rangle = \delta_{ij} + \sum_n \langle i|\mathscr{H}|n\rangle\langle n|\mathbf{G}|j\rangle \qquad [5.11]$$

where n runs over the sites which interact with $|i\rangle$. If we consider a Bethe lattice of coordination z, with a single orbital $|i\rangle$ per site which interacts only with its nearest neighbours with strength h, then generally:

$$E\langle n|\mathbf{G}|0\rangle = h\langle n-1|\mathbf{G}|0\rangle + (z-1)h\langle n+1|\mathbf{G}|0\rangle \qquad [5.12]$$

Equation 5.12 is solved successively to give a continued fraction expansion of $\langle 0|\mathbf{G}|0\rangle$, which of course gives the local density of states. Simple tight-binding Hamiltonians are used in this method, but complicated Hamiltonians describing additional interactions can be dealt with, although less easily than in the recursion method.

As an example of the use of the cluster-Bethe-lattice method, we consider another elemental system, arsenic. The common crystalline form has the rhombohedral A7 structure, with puckered double layers such that each As atom has three neighbours at 2.51 Å within the layer, and a further three at 3.15 Å in an adjacent layer – it is a semi-metal. The amorphous form is some 15% less dense, and experimental RDFs indicate that the threefold nearest-neighbour coordination is retained, although the correlation between layers is lost – this material is a semiconductor with a band gap ~ 1.2 eV. Pollard and Joannopoulos (1978a) considered the simple Hamiltonian which contains only interactions between s-orbitals $|s\rangle$ and between p-orbitals $|p\rangle$, but which neglects any s–p mixing, viz.:

$$\mathscr{H} = \sum_{ss'} V_{ss'}|s\rangle\langle s'| + \sum_{pp'} V_{pp'}|p\rangle\langle p'| \qquad [5.13]$$

where V_{ss} ($= -12$ eV) is the s-orbital binding energy, $V_{ss'}$ ($= -1$ eV) is the interaction between neighbouring s-orbitals, V_{pp}, the p-orbital binding energy is taken to be the zero of energy and $V_{pp'}$ is the interaction between p-orbitals via σ-bonding ($V_\sigma = 2.5$ eV) or via π-bonding ($V_\pi = -1.2$ eV). Figure 5.6(c) shows the result of such calculations for both amorphous and crystalline forms, together with experimental densities of states obtained from X-ray photoelectron spectroscopy (XPS) (Ley et al. 1973) (see sect. 5.2.2) and similar calculations using the recursion method (see Kelly 1980). The valence band is seen to be split into two distinct bands, a low-lying predominantly s-band and a higher-lying mainly p-band (Fig. 5.6(a)); the amorphous and crystalline densities of states are differentiated by the skewing to higher energies of the top of the p-band and the filling-in of the dip in the s-band, just

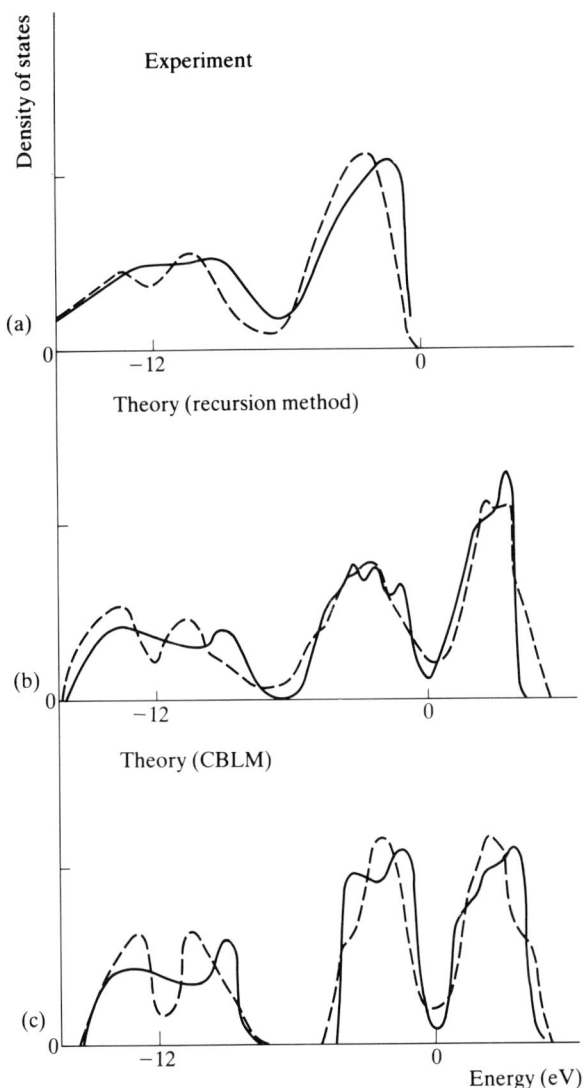

Fig. 5.6 Comparison of densities of states for amorphous (solid line) and crystalline thombohedral (dashed line) As. Experimental data are from XPS (Ley *et al.* 1973), and the theoretical curves are calculated using the recursion method (see Kelly 1980) and the cluster-Bethe-lattice method (CBLM) (Pollard and Joannopoulos 1978a).

as in the case of the tetrahedral materials, Si and Ge. Both sets of calculations, performed using the Greaves–Davis threefold coordinated CRN, reproduce these trends, although the recursion method calculations using a 'chemical pseudopotential' with atomic s- and p-orbitals which incorporate s–p hybridization, give the best agreement with the experimental density of states, in particular, and not surprisingly, in the region between the s- and p-bands. The differences in the densities of states between amorphous and crystalline forms is attributed to the presence of odd-membered rings as for Si, since the Greaves–Davis CRN contains a wide range of ring orders. Interestingly, the density of states of a *single* rhombohedral A7 layer calculated by the recursion method shows a gap at the Fermi level; the filling-in of this gap and the consequent metallic behaviour of the crystal is ascribed to interlayer interactions, which are much weaker in the amorphous form due to the loss of significant inter-layer correlations (Greaves *et al.* 1979).

The representation of the bonding in a-As as being mainly p-like (with a deep-lying non-bonding s-state), suggests an isomorphism between electronic and vibrational excitations (Pollard and Joannopoulos (1978(b)). This is so because if the bond angle is 90° and s–p mixing is neglected, the p-like orbitals transform as orthogonal vectors, and since the symmetry of the electronic Hamiltonian and the dynamical matrix as given by the Born model ([4.9]) are similar, the electronic and phonon densities of states become isomorphic, in that there is a one-to-one relationship between states in the p-like valence and conduction bands (Fig. 5.6(c)) and states in the acoustic and optical bands, respectively (Fig. 4.12). The bond stretching and bond bending force constants are analogous to the V_σ and V_π interactions in [5.13]. The importance of this observation is that local topology should affect electronic and vibrational states in a similar manner for this particular case.

Indeed, one of the principal uses and attractions of theoretical densities of states calculations is to act as a test of structural models for amorphous materials. If the electronic Hamiltonian (or dynamical matrix) is realistic (as may be ascertained by a comparison of calculated crystalline band structures (or dispersion relations) and densities of states with experiment), calculations using various amorphous structural models provide a sensitive probe of the topology, e.g. ring statistics, of such models. In fact, the sensitivity of the electronic or vibrational states to the ring statistics is significantly higher than is, say, the RDF.

5.2.2 Experimental determination

In this section we will be concerned with those experimental techniques which probe the gross features of the electronic structure and consequently give a measure of the density of states. Two such techniques are photoemission and UV and X-ray absorption. We will not discuss those experiments which probe the band-edge region for semiconductors, e.g. optical absorption at lower energies or electrical transport processes, since these will be dealt with later in this chapter.

Photoemission

The technique most widely used to explore the valence and conduction band densities of states is photoemission; as its name implies, this measures the energy distribution of photo-emitted electrons as a function of incident photon energies. The interested reader is referred to the two books edited by Cardona and Ley (1978). The commonest photon sources are the UV line spectra of rare gas discharge lamps, or X-ray anodes; these give rise, respectively, to two conventional branches of photoemission, namely 'ultra-violet photoemission spectroscopy' (UPS) and 'X-ray photoemission spectroscopy' (XPS). The UV photon energy range is 10–50 eV, whereas the K_α X-ray lines commonly employed are 1486.6 eV (Al) and 1253.6 eV (Mg), although this gap can be bridged by the use of synchrotron radiation, rendering the difference between the two techniques less distinct. The photo-emitted electrons can be energy analysed either by a retarding grid analyser or a dispersive electrostatic analyser. The experiments have to be conducted in ultra-high vacuum to minimize surface contamination, since the escape depth of the photo-electrons is very small (~ 5 Å for UPS and ~ 50 Å for XPS).

The photoemission process can be understood in terms of the 'three-step model', namely: (1) optical excitation of an electron; (2) its transport through the solid (including the possibility of inelastic scattering by other electrons; and (3) the escape through the sample surface into the vacuum, although this approach drastically approximates the many-body processes that take place. The 'energy distribution curves' (EDCs) of the photo-emitted electrons are given by:

$$I(E, \hbar\omega) = P(E, \hbar\omega)T(E)D(E) \qquad [5.14]$$

where $P(E, \hbar\omega)$ represents the distribution of photo-electrons of energy E excited by a photon of energy $\hbar\omega$, $T(E)$ is a transmission function (weakly and smoothly varying with E), and $D(E)$ is an escape function also a smooth function of E. Thus it is only $P(E, \hbar\omega)$ which contributes structure to the EDCs, and this may be written, when k-conservation is not important, as is the case for amorphous materials, as:

$$P(E, \hbar\omega) \propto N_c(E)N_v(E-\hbar\omega)M^2(E, E-\hbar\omega) \qquad [5.15]$$

where N_c and N_v are the conduction and valence band densities of states, respectively, and M is a matrix element which may be taken to be constant for limited photon energy ranges. Thus we see that the photoemission EDCs are determined by a *joint* density of states involving a convolution of occupied and unoccupied electron states.

The origin of $P(E, \hbar\omega)$ is illustrated schematically in Fig. 5.7, where the features in the EDCs can be traced back to structure in the densities of states of the bands. The valence band density of states has been raised by $\hbar\omega$ with respect to the conduction band to account for the term $N_v(E-\hbar\omega)$ and this is then convoluted with $N_c(E)$ and $T(E)$ to give $P(E, \hbar\omega)$ (assuming $D(E)$ and $M(E, E-\hbar\omega)$ to be constant). It can be seen from Fig. 5.7 that features in the EDCs obtained from UPS which *do* shift in position when $\hbar\omega$ is varied are to be associated with peaks in the *valence band* density of states, whereas conversely those features which do *not* shift in position are associated with maxima in the *conduction band* density of states, although conduction band states below the vacuum level of the semiconductor are inaccessible to the technique and can only be rendered accessible by means of a layer of caesium evaporated on to the surface to lower the work function. In this way the valence band and conduction band densities of states can be extracted independently from the EDCs. An illustration of this is given in Fig. 5.8(a) for the case of amorphous and crystalline Ge (Spicer 1974), where it is seen that the EDCs for both materials alter dramatically (in different ways) upon varying the UV photon energy. The densities of states deduced from these spectra are shown in Fig. 5.8(b) using [5.15] for the amorphous case (although not for c-Ge, where k is conserved and a different equation must be employed).

If photon energies greater than ~ 20 eV are employed, then the intensity

Electrons

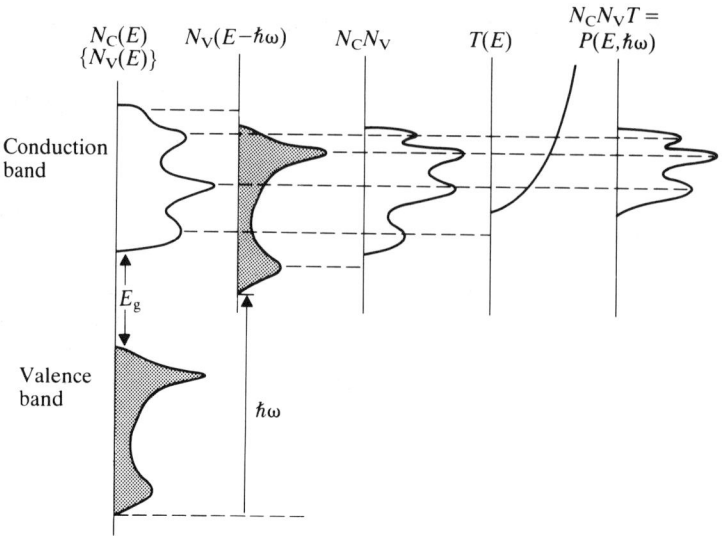

Fig. 5.7 Schematic illustration of the origin of the features which are observed in photoemission EDCs, $P(E, \hbar\omega)$ (after Mott and Davis 1979).

modulation by the final (conduction band) density of states rapidly becomes unimportant as the final state $N(E)$ approaches its featureless free-electron dependence $\propto E^{1/2}$. Thus soft X-ray photoemission probes only the valence band density of states, and the EDC is a direct reflection of this, modulated by slowly varying photo-ionization cross-sections. An example of the valence band density of states obtained from XPS data has already been given in Fig. 5.6 for the case of amorphous and crystalline As. Higher energy X-rays cause excitation from deep-lying core states, rather than the valence band, and produce narrow peaks in the EDCs at kinetic energies below the valence band spectra. The number of core levels, and their binding energies, are characteristic of a given element, and moreover the exact value of binding energy measured depends on the chemical environment of the element. This phenomenon forms the basis of the technique 'electron spectroscopy for chemical analysis' (ESCA) which is really only a surface probe, however, because of the short sampling depth (10–60 Å) of the XPS method. ESCA has been used to determine whether 'wrong', i.e. homopolar, bonds exist in amorphous semiconductors, e.g. in the III–V compounds, InP, GaAs, etc., but the results are somewhat inconclusive.

Ultra-violet and X-ray absorption

These techniques are analogous, and in a sense complementary, to photoemission. The absorption (or reflectivity) associated with electronic transitions from filled valence states to empty conduction states produced by photons having energies greater than the band-gap energy will be discussed here; excitation involving photon energies less than or equal to the band-gap energy, which probe states near the band edges of amorphous semiconductors, will be left until later.

The optical properties (for both UV and X-ray excitation) of amorphous and crystalline semiconductors are almost entirely determined by the imaginary part of the dielectric constant, where:

$$\varepsilon = \varepsilon_1 + i\varepsilon_2 \quad [5.16]$$

and $\varepsilon_2(\omega)$ is given in a one-electron expression (see, e.g. Connell 1979) as:

$$\varepsilon_2(\omega) = \frac{2}{V}\left(\frac{2\pi e}{m\omega}\right)^2 \sum_i \sum_f |\langle f|P|i\rangle|^2 \delta(E_f - E_i - \hbar\omega) \quad [5.17]$$

where V is the sample volume, P is the momentum operator, and the summations are over all initial valence states $|i\rangle$ and final conduction states $|f\rangle$. The wavefunctions $|i\rangle$ and $|f\rangle$ of the valence and conduction bands respectively, can be expanded in terms of a set of orthonormal, localized wavefunctions $|nv\rangle$ and $|nc\rangle$ centred on different atoms n. Thus

$$|i\rangle = \sum_n a_{inv}|nv\rangle \quad [5.18(a)]$$

$$|f\rangle = \sum_n a_{inf}|nc\rangle \quad [5.18(b)]$$

For a crystal, the a's are plane waves and $|nv\rangle$ and $|nc\rangle$ are Wannier functions. For an amorphous solid however, writing $a = Ae^{i\delta}$, the phases δ vary randomly from site to site if the uncertainty in $\mathbf{k}$ is of the order of the wavevector itself, $\delta k \simeq k$ (for strong scattering); the amplitude A may also vary randomly from atom to atom, although being of the same magnitude for extended states. This 'random-phase approxi-

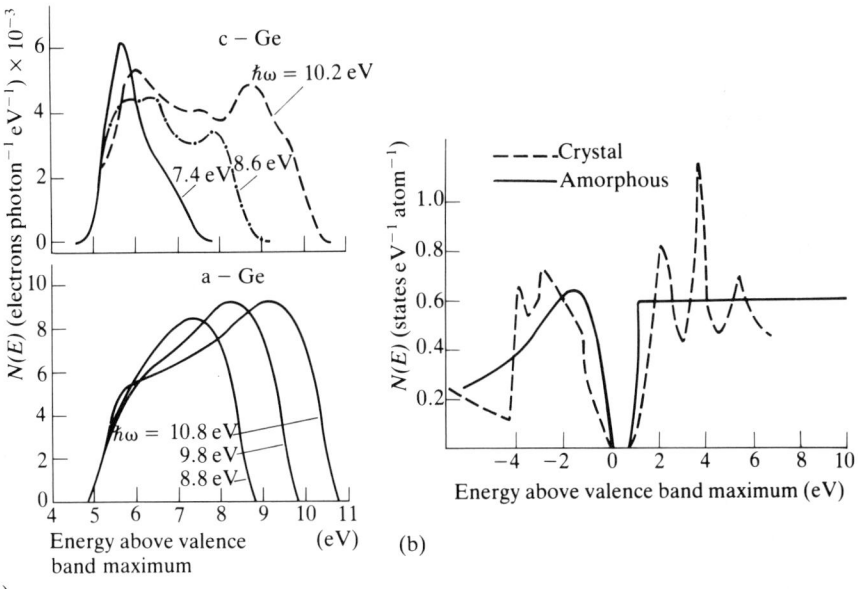

Fig. 5.8 Photoemission results for amorphous and crystalline Ge (Spicer 1974):
(a) EDCs for c-Ge and a-Ge for different photon energies;
(b) density of states for valence and conduction bands derived from the EDCs in (a). The profile for the conduction band for a-Ge is only approximate.

mation' (RPA) is an embodiment of the rule introduced by Ioffe and Regel that for an electron mean free path ℓ, $k\ell < 1$ is impossible, i.e. when the wavefunction loses phase memory from atom to atom, the mean free path cannot be less than the interatomic spacing. The RPA is not in general a realistic model since it does not take account of the considerable degree of short-range order exhibited by amorphous semiconductors (Ch. 3), although it is a useful starting point.

The momentum matrix element, averaged over an ensemble of random systems, can be evaluated for transitions between extended (delocalized) states:

$$\langle |\langle f|P|i\rangle|^2 \rangle_{\text{ensemble}} = \frac{a^3}{V} \left\langle \sum_{n'} |\langle n'c|P|nv\rangle|^2 \right\rangle_{\text{ensemble}} \qquad [5.19]$$

where a is the interatomic spacing. When $\langle nc|P|nv\rangle = P_{cv} \neq 0$, this term will dominate the sum, and so:

$$\langle |\langle f|P|i\rangle|^2 \rangle_{\text{ensemble}} = \frac{a^3}{V} P_{cv}^2 \equiv |P_{am}(\omega)|^2 \qquad [5.20]$$

where $P_{cv} \sim \hbar/a$.

Substituting this in [5.17] for $\varepsilon_2(\omega)$ gives for the amorphous case:

$$\varepsilon_2^{am}(\omega) = \frac{2}{V}\left(\frac{2\pi e}{m\omega}\right)^2 |P_{am}(\omega)|^2 \sum_i \sum_f \delta(E_f - E_i - \hbar\omega) \qquad [5.21(a)]$$

$$= 2\left(\frac{2\pi e}{m\omega}\right)^2 a^3 P_{cv}^2 \int_0^{\hbar\omega} dE\, g_v(-E)\, g_c(\hbar\omega - E) \qquad [5.21(b)]$$

where we have introduced the density of states as defined by [5.3], and the zero energy is taken at the top of the valence band edge.

We see from [5.21] that $\varepsilon_2(\omega)$ is a function of a *joint* density of valence and conduction band states, much as was the case for UPS, [5.15]. This is also true for the optical absorption α, since this is related to $\varepsilon_2(\omega)$ by the following relation:

$$\varepsilon_2(\omega) = \frac{n_0 c \alpha(\omega)}{\omega} \qquad [5.22]$$

where n_0 is the refractive index and c the speed of light. In practice, the absorption is often obtained by a Kramers-Krönig transformation of reflectivity data.

It is interesting to compare the equation for $\varepsilon_2(\omega)$ for the amorphous case ([5.21]) with that for the crystalline case. This is obtained from [5.17] by requiring that optical transitions conserve momentum (or k), (i.e. are 'vertical'), which leads to:

$$\varepsilon_2^{cr}(\omega) = \frac{2}{V}\left(\frac{2\pi e}{m\omega}\right)^2 |P_{cr}(\omega)|^2 \sum_i \sum_f \delta(E_f(k) - E_i(k) - \hbar\omega) \qquad [5.23(a)]$$

$$= 2\left(\frac{2\pi e}{m\omega}\right)^2 |P_{cr}(\omega)|^2 \frac{1}{(2\pi)^3} \int_B d^3k\, \delta(E_f(k) - E_i(k) - \hbar\omega) \qquad [5.23(b)]$$

where $|P_{cr}(\omega)|^2$ is independent of k, and the integral is taken over the Brillouin zone. Thus, *crystalline* spectra are expected to show sharp features (Van Hove singularities) when $\hbar\omega$ is in the vicinity of critical points in the *joint* density of states where $\nabla[E_i(k) - E_f(k)] = 0$, as well as when the joint density of states has structure, whereas *amorphous* solids should exhibit features in the spectra *only* at energies where the joint density of states has structure. This difference in ε_2 spectra is strikingly shown in Fig. 5.9(a) for amorphous and crystalline Ge (Spicer and Donovan 1970). Although the spectra are qualitatively similar, that for amorphous Ge is featureless with no trace of the sharp peak observed at 4.5 eV in crystalline Ge. This peak is known to be enhanced due to an Umklapp enhancement of the crystalline matrix element superimposed on an otherwise monotonically decreasing ω-dependence (see Fig. 5.9(b)) often taken as $|P_{am}(\omega)|^2 \propto \omega^{-2}$. This dependence is predicted by the simple isotropic 'Penn model' of a semiconductor, in which it is assumed that just two bands exist, and the Penn gap $\hbar\omega_g$ represents an *average* gap between valence and conduction bands in a real material, and is a measure of the cohesive energy. The Brillouin zone (or Jones zone for a heteropolar system) is assumed to be spherical, with a magnitude of the wavevector, k_F, at the Fermi level. The matrix element for this model is then:

$$|P_{Penn}(\omega)|^2 = \hbar k_F \left(\frac{\omega_g}{\omega}\right)^2 \qquad [5.24]$$

The density of states can be determined from ε_2 spectra using an estimate for the matrix element, either a suitably averaged crystalline value or from the Penn model ([5.24]). The *valence band* density of states can be obtained by using [5.21(b)], and by making the approximation that the conduction band density of states can be reasonably well represented as a step-function $g_c(E_c)$ at an energy E_c above the valence band edge (see Fig. 5.8(b) for an indication of the validity of this), the following expression for the valence band density of states $g_v(E)$ in terms of ε_2 is thus obtained:

$$g_v(E_c - \hbar\omega) \propto \frac{d}{d\omega}\left[\frac{\omega^2 \varepsilon_2(\omega)}{|P_{am}(\omega)|^2}\right] \qquad [5.25]$$

The valence band density of states for a-Ge obtained from ε_2 data using [5.25] is shown in Fig. 5.10(a), compared with results obtained from XPS. The agreement between the two methods is excellent, the only differences arising from the neglect of the width of the tail to the conduction band edge assumed in the derivation of [5.25].

The effective number, n_{eff}, of free electrons per atom contributing to the optical absorption up to an energy $\hbar\omega_0$ is given by the following plasma sum rule:

$$n_{eff} = \frac{m}{2\pi^2 N e^2} \int_0^{\omega_0} \omega \varepsilon_2(\omega) d\omega \qquad [5.26]$$

where N is the atomic density. n_{eff} is shown in Fig. 5.10(b) for the same samples of a-Ge as were measured to give Fig. 5.10(a), together with data for c-Ge for comparison. It can be deduced that n_{eff} reaches 1.85 electrons per atom at 4.5 eV, a value in accord with our theoretical understanding of the make-up of the upper part of the valence band (all that is shown in Fig. 5.10(a)), namely that it is a broadened δ-function derived from p-states and therefore contains 2 electrons per atom (i.e. the s^2p^2 electronic configuration characteristic of group IV elements).

Thus far, we have concentrated on optical transitions from valence states to conduction states, produced by UV excitation. If photons of higher energy, e.g. X-

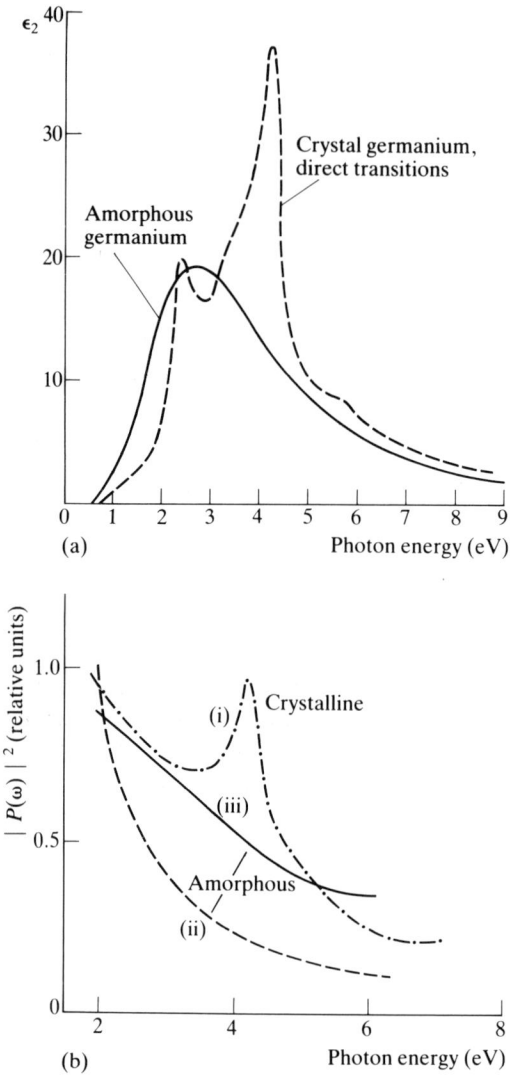

Fig. 5.9 (a) Spectral dependence of ε_2 for amorphous and crystalline Ge (Spicer and Donovan 1970). (b) Energy dependence of the optical transition matrix elements for Ge: (i) crystalline Ge; (ii) amorphous Ge (using convolution of crystalline density of states); (iii) amorphous Ge (using convolution of amorphous density of states (after Mott and Davis 1979).

rays, are used, the initial states become the narrow core levels situated at energies well below the valence band, and the technique becomes X-ray absorption spectroscopy. The advent of tunable X-ray synchrotron radiation sources has meant that this form of spectroscopy has become much more feasible. The *conduction band* density of states is explored by varying the X-ray photon energy and exciting from a given core level. This technique suffers, however, from two disadvantages. The first is that often the density of states near the edge (e.g. for a-Si) appears to be enhanced as compared with results obtained using UPS; this effect has been ascribed to core

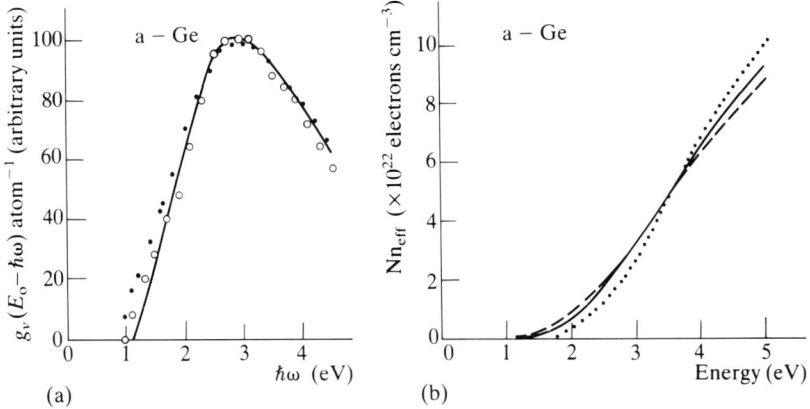

Fig. 5.10 (a) Valence band density of states of a-Ge measured from the conduction band edge. The points refer to data obtained from optical measurements (as in Fig. 5.9(a)) on sputtered material deposited on to substrates at 25 °C (solid circles) and 350 °C (open circles) (W. Paul, G. A. N. Connell and R. J. Temkin (1973) *Adv. Phys.* **22**, 529, together with data obtained by photoemission on evaporated material (L. Ley, S. Kowalczyk, R. Pollak and D. A. Shirley (1972) *Phys. Rev. Lett.* **29**, 1088).
(b) The effective number of electrons involved in transitions up to energy $\hbar\omega$ in a-Ge (dashed curve $T_d = 25°C$, solid curve $T_d = 350°C$; dotted curve crystalline Ge) (G. A. N. Connell, R. J. Temkin and W. Paul (1973) *Adv. Phys.* **22**, 643).

exciton formation. The second disadvantage is more serious in that it casts some doubt on the ability of the technique to give a true representation of the conduction band density of states. This possibility is seen dramatically in Fig. 5.11, where X-ray absorption spectra are shown for a-As, a-Se and the alloy a-As$_2$Se$_3$ (Bordas and West 1976). Disturbingly, it is apparent that the spectrum for the alloy is essentially identical to the sum of the individual spectra of the constituent elements, suggesting

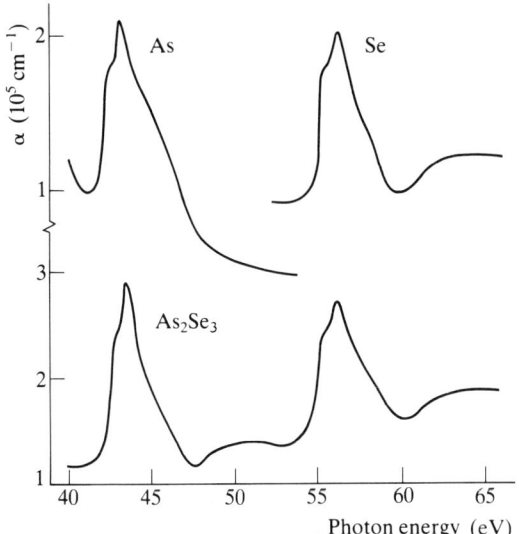

Fig. 5.11 X-ray absorption spectrum of a-As$_2$Se$_3$, compared with spectra for a-As and a-Se (Bordas and West 1976).

Table 5.1 Experimental techniques for density of states determination

Technique	Band probed
Ultra-violet photoemission spectroscopy (UPS)	Valence and/or conduction
X-ray photoemission spectroscopy (XPS)	Conduction
Ultra-violet absorption spectroscopy	Valence and conduction
X-ray absorption spectroscopy	Conduction (?)

that the conduction band of a-As_2Se_3 is *not* being probed, but instead a sum of final states associated with the excited atoms. Whether this is the case universally is not as yet certain.

We conclude this survey of experimental techniques which probe one or other of the bands of amorphous semiconductors by summarizing the capabilities of each method in Table 5.1.

5.3 Localization

5.3.1 Introduction

In the last section we considered the theoretical and experimental determination of the overall shape of the electronic density of states in amorphous solids, primarily semiconductors, yet paid no attention to the *nature* of such electron states, i.e. whether they are spatially extended throughout a sample or localized in the vicinity of a given atom. We remedy this omission in what follows, and show that the presence of disorder can greatly influence the nature of the electron wavefunctions. As before, we will consider only the effect of topological (or other) disorder in 'ideal' random structures and leave for later a discussion of the electron states associated with structural defects.

5.3.2 Effects of disorder

The presence of disorder manifests itself on the electron states in a variety of ways. We have seen that sharp features in the density of states – the Van Hove singularities – become smeared out if there is no long-range order. Furthermore, fluctuations in short-range order, such as bond-angle distortions, lead to 'tailing' of states into the gap at the band edges (Fig. 5.12(a)); this may be seen from a consideration of the Weaire–Thorpe Hamiltonian ([5.5]) in which variations of V_1 cause a broadening of the bands and consequent tailing of states into the gap. Band tails are even more pronounced if chemical disorder, such as the presence of 'wrong bonds', is present, as may be seen by reference to Fig. 5.5(b). The states at the top of the valence band in heteropolar systems such as a-GaAs derive primarily from cation-cation p-like bonding states which are extremely sensitive to the presence of cation like-bonds (Joannopoulos and Cohen 1976). Another cause of band tailing can be the occurrence of short-range disorder in *non-bonded* atom distances. This is particularly important for chalcogen atoms, which have one non-bonding p-like orbital (forming the top of the valence band), in addition to two predominantly p-

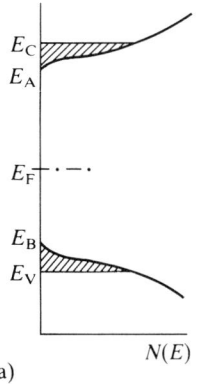 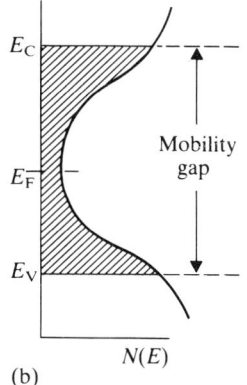

Fig. 5.12 Models for the density of states of an amorphous semiconductor.
(a) A true gap in the density of states with some band tailing.
(b) The Cohen–Fritzsche–Ovshinsky (1969) model of overlapping valence and conduction band tails (after Mott and Davis 1979).

like bonding orbitals per atom. If two such atoms are forced close to each other in the random network the interaction between the non-bonding lone-pair orbitals causes the energy of such states to be raised, again producing tailing of states from the valence band edge into the gap. The result of all these effects gives rise in the extreme case to the Cohen–Fritzsche–Ovshinsky (1969) (CFO) model for the density of states in the gap region of an amorphous multicomponent (chalcogenide) semiconductor (Fig. 5.12(b)). In this, the band tailing is so pronounced that the tails arising from the valence and conduction bands actually overlap in mid-gap; this model will be discussed in more detail in Chapter 6.

Another important consequence of disorder is that the electrons may become 'localized', i.e. spatially confined to the vicinity of predominantly a single atomic site. This occurrence is more probable the greater the degree of disorder in the potentials experienced by the electrons, and so is more likely in the band tails, since these arise in general from the most distorted sites. Criteria which dictate whether electron states should be localized or non-localized (i.e. extended) will be discussed later.

It should be stressed at the outset that localized states *per se* are not unique to the amorphous phase, but can also occur in crystalline materials, albeit generally containing a degree of disorder. For instance, if a single defect is introduced into an otherwise perfectly periodic lattice the Schrödinger equation becomes

$$[\mathcal{H}_0 + U(r)]\psi = E\psi \qquad [5.27]$$

where the solutions of the Hamiltonian for the perfect lattice $\mathcal{H}_0$ are Bloch functions, and $U(r)$ is the extra potential introduced by the defect, which may be positive or negative depending on whether an electron is repelled by or bound to the defect. The presence of the defect causes one state to be split off from the band under consideration: if $U(r)$ is positive, the uppermost state splits off, and if $U(r)$ is negative the lowermost state splits off. The states within the band are only marginally displaced in energy and their wavefunctions remain approximately Bloch-like. However, the wavefunction associated with the split-off state is spatially localized in the vicinity of the defect. (An analogous situation holds also for lattice vibrations where the impurity modes are also localized in the vicinity of the defect.)

In contrast, for systems having a considerable degree of disorder, e.g. amorphous semiconductors, *all* the states in a band may be localized if the disorder is sufficiently great. Although this statement must be qualified to specify the dimension of the system, the type of disorder and even the length scale of the sample under consideration, nevertheless it is true under certain conditions and is a very surprising observation for those versed in the physics of (crystalline) solids. However, in most circumstances for three-dimensional systems, the criteria for localization are not met for states throughout an entire band, and only states near the band edges are localized; hence critical energies (labelled E_v and E_c in Fig. 5.12) separate localized from extended states.

Criteria for localization must be established before any discussion concerning the amount of disorder necessary to induce it can be pursued. A variety have been suggested. The first such criterion was put forward by Anderson (1958) in his now classic paper on electron localization in random lattices, namely the absence of diffusion at $T=0$: if an electron is placed on an atomic site (or equivalently in a volume V such that an energy E is at the centre of the band) at time $t=0$, then the state (at energy E) is deemed to be localized if as $t \to \infty$ the electron has not diffused away and has a finite probability $\propto e^{-2\alpha r}$ of remaining at distance r within the volume; if there is a finite chance of diffusion at $T=0$ then the state is delocalized or extended. The quantity α^{-1} is known as the localization length. The exponentially decreasing nature of the localized wavefunction, $\psi \sim e^{-\alpha r}$, has suggested another criterion, namely that the energy levels of localized states should be insensitive to boundary conditions (see e.g. Thouless 1979). The *size* of the sample is thus seen to be crucial; if the localization length is much greater than the sample size, $\alpha^{-1} \ll L$, it becomes very difficult to determine whether the state is localized or extended. Another measure of the degree of localization that has been used is the 'participation ratio' P (or its inverse) that has already been mentioned in connection with vibrational modes ([4.17]), which is roughly speaking the number of sites over which a localized wavefunction has significant amplitude. Thus a localized state is characterized by the participation number $P \propto [\sum_i |\psi_i|^2]^2 / [\sum_i |\psi_i|^4]$ tending to zero, whereas an extended state has P finite. A final criterion for localization may be chosen to be the behaviour of the d.c. electrical conductivity as $T \to 0$; essentially this is related to the electron diffusion. States of energy E are deemed to be localized if the ensemble average of the d.c. conductivity $\langle \sigma_E \rangle$ is zero at $T=0$, whereas extended (metallic-like) states have a finite conductivity at $T=0$. (In other words, conduction between localized states can only take place by means of thermally assisted

Table 5.2 Criteria for distinguishing localized and extended electron states (after Kramer and Weaire 1979)

Name		Localized regime	Extended regime
Inverse localization length	α $\|\Psi\| \sim r^{-\beta} e^{-\alpha r}$	finite	0
Prefactor index	β	finite	∾
Participation ratio	$P \propto [\sum_i |\Psi_i|^4]^{-1}$	0	finite
Conductivity	$\langle \sigma_E \rangle$	0	finite

'hopping' as we shall see later.) The various criteria are summarized for convenience in Table 5.2.

We can now turn to the problem of how to obtain a numerical estimate for the degree of disorder necessary to localize, say, all states in a band, having at our disposal means of distinguishing between localized and extended states. However, in order to do this, we must address the question of dimensionality, since this profoundly influences the behaviour.

5.3.3 One dimension

We consider firstly one-dimensional systems for which the results are the most clear-cut. Put simply, any disorder, however small, is sufficient to localize *all* states as first shown by Mott and Twose (1960); extended states are only obtained if all the potentials in a 1-D chain are of equal magnitude and spacing (the Kronig–Penney model). This result can be proved rigorously for the case of one dimension; a review is given by Ishii (1973).

Much recent work has been devoted to the theoretical consideration of thin metal wires which Thouless (1977) first showed should exhibit *non-metallic* conduction, i.e. localized behaviour, if the *resistance* of the wire is greater than several tens of kilohms. This prediction, which has been borne out by experiment, is sufficiently startling and interesting that we shall examine it further here, for, although not concerning amorphous materials *per se*, it does shed some useful light on the localization process. The essence of the argument lies in the localization criterion that electron energy levels are sensitive to boundary conditions, since it can be shown that this sensitivity is closely related to the resistance of the system at that energy.

Consider a metal wire of length L and cross-sectional area A, which has sufficient disorder (impurities, etc.) that the bulk material has a finite conductivity σ at $T=0$. Now an electron wave-packet diffusing along the length of the wire is insensitive to the boundary for the time t it takes to travel the length of the wire, and so by the uncertainty principle, the energy levels can be shifted only by an amount ΔE where:

$$\Delta E = \frac{\hbar}{t} = \frac{\hbar D}{L^2} \qquad [5.28]$$

where D is the diffusion constant for the electron. This can be related to the conductivity, and the density of states dn/dE (per unit volume per unit energy interval) by the Einstein relation

$$\mu = \frac{eD}{n}\frac{dn}{dE}$$

and $\sigma = ne\mu$ i.e.:

$$\sigma = e^2 D \frac{dn}{dE} \qquad [5.29]$$

Electrons

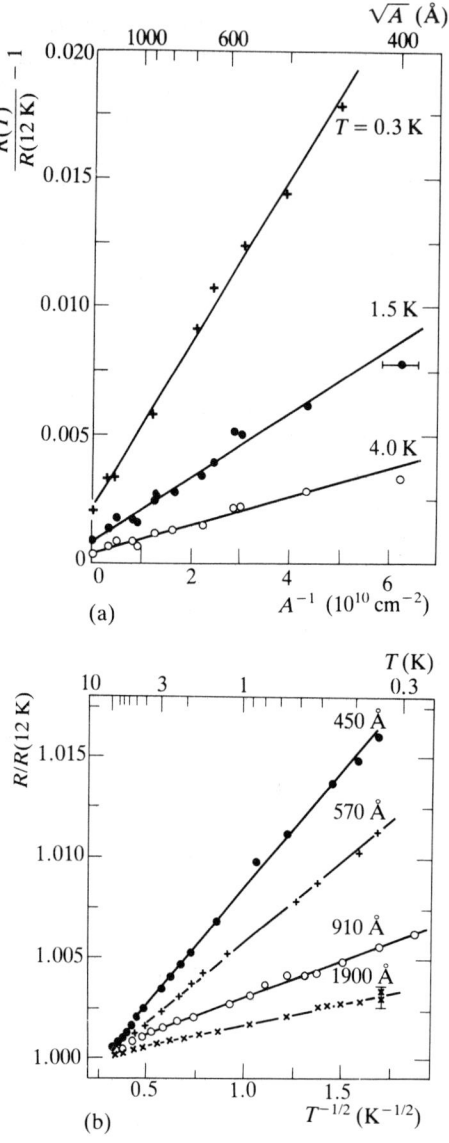

Fig. 5.13 Size effect for 'dirty' thin wires of $Au_{60}Pd_{40}$ (Giordano 1980).
(a) Plots of the resistance as a function of $1/A$ at several temperatures.
(b) Temperature dependence of the resistivity of wires of varying diameters.

Combining [5.28] and [5.29] yields:

$$\Delta E = \frac{\hbar}{e^2} \frac{\sigma}{L^2} \frac{dE}{dn} = \frac{\hbar}{e^2 R} \frac{dE}{dN} \qquad [5.30]$$

where R is the resistance of the wire and dE/dN is the average spacing between energy levels. If the energy shift, $\Delta E = \hbar/t$, due to a change in boundary conditions, is much less than the energy spacing, the criterion for localization of the states is satisfied, and the resistance should increase much faster with increasing length than the normal linear dependence. The condition necessary for this to occur is:

$$R_c \gg \hbar/e^2 \qquad [5.31]$$

or in other words the maximum resistance for metallic behaviour of a wire should be ~ 5 kΩ, and the localization length for such a system is such that it corresponds to a resistance of this amount. Later work, which considered that the time between inelastic collisions (rather than the diffusion time) limits the metallic behaviour, raised the critical resistance to 36.5 kΩ (Thouless 1980). Such effects should only be observable at low temperatures, since phonons will induce transitions between localized states at higher temperatures.

Anderson et al. (1980) further proposed that in one dimension the resistance should scale as the inverse localization length, α, according to:

$$R(L) = \frac{\pi \hbar}{e^2} \{\exp(\alpha L) - 1\} \qquad [5.32]$$

This scaling parameter can be rewritten in a form proportional to $\ln(1+R)$, and for small resistances the scaling reduces to R (i.e. normal additivity of resistances), but for large resistances, it reduces to $\ln R \propto \alpha L$, reflecting exponential localization. The length L is the length of the wire at $T=0$, or the distance between inelastic collisions at finite temperatures. If the electronic motion is diffusive this gives:

$$L \simeq \sqrt{(D\tau_i)} \qquad [5.33]$$

where τ_i is the inelastic scattering time. The localization length, α^{-1}, is related to the maximum metallic resistance:

$$R_c = \rho_0 \frac{\alpha^{-1}}{A} \qquad [5.34]$$

where ρ_0 is the bulk resistivity and A is the cross-sectional area, and is also related to the (large-scale) resistance R_0 of a wire of length L

$$\alpha^{-1} = \frac{\pi \hbar}{e^2} \frac{L}{R_0} \qquad [5.35]$$

Inserting these expressions into [5.32], expanded to the quadratic term, gives:

$$R \simeq R_0 \left(1 + \frac{\rho_0 \sqrt{(D\tau_i)}}{2 R_c A}\right) \qquad [5.36]$$

i.e. the extra resistance associated with localization should be inversely proportional to A. (NB This expression for the extra resistance differs by a factor of 4 from that given by Thouless 1980.) Wires of several hundred ångströms diameter and a few micrometres in length made from Au–Pd alloys have been studied, and typical results of the resistance as a function of A^{-1} are shown in Fig. 5.13(a) for different temperatures. The temperature dependence of the extra resistance is seen to be proportional to $T^{-1/2}$ (Fig. 5.13(b)); this behaviour can be understood if the scattering time is shorter and increases less rapidly with temperature than is

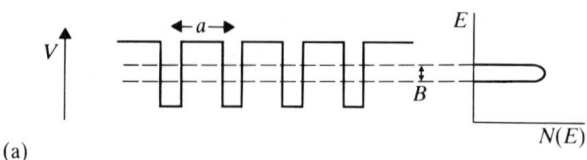

(a)

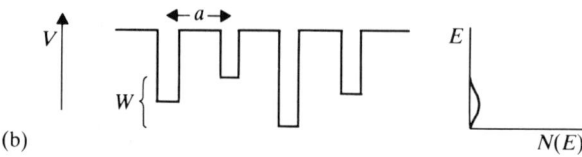

(b)

Fig. 5.14 (a) Representation of the potential wells for a crystalline lattice and the density of states expected for a tight-binding model.
(b) Representation of the potential wells in the Anderson model (in which there are random variations in the site energies) and the density of states expected for a tight-binding model (after Mott and Davis 1979).

expected for phonon processes, and it has been suggested that 'two-level systems' are responsible, for which $\tau_i \propto T^{-1}$. Thus, Thouless's ideas on size localization in one dimension appear to be borne out by experiment.

5.3.4 Two and three dimensions

The one-dimensional case is exceptional in that any amount of disorder, no matter how small, will localize the electron states. This is not the case for higher dimensions, for which we will see that a certain critical degree of disorder is necessary to induce localization in an entire band.

The question of localization in random lattices was first addressed by Anderson (1958), who considered the following simple tight-binding Hamiltonian:

$$\mathcal{H} = \sum_i \varepsilon_i |i\rangle\langle i| + V \sum_{i,j} |i\rangle\langle j| \qquad [5.37]$$

The basis states $|i\rangle$ are located on the sites of a *periodic* lattice and are coupled by constant nearest-neighbour interactions V. The disorder in this model occurs in the form of random site energies ε_i (or diagonal elements) termed 'diagonal disorder'. The ε_i are usually taken to be uniformly distributed over a certain energy width W, i.e.

$$p(\varepsilon) = 1/W \quad |\varepsilon| < W/2$$
$$= 0 \quad |\varepsilon| > W/2 \qquad [5.38]$$

and the strength of the disorder is then conveniently characterized by the ratio W/V. This model is illustrated schematically in Fig. 5.14. It can be seen that the problem is essentially a competition between the two terms in [5.37]; the first in isolation produces a localized (single) state, whereas the second produces an extended (Bloch) state, and the value of W/V is obviously crucial in deciding which of these two effects is dominant. Thus, for a given amount of disorder, all the states in the tails of a band having energies greater than or less than critical energies, E_c and E_c', will be localized,

Table 5.3 Critical strengths of disorder (W/V) for complete localization (after Kramer and Weaire 1979)

	Anderson (1958)	Ziman (1969)	Herbert and Jones (1971)	Licciardello and Economou (1975)
Square lattice	28	22	—	7.2
Diamond cubic lattice	32	22	—	8.2
Simple cubic lattice	62	32.4	40,24,21	14.5

References: P. W. Anderson (1958), *Phys. Rev.* **109**, 1492; J. M. Ziman (1969), *J. Phys.* **C2**, 1230; D. C. Herbert and R. Jones (1971), *J. Phys.* **C4**, 1145; D. C. Licciardello and E. N. Economou (1975), *Phys. Rev.* **B11**, 3697.

whereas states nearer the centre of the band will be delocalized (extended). As the degree of disorder increases, the energies E_c and E'_c move towards each other, until at a critical amount of disorder, *all* states in the band are Anderson localized. This value of W/V is the quantitative criterion by which the results of calculations using the various methods detailed later are compared. The localization criterion may be rewritten in the general form W/zV so that localization on lattices having different values of coordination number, z, may be compared directly. An alternative formulation (Mott and Davis 1979) is to use W/B, where the band width $B = 2zJ$ and J is the overlap integral, $J = \int \psi^*(r - R_n) \mathcal{H} \psi(r - R_{n+1}) \mathrm{d}^3 x$, equal to V in this case. Anderson's first estimate for the critical value of W/V was ~ 62 for a simple cubic lattice; this has since been reduced by a factor of ~ 4 by more sophisticated calculations. Table 5.3 lists the time evolution of estimates for the critical values of W/V to localize all states in a band for various lattices, both two- and three-dimensional.

A variety of techniques, both numerical and analytical, have been developed to study the problem of localization, and so to derive critical values of W/V. Matrix diagonalization techniques (see sect. 4.3) can be used to obtain eigenvalues (and hence determine their sensitivity to boundary conditions) or to obtain eigenvectors (and hence ascertain exponential character or calculate the participation ratio). An example of the latter method is shown in Fig. 5.15 where the eigenvectors have been calculated for the Anderson model on a square lattice; exponential behaviour is clearly seen for values of W/V in excess of 8.0 and extended states are observed for smaller values (Yoshino and Okazaki 1977). The analytical properties of the Green's function were studied in Anderson's original work, states being deemed to be localized if the perturbation expansion in V of G converged, G being therefore real.

The Anderson model is highly simplistic in considering only *diagonal* disorder, i.e. a random variation in the site energies ε_i. Perhaps a more realistic model Hamiltonian for amorphous systems is one which incorporates *off-diagonal* disorder, i.e. random variations in the inter-site interaction V (now becoming V_{ij}) caused by a random spatial distribution of centres. This form of disorder, however, is found to be much less effective in inducing Anderson localization; indeed there are some doubts whether all states in a band can be localized in this manner, although band-tail states certainly become localized.

An obvious analogy exists between the localization transition and a phase transition, and all the theoretical machinery developed to study phase transitions has been brought to bear on the localization problem too. The critical behaviour on

Electrons

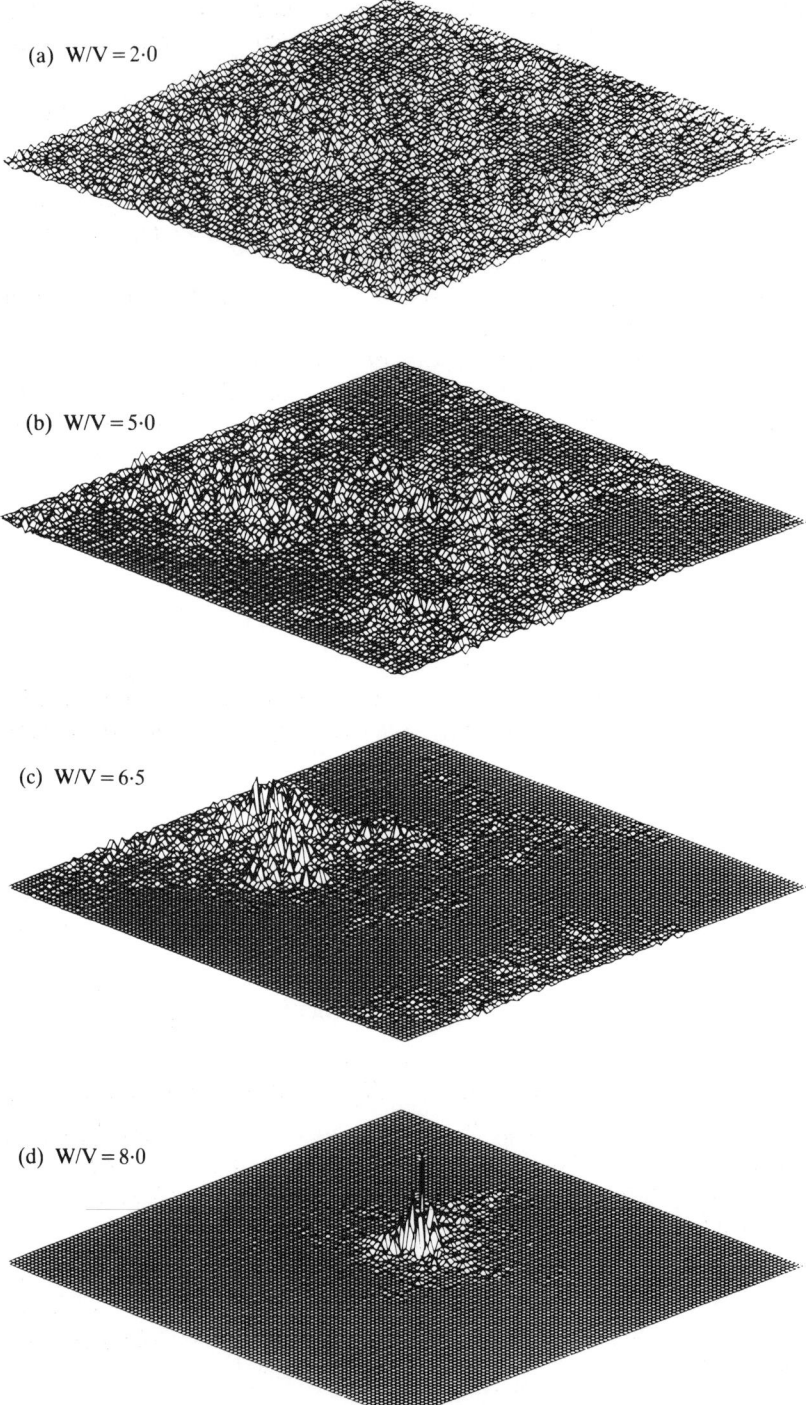

(a) W/V = 2·0

(b) W/V = 5·0

(c) W/V = 6·5

(d) W/V = 8·0

Fig. 5.15 Wavefunction amplitudes for the Anderson model on a square lattice, showing (a) extended and (d) localized states (Yoshino and Okazaki 1977).

the localized side of the transition can be expressed in terms of the divergence of a scaling length, in this case the localization length α^{-1}. This may be written in the form

$$\alpha \propto (E - E_c)^s \qquad [5.39]$$

where the critical exponent has been variously calculated to be 0.6 or $\frac{2}{3}$ in three dimensions, and $\frac{3}{4}$ in two dimensions (Mott 1981). The exact value of s in 3-D will be seen to be crucially important in deciding the behaviour of the conductivity at the critical energy separating localized and extended states.

Scaling theories for the localization transition have recently been developed which give results for the case of *two* dimensions (Abrahams et al. 1979) in marked contrast to those obtained by other methods. The approach is based on the ideas of Thouless that have already been discussed in the section on 1-D systems, and on the renormalization-group scheme. Small blocks of material of size L^d (where d is the dimensionality and $L > \lambda$, the mean free path) are coupled together to form larger blocks, and the strength of this coupling, g, is taken to be the ratio of the fluctuation in energy levels ΔE (resulting from a change in boundary conditions) to the mean spacing of the levels dE/dN, i.e.

$$g(L) = \frac{\Delta E(L)}{dE(L)/dN} \qquad [5.40]$$

This parameter g is a dimensionless conductance (related to the real conductance by $g = 2\hbar G/e^2$) and is related to the conductivity σ by an expression of the form already given in [5.30]

$$g(L) = \frac{2\hbar}{e^2} \sigma L^{d-2} \qquad [5.41]$$

The fundamental assumption made is that the coupling strength of the *larger* blocks, i.e. the scaled system, should only depend on the *original* g parameter. Put another way, in continuous terms:

$$\frac{d(\ln g(L))}{d(\ln L)} = \beta(g(L)) \qquad [5.42]$$

Thus, the quantity β is all that is required to ascertain how the coupling depends on L, and hence whether the states are localized or not. The behaviour of β with g can be ascertained from a consideration of its asymptotic behaviour as $g \to 0$ and ∞. For large g, macroscopic transport theory is valid (via extended states), so that $G(L) = \sigma L^{d-2}$ and hence

$$\lim_{g \to \infty} \beta(g) = d - 2$$

In fact it can be shown that this is the zeroth order term in a perturbation series, which to first order is:

$$\lim_{g \to \infty} \beta(g) = (d - 2 - a/g + \cdots) \qquad [5.43]$$

In 2-D, $a = \pi^{-2}$ and in 3-D, $a = 2^{1/2}4\pi^2$ (see Mott and Kaveh 1981). In the limit of

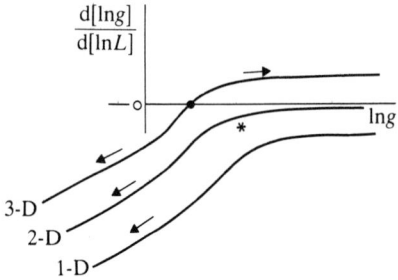

Fig. 5.16 Scaling trajectories for various dimensions according to Abrahams *et al.* (1979). In 3-D there is a fixed point ● and hence an Anderson transition for non-zero disorder strength. In 2-D there is only a smooth cross-over ∗ from weak to strong localization and hence no Anderson transition.

small coupling, the states are exponentially localized, and Abrahams *et al.* (1979) take $g(L) \sim e^{-\alpha L}$ for both 2-D and 3-D in this regime (although Mott and Kaveh (1981) take $g(L) \sim \alpha L e^{-\alpha L}$ for 3-D). Thus:

$$\lim_{g \to 0} \beta(g) \sim \ln g \qquad [5.44]$$

Hence, the overall form of $\beta(g)$ can be deduced to be as in Fig. 5.16. It can be seen that all states are predicted to be localized in 1-D, in accord with our previous discussion. However, the scaling theory predicts that *all* states in 2-D should be localized too, albeit only *weakly* localized below a critical amount of disorder. Integration of [5.42] together with the form of $g(L)$ appropriate to 2-D yields:

$$g = g_0 - A \ln L \qquad [5.45]$$

and the logarithmic size correction has been observed in many 2-D systems, including very thin metal films and 'inversion layers'. In contrast to previous theories, it is only in 3-D that an 'Anderson transition' between extended and localized states should occur as the disorder is increased, according to the scaling model.

5.3.5 Minimum metallic conductivity and the mobility edge

A central theme of the theory of electrons in amorphous solids has been the concept of the so-called 'mobility edge' separating localized and extended states, developed by Mott and expounded in the book by Mott and Davis (1979). This hypothesis asserts that at $T = 0$ the electron mobility, and hence the d.c. electrical conductivity, should be finite on the delocalized (metallic) side of the Anderson transition, but *zero* for localized states. Thus, it is proposed that the conductivity or mobility changes discontinuously at the critical energies separating localized and extended states (E_v and E_c in Fig. 5.12), even though the density of states is continuous there. This behaviour is shown schematically in Fig. 5.17 (together with a power-law behaviour that has recently been suggested (Abrahams *et al.* 1979). Note that if the Fermi level can be moved in some way through E_c or E_v, a metal–insulator transition occurs; the situation when E_F lies among a series of localized states on the localized side of the critical energy has been termed a 'Fermi glass'.

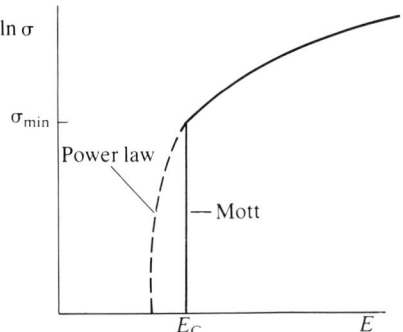

Fig. 5.17 Behaviour of the d.c. conductivity near a mobility edge as suggested by Mott and scaling theories (Abrahams *et al.* 1979) for a three-dimensional system.

Consider first extended states for which the energy lies deep in the band, well away from the mobility edge. The form of the wavefunctions is as shown schematically in Fig. 5.18(a), and the mean free path ℓ is long ($k\ell \gg 1$) although not infinite as for a perfect crystal. The conductivity σ_E can be evaluated from the Kubo formula:

$$\sigma(\omega) = \frac{e^2 \pi}{m^2 \omega \Omega} \sum_\alpha \sum_{\beta \neq \alpha} |\hat{p}_{\alpha\beta}|^2 (f_\beta - f_\alpha) \delta(E_\alpha - E_\beta - \hbar\omega) \qquad [5.46]$$

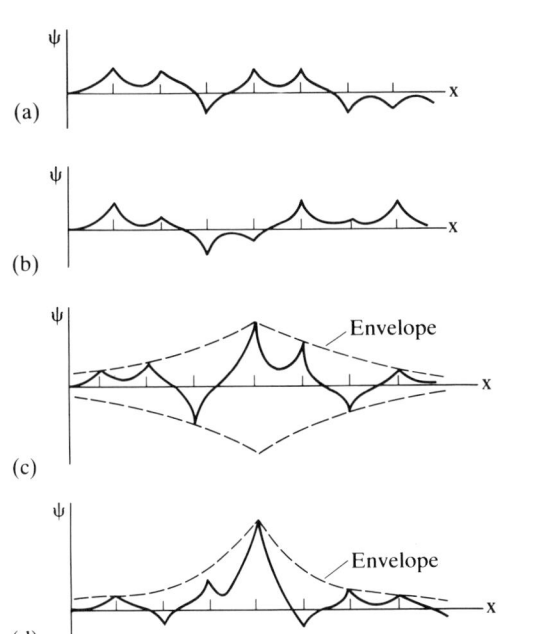

Fig. 5.18 Schematic illustration of the form of the electron wavefunction in the Anderson model (after Mott and Davis 1979).
(a) when $\ell \sim a$;
(b) when states are just non-localized ($E \gtrsim E_c$);
(c) when states are just localized ($E \lesssim E_c$);
(d) strong localization.

where $\hat{p}_{\alpha\beta}$ is the matrix element of the momentum operator between states of energy E_α and E_β, whose probabilities of occupation in thermal equilibrium are given by the Fermi–Dirac distributions f_α and f_β, and Ω is the volume of the system. For low temperatures, this may be written as:

$$\sigma(\omega) = \frac{\pi e^2 \hbar \Omega}{2m^2} \int_{E_F - \hbar\omega}^{E_F} \frac{|\hat{p}|^2_{av} N(E) N(E + \hbar\omega) \, dE}{\hbar\omega} \quad [5.47]$$

In the random-phase approximation, in which the wavefunctions are written as a sum of plane waves, with random amplitudes distributed in a gaussian fashion, $|\alpha\rangle = \sum a_k^\alpha |k\rangle$ the conductivity can be evaluated to be (see problem 5.3 and Thouless 1979):

$$\sigma(\omega) = \frac{k_F^3 e^2 \tau}{3\pi^2 m} \frac{1}{(1 + \omega^2 \tau^2)} \quad [5.48]$$

This is of the Drude form, and the relaxation time τ is given by $(m\ell/\hbar k_F)$. For zero frequency, and a spherical Fermi surface of wavevector k_F and area S_F, this reduces to the formula that can be obtained also by the Boltzmann formulation, i.e.:

$$\sigma_E = \frac{e^2 k_F^2 \ell}{3\pi^2 \hbar} = \frac{e^2 \ell S_F}{12\pi^3 \hbar} \quad [5.49]$$

As the amount of disorder is increased, the mean free path decreases until it reaches the Ioffe–Regel minimum, such that $k\ell \sim 1$, where $\ell \sim a$, the interatomic spacing (or a_E, the distance in which the wavefunction loses phase memory, whichever is the shorter). The conductivity at this point is given approximately, therefore, by substituting these limits into [5.49], i.e. $\sigma_E \sim e^2/3\hbar a$. Any further increase in disorder only serves to broaden the bands, thereby lowering $N(E)$ from its free electron value by a factor Γ, rather than decreasing ℓ further. Thus, the 'minimum metallic conductivity', σ_{min}, is the minimum conductivity at the mobility edge on the delocalized side before all states become localized and the conductivity $\langle \sigma_E \rangle$ drops precipitously to zero (Mott and Davis 1979, Mott 1981); for three dimensions it can be written as:

$$\sigma_{min}^{3-D} \simeq \left(\frac{e^2}{3\hbar a_E}\right) \Gamma^2 \quad [5.50]$$

where $\Gamma \sim \frac{1}{3}$ but depends on the coordination number z. Thus, it is expected that σ_{min}^{3-D} should lie in the range 300–600 Ω^{-1} cm^{-1}, if a_E is taken as ~ 3 Å. In 2-D, a similar analysis shows that σ_{min}^{2-D} is independent of a_E:

$$\sigma_{min}^{2-D} \simeq 0.1 e^2/\hbar \quad [5.51]$$

The constant of proportionality is predicted to be independent of the type of disorder.

However, recent scaling theories of the localization transition have cast doubts on the concept of a minimum metallic conductivity, certainly in 2-D and possibly in 3-D. As we have seen, scaling arguments predict that in two dimensions all states are (power-law) localized, albeit weakly, if any disorder is present, and strongly localized (exponentially) above a certain degree of disorder. Thus, there can be no minimum metallic conductivity in 2-D. However, Abrahams et al. (1979) also assert that there should be no minimum metallic conductivity, either, in 3-D; instead, they predict that in 3-D the conductivity near a mobility edge behaves like:

$$\sigma_E \simeq \sigma_0 \left[\frac{(E-E_c)}{B} \right]^v \qquad [5.52]$$

where B is the bandwidth and v is a constant of the order of unity. This expression is obtained by assuming β ([5.42]) varies as $\beta \simeq 1/v \ln (g/g_c) \simeq (g - g_c)/v g_c$, where g_c is the value of g where $\beta = 0$; integration of [5.42] then gives [5.52]. Mott and Kaveh (1981), however, argue that the absolute value of σ_E at E_c cannot be obtained in this way by a scaling argument because of the undetermined value of the constant of integration which results; all that can be deduced is the size dependence of the conductivity.

Mott (1981) has argued recently that whether or not σ_{min} exists in 3-D depends crucially on the value of the exponent, s, in the power-law behaviour of the energy dependence of the inverse localization length, α ([5.39]). This is because the derivation of σ_{min}^{3-D} given above depends for its validity on the absence of significant fluctuations in the random coefficients a_n of the wavefunction $\Psi = \sum a_n \exp(i\phi_n) \psi_n(r)$, where ϕ_n are random phases and ψ_n are atomic (s-like) wavefunctions. The behaviour of the conductivity at E_c can be understood from a consideration of the Kubo formula, [5.46]. We assume that the only relevant factors are the matrix element of the momentum operator and the square of the density of states, viz.

$$\left\langle \left| \int \psi_1 \frac{\partial}{\partial x} \psi_2 \, d^3x \right|^2 \right\rangle \{N(E_F)\}^2.$$

$\sigma(E < E_c)$ vanishes because two overlapping localized states with the same energy E cannot exist. For energies at and above the mobility edge, $E \geqslant E_c$, $\sigma \geqslant \sigma_{min}$ if the states are just extended, but $\sigma \geqslant 0$ if significant fluctuations in a_n exist such that the elements of the matrix element vanish because the wavefunctions have maxima and minima at different spatial positions. Significant fluctuations are absent, and hence σ_{min} is finite, if $s > \frac{2}{3}$ in 3-D; if $s < \frac{2}{3}$ there is no discontinuity in σ_E across E_c, although it does increase very rapidly.

Consider a volume containing N^3 potential wells (in the Anderson model); then a mean fluctuation $\delta V/V \simeq N^{-3/2}$ is expected for an ensemble of such volumes. For an electron with $E = E_c$, half the volumes will have a mean potential greater than V by δV, and for these the wavefunction there will behave in an exponential manner $\sim \exp(\pm \alpha r)$, where (from [5.39]):

$$\alpha a \simeq (\partial V/V)^s \qquad [5.53]$$

For a block of side Na, [5.53] is therefore proportional to $N^{-3s/2}$, and thus

$$\alpha r = \alpha Na \sim N^{1-3s/2} \qquad [5.54]$$

Hence, if $s \geqslant \frac{2}{3}$, [5.54] is well behaved, but if $s < \frac{2}{3}$, it diverges as $N \to \infty$, leading to fluctuations in Ψ. The fluctuations in V between blocks are only significant if $\delta V \simeq \delta E = E - E_c$. Thus, the wavefunctions Ψ will have fluctuations of order

Electrons

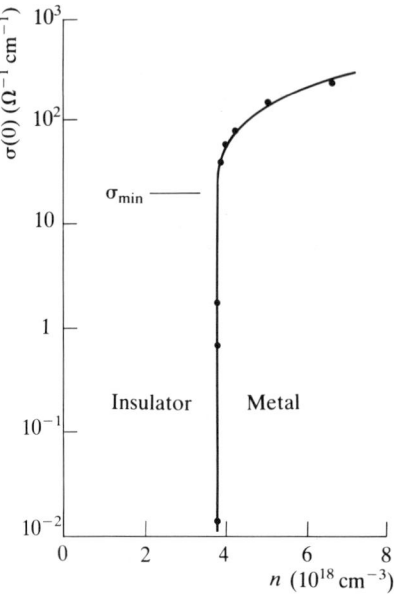

Fig. 5.19 Conductivity in the limit as $T \to 0$ of Si:P as a function of phosphorus concentration (Rosenbaum et al. 1980).

$\exp\{\pm c(E/\Delta E)^{(2/3)-s}\}$, where c is a constant, and hence the matrix element in the Kubo formula should behave as $\exp\{-c(E/\Delta E)^{(2/3)-s}+\eta\}$, where η is the phase difference between wavefunctions. Integration over η from 0 to π suggests that $\langle |\int \psi_1 (\partial/\partial x)\psi_2 \, d^3x|^2 \rangle \simeq (\Delta E/E)^{(2/3)-s}$, and so for $s < \frac{2}{3}$ the conductivity should behave as:

$$\sigma_E \simeq \sigma_{\min}^{3\text{-D}}(\Delta E/E)^{(2/3)-s} \qquad [5.55]$$

Figure 5.19 qualitatively shows this behaviour in the system crystalline Si:P (in which the P donors produce an impurity band having a random array of potentials much as in the Anderson model). Note the several experimental points for which the conductivity is *less* than $\sigma_{\min}$. However, the presence of compositional or other fluctuations might also lead to the absence of both $\sigma_{\min}$ and a discontinuous rise in conductivity at E_c due to the existence of percolation channels, which will be discussed next. It should be admitted in conclusion that the question of the existence of a minimum metallic conductivity is still controversial.

5.3.6 Percolation

We have seen in Chapter 3 that structural and compositional fluctuations may exist in an amorphous solid. If the length scale of these fluctuations is of the order of hundreds of ångströms, then tunnelling of electrons between such regions is essentially precluded, and a *classical* approach may be adopted. The long-range potential fluctuations we are considering are shown schematically in Fig. 5.20 (Fritzsche 1973); note that these represent only *symmetric* fluctuations due to electrostatic charges, and the antisymmetric potential fluctuations due to density

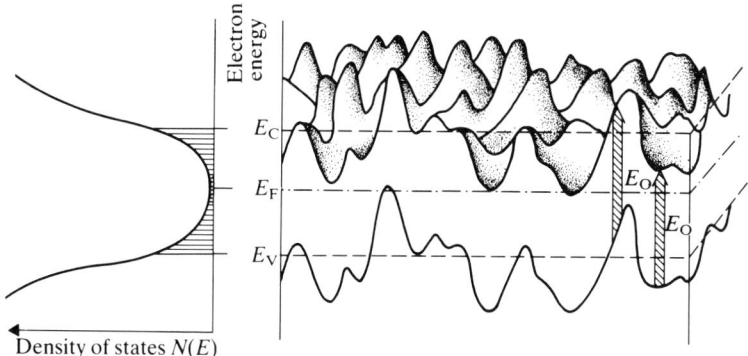

Density of states $N(E)$

Fig. 5.20 Schematic illustration of the symmetric part of long wavelength potential fluctuations. Short-range fluctuations, as well as antisymmetric fluctuations which shift valence and conduction band states in opposite directions, are omitted for clarity. E_o corresponds to an average optical gap, E_c and E_v are percolation thresholds or mobility edges. The density of states is shown on the left-hand side (Fritzsche 1973).

variations which shift the valence and conduction band states in different directions, together with short-range fluctuations, are omitted for clarity. Electrons of a given energy E are only able to reside in those regions where the potential energy is lower than E. At low energies, the allowed regions would form isolated pockets in space, and as E is increased these regions would grow and merge until, at a critical energy, an allowed channel would be formed throughout the volume of the sample and the electron could therefore move anywhere.

This occurrence is termed 'percolation' and has applications in many disparate fields; a review of the application to the theory of electrons in inhomogeneous media has been given by Cohen *et al.* (1978). There are essentially two types of percolation problem: *site* percolation in which if a site has a probability p of being occupied then the percolation probability function $P(p)$ is the probability that a given atom belongs to an infinite cluster, and *bond* percolation in which if a bond connecting two sites has a probability p of being 'open' then $P(p)$ is the probability that such a favourable bond is part of an infinite cluster linked by such bonds. $P(p)$ is thus the probability that a particle can percolate an infinite distance from a site; there is a threshold or critical value p_c such that $P(p<p_c)=0$, which is different for different types of lattice and is generally different for bond and site percolation on the same lattice. For example, $p_c \simeq 0.25$ for bond percolation on a 3-D simple cubic lattice.

Thus applying these ideas to the problem of long-range fluctuations, we might expect the conductivity to behave as:

$$\sigma_E = \sigma_0 P(x) \qquad [5.56a]$$

where $x(E)$ is the allowed volume fraction at a given energy E and σ_0 is the energy-independent value of conductivity in the allowed regions of extended states. The form of [5.56a] is expected to be like curve (ii) in Fig. 5.21, in contrast to the discontinuous Mott mobility edge (curve (i)). Note that the *percolation* threshold at E_c marks the boundary between localized and extended states in this model. In fact, we do not expect $\sigma(E)$ to follow $P(E)$ just beyond the threshold, since, although the fraction of allowed channel volume does increase sharply above x_c or E_c, the first

Electrons

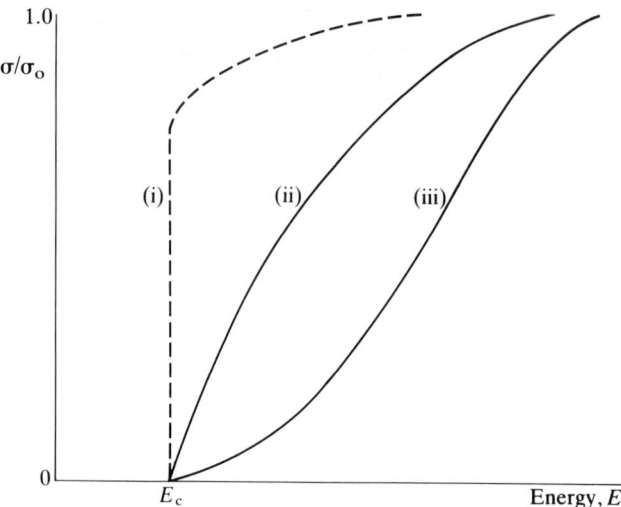

Fig. 5.21 Behaviour of conductivity as a function of energy. Curve (i) is the sharp Mott mobility edge for homogeneous material, (ii) is the prediction of percolation theory and (iii) is the prediction of percolation theory for an inhomogeneous material.

percolation channels will be very narrow and convoluted in shape and contribute little to conduction. Thus, $\sigma(E)$ will increase less rapidly than $P(E)$, and in fact is predicted to follow the power law (see Fritzsche 1973):

$$\sigma_E = \sigma_0 (E - E_c)^{1.6} \qquad [5.56b]$$

This behaviour is shown in curve (iii) of Fig. 5.21. The proposal by Cohen *et al.* that long-range potential fluctuations in the band tails near a mobility edge should always give rise to percolation behaviour and hence to a conductivity that decreases *continuously* to zero has been criticized (see e.g. Mott and Davis 1979). The classical model is only valid if tunnelling is precluded. However if the potential fluctuations are such that tunnelling *is* allowed, Mott has argued that a sharp mobility edge (at a higher energy than the percolation edge) should still exist, although the jump in σ may be less than σ_{min} ([5.50]), because localized and extended states cannot coexist at the same energy.

5.4 Transport properties

We turn our attention now to the electrical transport properties of *ideal* amorphous semiconductors (leaving until Ch. 7 a discussion of the electrical properties of glassy metals). We regard materials as 'ideal' if they are free of structural defects, such as dangling bonds, which give rise to electron states lying deep in the energy gap, and hence states which *do* exist in the gap arise only from band tails. Transport properties such as d.c. electrical conductivity, thermopower and Hall effect will be discussed in the light of the models for electron states in amorphous materials that we have developed in the preceding sections. Extensive reviews of these subjects may be found in Mott and Davis (1979) and Nagels (1979). Optical properties involving transitions of energies comparable to the band gap form the subject of the next section.

5.4.1 Direct current electrical conductivity

The total conductivity at a finite temperature can be written as:

$$\sigma = -\int \sigma_E \frac{\partial f(E)}{\partial E} dE \qquad [5.57]$$

where $f(E) = (1 + \exp(E - E_F)/k_B T)^{-1}$, the Fermi–Dirac distribution function and hence $\partial f(E)/\partial E = -f(E)[1 - f(E)]/k_B T$. Conduction involving localized states can only occur by phonon-assisted hopping, and this will be left for consideration until later. Thus, the only current-carrying states in this model are the extended states beyond E_v and E_c in the valence and conduction bands, respectively. Equation 5.57 can be rewritten in the form

$$\sigma = e \int N(E) \mu(E) f(E)[1 - f(E)] dE \qquad [5.58]$$

where $\mu(E)$ and $N(E)$ are the mobility and density of states at an energy E. Assuming Mott's model for a sharp mobility edge at E_v and E_c is valid, and if conduction takes place far enough away from E_F that $f(E)[1 - f(E)] \simeq \exp(-(E - E_F)/k_B T)$ [5.58] may be integrated to give (for electrons):

$$\sigma_{ext} = eN(E_c)\mu_e k_B T \exp[-(E_c - E_F)/k_B T] \qquad [5.59]$$

where μ has been taken to be zero in the mobility gap, and to be some average value μ_e in the band. The conductivity in the Mott model due to extended states can also be written (for electrons) as:

$$\sigma_{ext} = \sigma_{min} \exp[-(E_c - E_F)/k_B T] \qquad [5.60]$$

Thus, we may write $\mu_e = \sigma_{min}/ek_B T N(E_c)$, and if the mobility edges lie far from the centre of the band (i.e. $V_0 \ll B$), then we may make the approximation (Mott and Davis 1979):

$$N(E_c) \simeq 1/Ba_E^3 \qquad [5.61]$$

where a_E as before is the distance over which the wavefunction loses phase memory. Hence, using [5.50] for the minimum metallic conductivity, an estimate for the mobility is:

$$\mu_e \simeq \frac{ea_E^2 B\Gamma^2}{3\hbar k_B T} \qquad [5.62]$$

Taking $\Gamma^2 = 1/10$, $a_E = 3$ Å and $B = 5$ eV, gives $\mu_e \simeq 9$ cm^2 V^{-1} s^{-1} at room temperature; for comparison, the electron mobility in, say, crystalline GaAs is $\simeq 9 \times 10^3$ cm^2 V^{-1} s^{-1}. A similar estimate has been obtained by Cohen by assuming that the transport is diffusive (i.e. by Brownian motion) when the mean free path is very short; in this case $\mu_e = eD/k_B T$, where the diffusion constant is given by

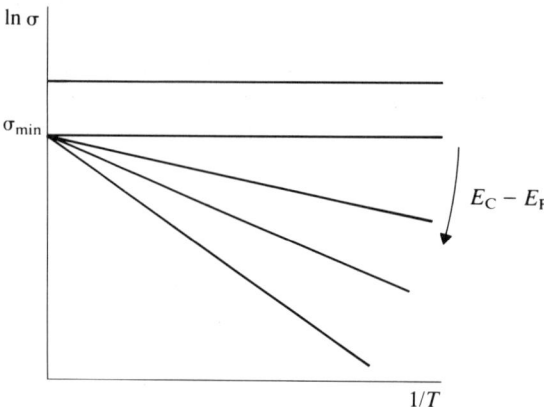

Fig. 5.22 Schematic representation of the temperature dependence of the conductivity as the Fermi level E_F is moved away from E_c into the gap in the direction indicated.

$D = (1/6)v_{el}a^2$, and v_{el} is an electronic frequency ($v_{el} \simeq \hbar/m_e a^2 \simeq 10^{15}$ s^{-1}). Hence (see e.g. Mott and Davis 1979):

$$\mu_e^{BM} \simeq \frac{e\hbar}{6m_e k_B T} \qquad [5.63]$$

which gives a value $\mu_e \sim 7.5$ cm^2 V^{-1} s^{-1} at room temperature. The frequency v_{el} may also be written in the form $\sim JF/\hbar$, where J is an electronic transfer integral and F is a factor accounting for the fact that increasing disorder decreases the carrier diffusion (Emin 1976). In the random phase model, the role of F is played by the factor $[a^3(2zJ)N(E_c)]$, so that the mobility can be written:

$$\mu_e^{RPA} = \frac{2\pi e a^5 z J^2 N(E_c)}{3\hbar k_B T} \qquad [5.64]$$

It can be seen from [5.59] or [5.60] that extended state conduction should be activated. If, in some manner, E_F may be moved from its usual position in mid-gap through the mobility edge, e.g. by doping in a-Si:H or by alteration of the gate voltage in a MOSFET 2-D system, one expects the behaviour shown in Fig. 5.22; the physical meaning of σ_{min} is immediately apparent, being the limiting metallic conductivity, and also (on this model) the focus of the extrapolated $\ln \sigma$ v.s. $1/T$ curves. It is important to understand, however, that the observed activation energy for conduction, ΔE_σ, is *not* $E_c - E_F$, because the band gap (or mobility gap) is itself a function of temperature, decreasing with increasing temperature, since the phonons contribute a term to the electron self-energy. Optical measurements show that at room temperature, the dependence is nearly linear, i.e.:

$$E_c - E_F \simeq \Delta E_\sigma - \gamma T \qquad [5.65]$$

but at low temperatures ($\lesssim 77$ K) the gap becomes almost temperature independent; this behaviour is shown schematically in Fig. 5.23. Since optical studies give the temperature coefficient of the optical gap in most materials to be ~ 4–8×10^{-4} eV K^{-1}, and E_F often lies near mid-gap, we expect γ to have a value roughly half this amount. Thus we conclude that the *observed* activation energy, ΔE_σ, has no physical significance in itself, simply being the $T=0$ intercept of the curve of $E_c - E_F$ v.s. T: the energy interval between the Fermi level and the mobility edge can only be obtained if γ is known from complementary optical or thermopower studies. The result of this is that the pre-exponential in [5.59] becomes (for electrons):

$$C_0 = eN(E_c)k_B T \mu_e \exp(\gamma/k) \qquad [5.66]$$

where, because $\mu_e \propto 1/T$ ([5.62] and [5.63]), C_0 is temperature independent and, in general, values for C_0 lie in the range $\sim 10^2 - 10^4 \, \Omega^{-1} \, \text{cm}^{-1}$.

Thus far, we have developed the theory in the framework of Mott's model of a discontinuous mobility edge. One might ask whether a constant activation energy for conduction would still be observed if instead the mobility edge were more gradual, e.g. at a percolation edge. However, it can be shown that if the mobility increases as $(E-E_c)^m$, integration of [5.58] introduces a term $(k_B T)^m$ into the pre-exponential, leaving a constant activation energy as before. Long-range fluctuations, such as might also give a gradual mobility edge, have been invoked to account for the interesting correlation between C_0 and ΔE_σ (or the difference between conductivity and thermopower activation energies) that has been observed in a-Si:H samples; we will consider this and other alternative explanations, when we discuss thermopower in the next section.

Several other features concerning conduction arising from carriers excited to the mobility edge merit discussion. Most (undoped) amorphous semiconductors have an electrical activation energy which is approximately half the optical gap, $\Delta E_\sigma \leq \frac{1}{2} E_{\text{opt}}$, so the Fermi level lies near the centre of the gap. This fact, and the observation that the Fermi level is often 'pinned', i.e. the zero-temperature Fermi-level position does not shift appreciably with the introduction of a few shallow

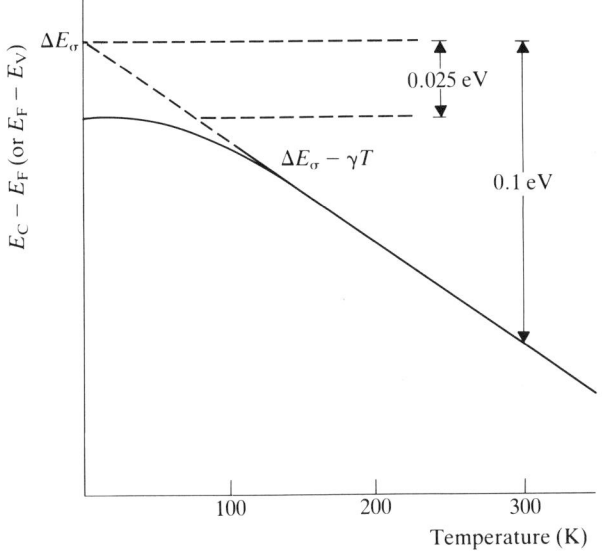

Fig. 5.23 Schematic illustration of the temperature dependence of $E_c - E_F$ (or $E_F - E_v$) with $\gamma = 3 \times 10^{-4}$ eV K^{-1}. ΔE_σ is the experimental slope of the $\ln \sigma$ v.s. $1/T$ curve measured in the range $200 < T < 400$ K (after Fritzsche 1973).

donors or acceptors, cannot readily be understood on the basis of the present model unless the band tails completely overlap in mid-gap, as in the CFO model (Fig. 5.12(b)). Indeed, we shall see in the next chapter that it is often the presence of certain structural defects, and their resulting states in the gap, which dictate this behaviour. Hence, we cannot write $2\Delta E_\sigma =$ band gap, as if the material were *intrinsic*. This fact can also be illustrated in a different way (Fritzsche 1973): in intrinsic crystalline semiconductors, all the electrons in the conduction band are excited from the valence band, and so electrical neutrality demands that the number of electrons and holes must be equal ($n = p$) and this dictates the position of E_F, so that:

$$\sigma^{cr}(\text{intrinsic}) \propto \exp\left[-\frac{(E_c - E_v)}{2k_B T}\right] \qquad [5.67]$$

This is not the case for amorphous semiconductors, where $n \neq p$ because of the much larger number of electrons and holes in localized tail states (or gap states) than are free and produced by thermal excitation (certainly at low temperatures where the number of intrinsic carriers is very small), and hence the electrical properties must in general be controlled extrinsically.

This is perhaps the place to discuss one of the most significant recent developments in the field of amorphous semiconductors, namely the discovery that certain non-crystalline materials, notably a-Si:H, can be systematically doped n- or p-type. The dopability of conventional crystalline semiconductors, e.g. Si, has led to their use in countless electronic device applications and to their forming the foundation of the present-day electronics industry. Doping is achieved in these materials by the substitutional replacement of, say, the Si by either group V elements, e.g. P or As, to give n-type doping, or by group III elements, e.g. B, to give p-type doping; in both cases electrical doping occurs because the dopant atoms are forced into chemically unfavourable tetrahedral substitutional sites and therefore donate or accept an electron, respectively.

It was thought for a long time that amorphous semiconductors could not be doped because of two factors: (1) it was believed that any foreign atom introduced into an amorphous matrix would take up its normal chemical valence because of the supposed flexibility of the random network and would not be constrained into occupying a tetrahedral site as must happen in a crystalline (Si) matrix, e.g. a P atom would be threefold, rather than fourfold, coordinated in accordance with its normal valence; and (2) the density of states in the gap of most amorphous semiconductors due to band tailing or structural defect states (see sects 5.3.2 and 6.2.1) is sufficiently high to preclude displacement of the Fermi level by the donation of electrons or holes by impurity atoms, because any surplus carriers so introduced are simply trapped by these gap states and consequently do not shift the Fermi level appreciably. The discovery that *hydrogenated* amorphous Si, prepared either by the glow-discharge decomposition of silane or by sputtering in an Ar/H_2 atmosphere, possesses a relatively low density of states in the gap ($N(E_F) \lesssim 10^{16}$ cm^{-3} eV^{-1}), much lower than non-hydrogenated (e.g. evaporated) a-Si ($N(E_F) \simeq 10^{20}$ cm^{-3} eV^{-1}), immediately suggested that a-Si:H might be a suitable candidate for doping. The first results on systematic doping in a-Si:H were reported by Spear and LeComber (1976), and are shown in Fig. 5.24(a). It can be seen that the addition of P (in the form of PH_3) incorporated into the SiH_4 gas stream) increases the d.c.

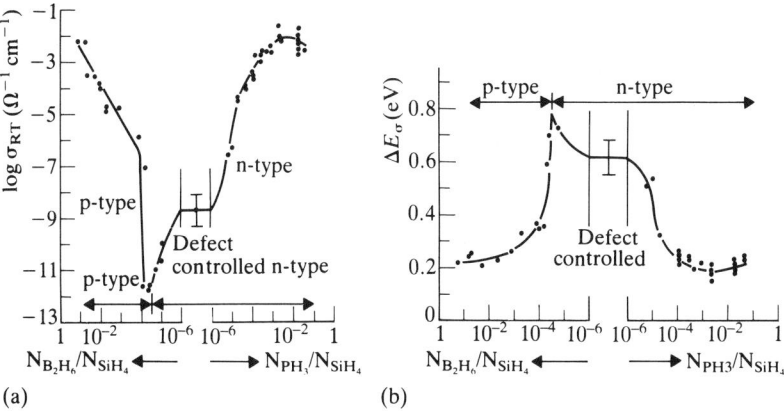

Fig. 5.24 Substitutional doping in GD a-Si:H (Spear and Le Comber 1976).
(a) Changes in room temperature d.c. conductivity by doping with P and B. The data point labelled 'defect-controlled' refers to intrinsic, i.e. undoped, material.
(b) Changes in the d.c. conductivity activation energy ΔE_σ by doping with P and B.

conductivity by about 7 orders of magnitude; the addition of B (in the form of B_2H_6) at first *decreases* the conductivity (because E_F is moved down from its position slightly above mid-gap in intrinsic material to exactly mid-gap) and thereafter the conductivity rapidly increases by about 10 orders of magnitude as E_F is progressively moved nearer the valence band edge. Note that at high doping levels for either p- or n-type dopants, the rate of increase of the conductivity slows down markedly; the reason for this is that movement of the Fermi level becomes progressively more difficult as E_F moves through the rapidly increasing density of tail states at the band edges. This change in doping efficiency with dopant concentration is also clearly illustrated by the behaviour of the d.c. conductivity activation energy ΔE_σ shown in Fig. 5.24(b), since ΔE_σ is related to $E_c - E_F$ (or $E_F - E_V$) by [5.65]; the conductivity changes shown in Fig. 5.26(a) are brought about by a total movement of the Fermi level through the gap $\simeq 1.2$ eV. However, not all the dopant atoms which are present are electrically active (i.e. in tetrahedral sites). The fraction which is active can be estimated by deducing the number of dopant centres required to shift the Fermi level by a given amount through the density of gap states, known from, say, field-effect measurements (see Fig. 6.3); however, the magnitude of the density of states deduced from field-effect measurements is notoriously sensitive to the presence of surface states at the film interfaces, rendering a knowledge of the bulk density of states uncertain in many cases. This uncertainty therefore causes an equal uncertainty in the estimation of the fraction of active dopant atoms, and values range between 0.3 and 0.01.

Introduction of dopant gases into the gas stream of a glow-discharge deposition system is not the only way of doping a-Si:H; r.f. sputtering of a Si target using an Ar/H_2 plasma into which PH_3 or B_2H_6 has been added has been found to produce doped films (Paul *et al.* 1976), as has ion-implantation into glow-discharge a-Si:H (Spear *et al.* 1979, Kalbitzer *et al.* 1980). Reviews of doping studies in a-Si:H have been given by Spear (1977) and Le Comber and Spear (1979), and the interested reader is referred to these for more details.

The discovery that amorphous semiconductors can be doped is not just a matter for scientific curiosity, but has led to considerable technological interest. The advantage that the amorphous films have over conventional crystalline semiconductors lies chiefly in the ease and relative cheapness with which they may be manufactured in the form of large areas. Where large area devices are concerned, such as in solar cells, or thin-film transistors (TFT) acting as switching arrays for large area liquid crystal displays for example, a-Si:H films are obviously potentially more attractive than expensive single crystal Si devices, provided that the efficiency is not so low as to negate the advantages that such films have by virtue of their cheapness and ease of fabrication. Reviews of progress in thin-film solar cells, particularly with regard to a-Si:H have been given by Debney and Knight (1978), Wilson et al. (1978) and Hamakawa (1981); a similar review of TFT applications has been given by Le Comber et al. (1981).

The essential function of a solar cell is the creation of electron-hole pairs by incident photons and their subsequent separation by some form of rectification action such that the carriers can produce a current in an external circuit; rectifying structures can be p–n homo- or hetero-junctions, p–i–n structures or Schottky barriers. The power conversion efficiency of such a device can be expressed as $\eta_c = V_{oc} J_{sc} f / P_{in}$, where V_{oc} is the open-circuit voltage, J_{sc} is the short-circuit current density, f is the so-called 'fill factor' (which is a measure of the degree to which the current–voltage curve is rectangular, and is unity in the ideal case) and P_{in} is the power density incident on the device. Efficiencies for devices fabricated from a-Si:H in a p–i–n geometry have now reached $\eta_c \simeq 8\%$; this is considerably smaller than the $\simeq 15\%$ efficiencies that are achieved by conventional crystalline (say Si) solar cells, but is still sufficiently high to warrant their exploitation in certain small-scale applications. Amorphous hydrogenated silicon is in fact superior to crystalline Si in respect of its optical properties in relation to the sun's spectrum at the earth's surface. The solar illumination intensity, P_{in}, depends significantly on the atmospheric conditions; on a clear sunny day with the sun at the zenith, a power density of 100 mW cm^{-2} is received at sea-level and this is termed AM1 ('air mass 1') illumination, whereas the intensity received outside the atmosphere is termed AM0 ($P_{in} = 135$ mW cm^{-2}). The AM1 spectrum peaks at $\simeq 0.5$ μm which matches the absorption edge of a-Si:H better than c-Si; furthermore, because the absorption edge of c-Si is due to indirect optical transitions it lies considerably below that of a-Si:H (Fig. 5.25) which means that a film of a-Si:H only a few microns in thickness is needed to absorb all the useful sunlight, whereas several hundred microns of c-Si are required to do the same.

We conclude this section by considering the electrical conductivity in a 'perfect' amorphous semiconductor resulting from excitation of carriers to the localized tail states below the mobility edge, and the conduction processes that take place there. If the states are localized, so that $\sigma_E = 0$, conduction can only take place by *thermally activated hopping* between sites, in which energy is exchanged with a phonon (see sect. 6.3.3 for a fuller discussion). Thus, it is expected that the mobility is now activated:

$$\mu_{hop} = \mu_0 \exp\left[-\Delta(E)/k_B T\right] \qquad [5.68]$$

where $\Delta(E)$ is the activation energy for hopping and where by analogy with our

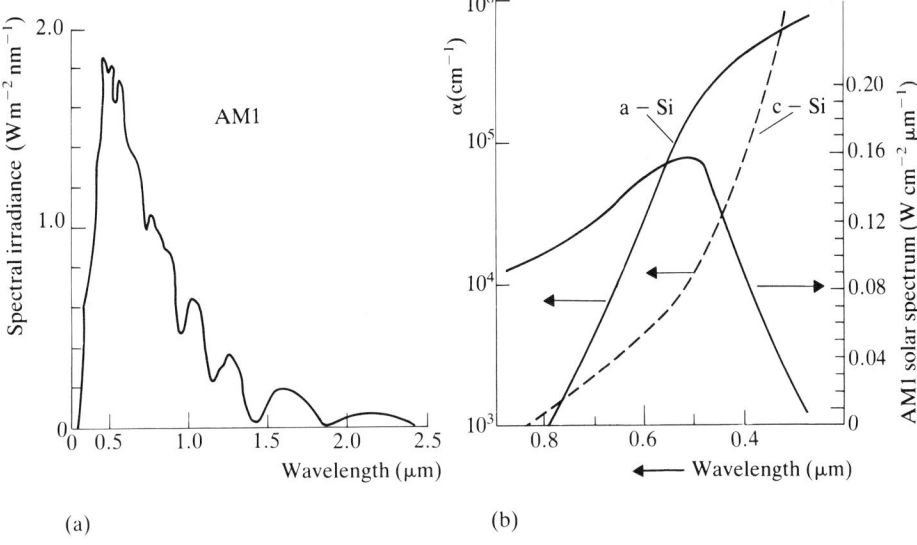

Fig. 5.25 (a) Solar energy spectrum (AM1) (Treble, F., *Proc. Photovoltaic Solar Energy Conference* (Reidel: 1977), p. 732).
(b) Comparison of the spectral dependence of the absorption coefficient of amorphous and crystalline Si with the AM1 solar spectrum (R. A. Gibson, P. G. Le Comber and W. E. Spear (1978) *IEE: J. Solid-State and Electron Devices*, **2**, S3.

previous expression for μ_e^{BM}, μ_0 may be written:

$$\mu_0 = \frac{v_0 e R^2}{6 k_B T} \quad [5.69]$$

where v_0 can approximate to a typical phonon frequency ($\simeq 10^{13}$ s^{-1}) and R is the hopping distance; if R is taken to be ~ 2 Å, and $\Delta \sim k_B T$, then [5.69] gives $\mu_{hop} \simeq 10^{-2}$ cm^2 V^{-1} s^{-1} at room temperature. Hence, the three decade increase in mobility at E_c or E_V going from μ_{hop} to μ_e justifies the term 'mobility edge'.

The density of states naturally increases rather sharply in the tail-state region, and if a power-law behaviour is assumed:

$$N(E) = \frac{N(E_c)}{(E_c - E_A)^s}(E - E_A)^s \quad [5.70]$$

then σ_{hop} may be calculated from [5.58] assuming that hopping is only between nearest-neighbour states (Nagels 1979):

$$\sigma_{hop} = \sigma_{hop}^0 \left(\frac{k_B T}{E_c - E_A}\right)^s C \exp[-(E_A - E_F + \Delta)/k_B T] \quad [5.71]$$

where $\sigma_{hop}^0 = e k_B T N(E_c) \mu_0$, and C is given by:

$$C = s! - \left(\frac{E_c - E_A}{k_B T}\right)^s \exp\left[-\frac{(E_c - E_A)}{k_B T}\right]\left[1 + s\left(\frac{k_B T}{E_c - E_A}\right)\right.$$

$$\left. + s(s-1)\left(\frac{k_B T}{E_c - E_A}\right)^2 + \cdots\right] \quad [5.72]$$

Electrons

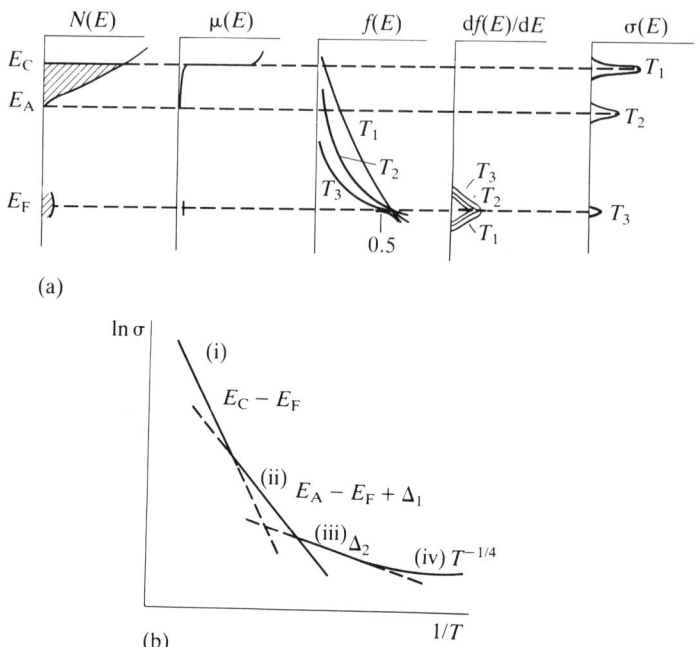

Fig. 5.26 (a) Schematic illustration of the change in mode of conduction with temperature ($T_1 > T_2 > T_3$). (b) Schematic illustration of the temperature dependence of conductivity expected for the model shown in (a). The activation energies found in the various regimes are indicated (after Mott and Davis 1979).

Thus, the predominant temperature depencence is from the exponential term with exponent ($E_A - E_F + \Delta$), and consequently this behaviour is expected to be observed at temperatures lower than that at which extended state conduction occurs. The origin of this change of conduction mechanism with temperature (two-channel model) is shown diagrammatically in Fig. 5.26(a), and the temperature dependence of σ in Fig. 5.26(b). Note that conduction among states in mid-gap has been included for completeness in Fig. 5.26; this will be discussed in detail in section 6.3.3. However, if the density of gap states is high, the conductivity due to band-tail hopping (region (ii) in Fig. 5.26(b)) may not be observed at all, a direct transition from mechanism (i) to (iii) or (iv) occurring instead.

If band-tail hopping conduction occurs at all, it is likely that *nearest-neighbour* hopping (assumed in the derivation of [5.71]) does not take place. Instead, 'variable-range hopping' will probably occur, in which an electron will hop to a more distant neighbour for which the energy difference between the states is less. The transition probability can be written as $\gamma' = \exp(-2\alpha r)\exp(-\Delta/k_B T)$, the first term reflecting the overlap between exponentially localized states separated by a distance, r, and the second the Boltzmann term for activation to overcome the energy difference, Δ, between sites. The transition probability is maximized by optimization of the exponent ($2\alpha r + \Delta/k_B T$), and the procedures for doing this will be discussed in more detail in the next chapter dealing with defects (sect. 6.3.3), since variable-range hopping conduction is much more prevalent among localized defect gap states. If E_m is the energy in the tail above the band edge, E_A, where the number of carriers is a

Table 5.4 Electrical and optical parameters for some amorphous semiconductors

Material	Preparation technique	ΔE_σ (eV)	σ(300 K) Ω^{-1} cm^{-1}	E_o(300 K) (eV)	Comments
Se	LQ	1.10[1]	$\simeq 10^{-16}$[1]	2.05[2]	—
As	VT	0.81[3]	4×10^{-9}[3]	1.4[3]	Bulk (H-transported)
P	S	1.03[4]	$\simeq 10^{-13}$[4]	1.72[4]	Annealed
Si	E	†	2×10^{-3}[5]	1.08[5]	Oxygen-free
Si:H	GD	0.6[6]	2.5×10^{-9}[6]	1.64[7]	$T_d = 600$ K. Intrinsic
Ge	E	†	$\simeq 10^{-5}$[8]	1.00[9]	Annealed at 670 K
Ge:H	GD	0.43[10]	$\simeq 10^{-4}$[10]	0.92[11]	$T_d = 500$ K. Intrinsic
As$_2$S$_3$	LQ	1.14[12]	$\simeq 10^{-17}$	2.32[13]	—
As$_2$Se$_3$	LQ	0.91[12]	$\simeq 10^{-12}$[12]	1.76[13]	—
As$_2$Te$_3$	LQ	0.42[12]	$\simeq 10^{-4}$[12]	0.83[14]	—

Key: † = $T^{-1/4}$ conductivity; VT = vapour transport; S = sputtered; E = evaporated; GD = glow discharge; LQ = liquid quench.
References: (1) J. L. Hartke (1962), *Phys. Rev.* **125**, 1177; (2) E. A. Davis (1970), *J. Non-Cryst. Sol.* **4**, 107; (3) G. N. Greaves, S. R. Elliott and E. A. Davis (1979), *Adv. Phys.* **28**, 49; (4) P. Extance and E. A. Davis (to be published); (5) S. K. Bahl and S. M. Bhagat (1975), *J. Non-Cryst. Sol.* **17**, 409; (6) W. E. Spear and P. G. Le Comber (1976), *Phil. Mag.* **33**, 935; (7) R. J. Loveland, W. E. Spear and A. Al-Sharbaty (1973/4), *J. Non-Cryst. Sol.* **13**, 55; (8) H. Mell (1974), *Proc. 5th Int. Conf. on Amorphous and Liquid Semiconductors*, J. Stuke and W. Brenig (eds) (Taylor and Francis) p. 203; (9) M. L. Theye (1974), *Proc. 5th Int. Conf. on Amorphous and Liquid Semiconductors*, J. Stuke and W. Brenig (eds) (Taylor and Francis) p. 479; (10) D. I. Jones, W. E. Spear and P. G. Le Comber (1976), *J. Non-Cryst. Sol.* **20**, 259; (11) D. I. Jones – private communication; (12) C. H. Seager and R. K. Quinn, *J. Non-Cryst. Sol.* **17**, 386 (1975); (13) N. F. Mott and E. A. Davis (1979), *Electronic Processes in Non-Crystalline Materials* (OUP); (14) K. Weiser and M. H. Brodsky (1970), *Phys. Rev.* **B1**, 791.

maximum, it can be shown that $(E_m - E_A) = sk_B T$ and from [5.70] $N(E_m) = N(E_c)(sk_B T/(E_c - E_A))^s$. The hopping mobility can then be shown to be:

$$\mu_{hop} \propto v_0 T^{-(s+3)/2} \exp(-BT^{-(s+1)/4}) \qquad [5.73]$$

where B is a constant depending on $(E_c - E_A)$ and $N(E_c)$ (Grant and Davis 1974).

Finally, we include in Table 5.4 a selection of values for the d.c. conductivity activation energy, ΔE_σ, for those materials in which extended state conduction is observed, together with the optical gap and other parameters.

5.4.2 Thermopower

Thermoelectric power (or thermopower for short) is another transport property, closely related to σ, which gives a great deal of information about transport mechanisms in amorphous semiconductors. The thermopower or Seebeck coefficient, denoted by S, is the constant of proportionality between the voltage (in the absence of a current) and the temperature gradient which causes it. The Peltier coefficient, Π, is the constant of proportionality between the heat flux transported by the electrons and the current density, and S and Π are simply related via:

$$S = \Pi/T \qquad [5.74]$$

The thermopower may be written as (Mott and Davis 1979):

$$S = -\frac{k_B}{|e|\sigma} \int \sigma_E \frac{(E - E_F)}{k_B T} \frac{\partial f(E)}{\partial E} dE \qquad [5.75]$$

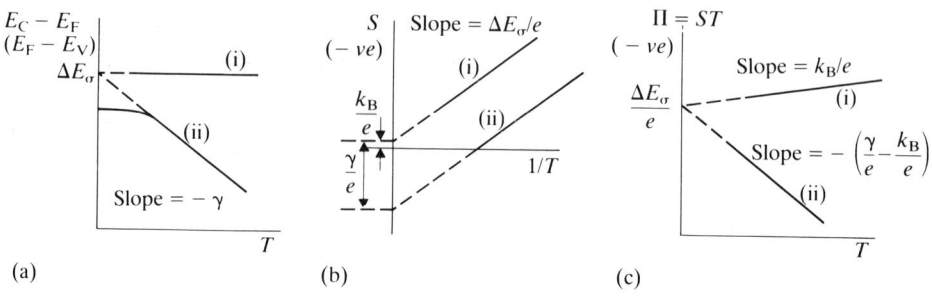

Fig. 5.27 Schematic illustration of methods of obtaining the temperature coefficient of the gap (curves (ii)); curves (i) arise if $\gamma=0$.
(a) Electrical conductivity, $E_c - E_F = \Delta E_\sigma - \gamma T$
(b) Seebeck coefficient $S = \dfrac{-k_B}{|e|}\left(\dfrac{\Delta E_\sigma}{k_B T} - \dfrac{\gamma}{k_B} + 1\right)$
(c) $\Pi = -\dfrac{\Delta E_\sigma}{|e|} + \left(\dfrac{\gamma}{|e|} - \dfrac{k_B}{|e|}\right)T$
(after Mott and Davis 1979).

where $|e|$ is the value of the electronic charge and where $\sigma_E = eN(E)\mu(E)k_B T$, as before; substituting this and [5.58] for σ gives, in the non-degenerate limit when the current path is far from E_F:

$$S = -\frac{k_B}{|e|} \frac{\int \mu(E)N(E)[(E-E_F)/k_B T]\exp[-(E-E_F)/k_B T]\,dE}{\int \mu(E)N(E)\exp[-(E-E_F)/k_B T]\,dE} \quad [5.76]$$

For the case of band conduction of electrons in extended states at and beyond E_c, assuming that $N(E)$ and $\mu(E)$ are *constant*, [5.76] can be readily integrated to give:

$$S_{\text{ext}} = -\frac{k_B}{|e|}\left(\frac{E_c - E_F}{k_B T} + A\right) \quad [5.77]$$

where $Ak_B T$ is the average energy of the transported electrons with respect to E_c; for amorphous semiconductors, and conduction in extended states, $A=1$. Note that if $(E_c - E_F)$ is temperature dependent ([5.65]), then:

$$S_{\text{ext}} = -\frac{k_B}{|e|}\left(\frac{\Delta E_\sigma}{k_B T} - \frac{\gamma}{k_B} + 1\right) \quad [5.78]$$

and it is immediately apparent that the slopes of the curves $\ln \sigma$ v.s. $1/T$ and S v.s. $1/T$ should be the *same* if conduction is in *extended* states. If both electrons and holes contribute to the conductivity, the thermopower consists of the weighted sum:

$$S = \frac{S_e \sigma_e + S_h \sigma_h}{\sigma_e + \sigma_h} \quad [5.79]$$

A most important point to note is that the sign of the thermopower is reversed if holes carry the current, and thus the thermopower is a reliable guide to the nature of the dominant charge type, whether n- or p-type. For instance, most chalcogenide glasses exhibit a positive thermopower, indicating p-type conduction. Note that, unlike the case for crystalline materials, the Hall effect *cannot* be used in this way, as we will see in the next section.

Measurements of the thermopower can also be used to obtain a value for γ, the temperature coefficient of $(E_c - E_F)$, from the *intercept* of S v.s. $1/T$ or the slope of Π v.s. T (Fig. 5.27).

A thermopower also results if current is carried by carriers hopping between localized tail states. If the density of states in this region is again assumed to vary as $N(E) \propto (E - E_A)^s$, the thermopower takes the form:

$$S = -\frac{k_B}{|e|}\left(\frac{E_A - E_F}{k_B T} + K\right) \qquad [5.80]$$

where K is a constant different from A in [5.77]. Comparison of [5.71] and [5.80] shows that the activation energy for conduction and the slope of S v.s. $1/T$ *differ* by the hopping energy Δ if conduction takes place in the band tails.

In the 'two-channel model' for transport (see e.g. Nagels 1979), where the conduction path changes from extended states to localized tail states with decreasing temperature, S is expected to change over from a slope $(E_c - E_F)$ to $(E_A - E_F)$ sharply at the same temperature that the conductivity shows a kink plotted as $\ln \sigma$ v.s. $1/T$. This behaviour has been observed in a-As, an n-type material, and the experimental curves for σ and S (as well as the Hall mobility) as a function of temperature are shown in Fig. 5.28.

This two-channel model of conduction has been called into question recently, at least for the case of a-Si:H, which comes closest to the realization of an 'ideal' amorphous semiconductor. Overhof and Beyer (1981) pointed out that the effects of any shifts in energy of E_F, E_c or E_A with temperature or doping are cancelled if the function $Q(T)$ is considered:

$$Q(T) = \ln \sigma(T) + eS(T)/k_B \qquad [5.81]$$

where $\sigma = \sigma_0 \exp[-(E_\sigma - E_F)/k_B T]$ and

$$S = -\frac{k_B}{|e|}\left[\frac{(E_s - E_F)}{k_B T} + A\right]$$

The two-channel model would predict that $Q(T)$ v.s. $1/T$ would be 'S-shaped' whereas Overhof and Beyer report linear behaviour, viz.

$$Q(T) = C - E_Q/k_B T \qquad [5.82]$$

where $C = \ln \sigma_0 + A$ and $E_Q = E_\sigma - E_s$ (0.05 eV $< E_Q <$ 0.25 eV). They ascribe this behaviour to local modulation of the mobility edge, E_c, by the long-range screened electrostatic potential of randomly distributed charged centres. A percolation treatment is used to argue that E_Q should be roughly equal to the disorder introduced by the charge centres, and assuming transport occurs only in extended states, $Q(T)$ can be calculated. This model of long-range fluctuations is then able to account qualitatively for the puzzling correlation between conductivity pre-factor σ_0 and activation energy ΔE_σ that has been observed for undoped a-Si:H films prepared by glow-discharge decomposition (see Fig. 5.29). Spear *et al.* (1980), on the

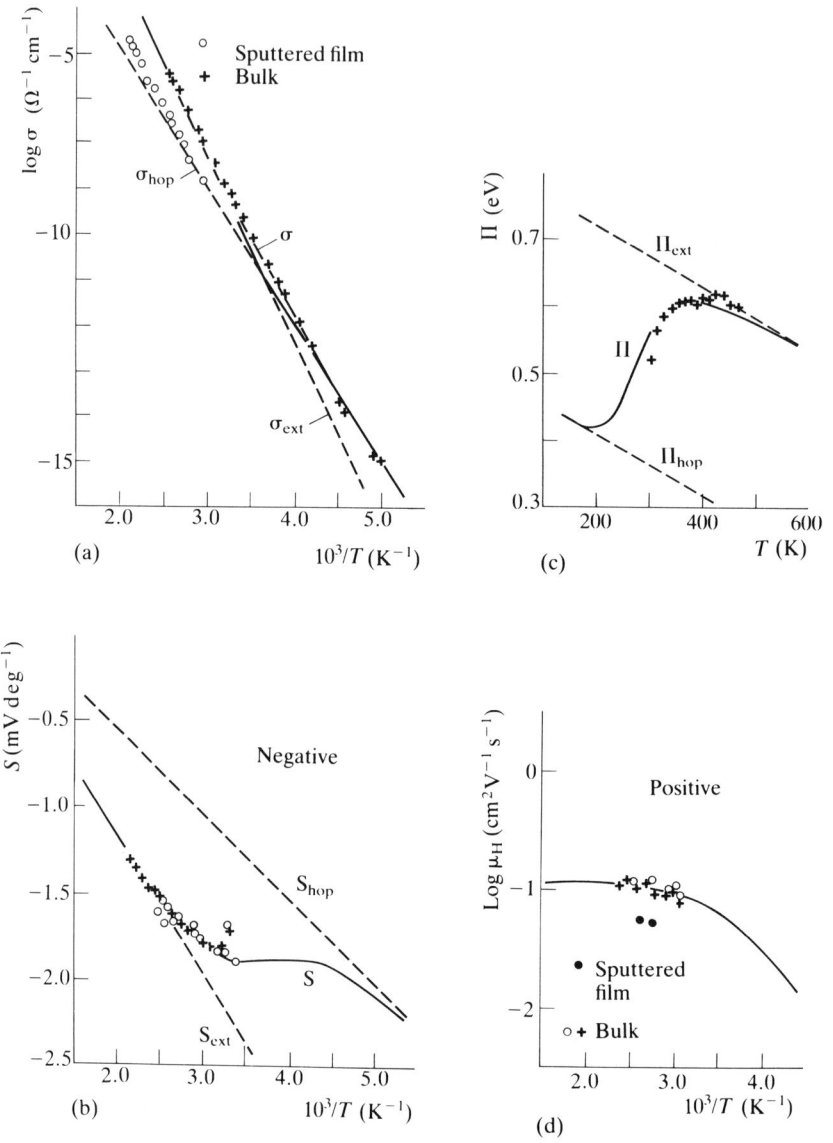

Fig. 5.28 Conductivity (σ), thermopower (S), Peltier coefficient (Π) and Hall mobility (μ_H) as a function of temperature for amorphous arsenic. The solid lines are theoretical fits assuming two conduction paths, at E_c and at E_A (Mytilineou and Davis 1977).

other hand, ascribe this correlation to a shift in the mobility edge with temperature, $E_c(T)$. They suppose this occurs because the overlap integral, J, increases with temperature due to an increase in thermal fluctuations, and hence E_c decreases to keep the localization condition, $W/2zJ$, constant. This model leads to a *linear* dependence of $\ln \sigma_0$ on ΔE_σ, with the constant of proportionality $G \simeq \delta_c/k_B(E_c(0) - E_A)$ where δ_c is the linear temperature coefficient of the shift of E_c from its position at $T=0$, $E_c(0)$. This appears on present evidence to be a better fit to

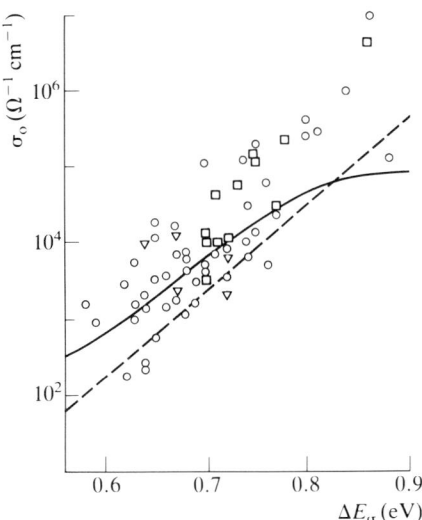

Fig. 5.29 Plot of prefactor versus slope for undoped GD a-Si:H films (Spear *et al.* 1980). The solid line is the fit to the theory of Overhof and Beyer (1981) and the dashed line is the fit to the theory of Spear *et al.* (1980) and has a slope $G = 27\,\text{eV}^{-1}$.

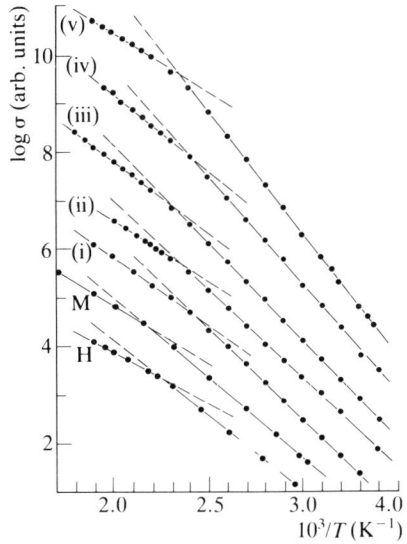

Fig. 5.30 Temperature dependence of the d.c. electrical conductivity of a-Si:H for various samples showing the 'kink' at high temperatures. The curves have been shifted vertically to facilitate comparison. Curves (i)–(v) are GD a-Si:H from Dundee, M is from Marburg, and H is a sputtered a-Ge:H sample from Harvard Laboratories, respectively (Spear *et al.* 1980).

the data (Fig. 5.29), although both models are at an early stage in their development and it is difficult to see which is the more correct. The Spear theory has the added advantage that it can also explain the kink in the *high* temperature conductivity of undoped a-Si:H (Fig. 5.30), which is in the *opposite* sense to that in Fig. 5.26(b) associated with the transition between extended-state and tail-state conduction. They ascribe the kink in Fig. 5.30 to the point at which the mobility edge has decreased to the band edge at E_A, and the slope is predicted to change from ΔE_σ ($= E_c(0) - E_F(0)$) to $\Delta E_\sigma - [E_c(0) - E_A]$.

5.4.3 Hall effect

Measurements of the Hall voltage are a natural complement to conductivity measurements in transport studies of *crystalline* semiconductors since, if unipolar conduction occurs, the Hall coefficient provides a reliable estimate of the carrier concentration n and the carrier type:

$$R_H = r/ne \qquad [5.83]$$

where r is the 'scattering factor' (of the order unity) equal to the ratio of the Hall and conductivity mobilities, μ_H/μ_e; R_H is negative for electrons and positive for holes. Equation 5.83 is valid only if the mean free path is sufficiently long for Boltzmann transport theory to apply; it is *not* appropriate for amorphous semiconductors. It is found, moreover, that the sign of R_H is always *opposite* to that of the thermopower in all the amorphous semiconductors examined so far except for certain compositions in the alloy system, $CdGe_xAs_2$, in which both n-type S and R_H are observed. This 'sign anomaly' is especially dramatic in a-Si:H where it is found that on changing the dopant from B (p-type) to P (n-type), the sign of the Hall effect changes from negative to positive in opposition to the change in sign of the thermopower (Le Comber et al. 1977). The sign of R_H for P-doped a-Si:H reverts however to its 'proper' negative value on recrystallization (Reilly and Spear 1978). This behaviour, completely at variance with that observed in crystals, is a major challenge to any theory of the Hall effect.

Consider first the case of carriers excited to a mobility edge where, although the states are extended, the scattering is very strong and the mean free path is very short. It is a central feature of all calculations that transitions between *two* sites do *not* contribute to the Hall voltage; instead a minimum of *three* sites is necessary, for which the carrier is given a choice of which way to proceed in the direction perpendicular to both the applied electric and magnetic fields, **E** and **H**, either moving directly between initial and final states, or alternatively via occupation of one or more other sites as virtual states. The three sites contributing to the Hall effect may be three of the atoms (or their bonds) forming a ring in the structure, or alternatively for group IV or V elements, they may be three of the bonds emanating from a single given atom (see Fig. 5.32).

Using the random phase model, Friedman (1971) first showed that in this regime the Hall mobility should be

$$\mu_H \simeq 2\pi \left(\frac{ea^2}{\hbar}\right)\left(\frac{B_0}{B}\right)\left(\frac{\eta \bar{z}}{z^2}\right)$$

Here B_0 is the bandwidth in the absence of disorder ($=2zJ$), where z is the coordination number and J the overlap integral, B is the bandwidth with disorder ($\simeq (a^3 N(E_c))^{-1}$), $\bar{z}$ is the average number of closed three-site paths about an arbitrarily chosen site, and η is the projection of the area of a three-site path in the direction perpendicular to the field ($\sim 1/3$ for 3-D). Putting all these factors together gives:

$$\mu_H \simeq \frac{4\pi e a^5}{3\hbar} JN(E_c)\left(\frac{\bar{z}}{z}\right) \qquad [5.84]$$

It has been assumed that the phase correlation of wavefunctions extends to nearest neighbours, but that there is no correlation between the phase differences of distinct pairs of sites, which is a somewhat less stringent condition than that used in the original random phase model, for which the phases of individual site wavefunctions are uncorrelated and hence J would vanish on averaging over the random phases. Note that [5.84] predicts that μ_H should be *temperature independent* (in contrast to $\mu_e \propto 1/T$, [5.62–5.64]), and if it is assumed that $\bar{z} \simeq z$ and $a = 3$ Å, $\mu_H \simeq 10^{-1}$ cm^2 V^{-1} s^{-1} (cf. $\mu_e \simeq 10$ cm^2 V^{-1} s^{-1}). Comparison with the random phase model prediction for μ_e ([5.64]), indicates that:

$$\frac{\mu_H}{\mu_e} \simeq \frac{2 k_B T}{z\; J} \qquad [5.85]$$

where generally $J \gg k_B T$.

The Hall mobility is commonly found to decrease with decreasing temperature from a temperature-independent value at high temperatures. In the 'two-channel' model, this observation is reconciled with the random phase model results by assuming that, at high temperatures, carriers are excited to the mobility edge, [5.84] should be obeyed, and μ_H should be temperature independent. It was tacitly assumed for a long time that the Hall mobility for carriers hopping among localized states (in the band tails) would be negligibly small, and hence the decrease in μ_H at low temperatures (see Fig. 5.28(d)) is simply a reflection of the decreasing number of carriers excited to the mobility edge. The total Hall mobility, like the thermopower ([5.79]) is a sum of various contributions, weighted by the conductivity:

$$\mu_H^{Tot} = \frac{\mu_H^{ext} \sigma_{ext} + \mu_H^{hop} \sigma_{hop}}{\sigma_{ext} + \sigma_{hop}} \qquad [5.86]$$

where, if $\mu_H^{hop} \ll \mu_H^{ext}$, and considering only electron conduction (Nagels 1979):

$$\mu_H^{Tot} \simeq \frac{\mu_H^{ext}}{\left[1 + \dfrac{\sigma_0^{hop}}{\sigma_0} \exp(E_c - E_A - W)/k_B T\right]} \qquad [5.87]$$

There has been much recent work on the size of the Hall effect expected for carriers hopping among disorder-localized states. The consensus of opinion seems to be that μ_H^{hop} is very small, but not negligible ($\sim 10^{-4}$ cm^2 V^{-1} s^{-1} at room temperature), although various authors differ as to the precise temperature dependence to be expected. An illustrative argument as to why μ_H should be small for carriers hopping among disorder-localized states is indicated in Fig. 5.31; a microscopic Hall voltage will only develop at *junctions* in the percolation cluster, i.e. where percolation paths intersect (Mott et al. 1975a), and it is expected that these are few for the case of hopping conduction.

We turn our attention now to the intriguing problem of the sign anomaly. It

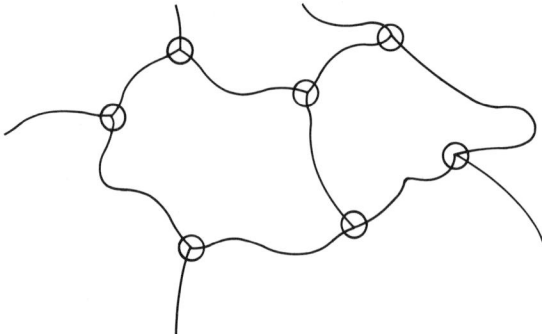

Fig. 5.31 Percolation cluster showing three-site junctions.

was originally believed (Friedmann 1971, Holstein 1973) that the sign of the Hall effect should be *negative* for *both* electrons and holes for a three-site process. The ratio of the signs of R_H for electrons and holes moving in the same band was shown to be equal to $(-1)^{n+1}$, where n is the number of transfer integrals involved in the closed loop process; for three-site interference $n=3$ and hence $(-1)^{n+1} = +1$, and both holes and electrons should have the same sign of R_H, namely negative. This is obviously contrary to experimental observation, in which several materials exhibit a positive R_H.

Emin (1977) showed that the absolute sign of R_H is given by:

$$\text{Sign}(R_H) = \text{Sign}\left[\varepsilon^{n+1} \prod_{i=1}^{n} J_{i,i+1}\right] \quad [5.88]$$

where $\varepsilon = +1$ if the carrier is an electron moving in conduction band states or $\varepsilon = -1$ if the carrier is a hole in valence band states, and $J_{i,i+1}$ denotes the transfer integral connecting orbitals i and $i+1$. $J_{i,i+1}$ is *negative* for interactions between bonding orbitals (since the energy entering into the transfer integral, the local atomic potential, is negative), but *positive* for interactions between antibonding orbitals. This approach is powerful in that the sign of R_H depends on the local geometry, as well as on the nature and relative orientation of the local orbitals between which the carrier moves. We consider several cases, shown schematically in Fig. 5.32. For atoms comprising a threefold ring, connected by s- or p-like (or hybrid) σ bonding orbitals, the sign of all the orbitals is the same, say positive (Fig. 5.32(a)), and hence J is negative, and if holes are the carriers ($\varepsilon = -1$), [5.88] gives a *negative* R_H (i.e. n-type), although the thermopower would be positive. For the case of electron transport between conduction (antibonding) states in a threefold ring (Fig. 5.32(b)), J is positive, $\varepsilon = +1$, and R_H is predicted to be *positive*, i.e. p-type. Chalcogenide glasses almost invariably exhibit p-type thermopower, but an n-type Hall coefficient. This might appear to an example of case (a), but the top of the valence band in these materials is composed of non-bonding, lone-pair orbitals, rather than bonding orbitals. This apparent contradiction can be understood if the atoms do form a threefold ring formed by p-like σ-bonding orbitals, but with the non-bonding p-orbitals forming π-bonds in addition; in this case, situation (a) is recovered. Alternatively, the three-site interaction may take place between three directed

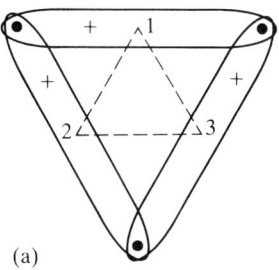 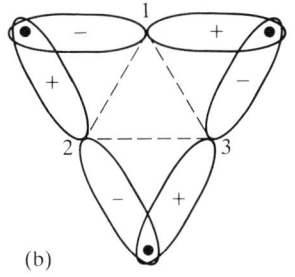 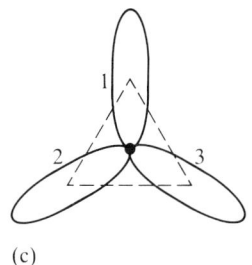

Fig. 5.32 Schematic representation of three-site interactions which can give rise to a Hall effect. Rings between three atoms comprised of
(a) bonding orbitals and
(b) antibonding orbitals;
(c) three directed orbitals emanating from a single atom.

orbitals emanating from the *same* atom (Fig. 5.32(c)), for which J is negative, and if holes are the carriers ($\varepsilon = -1$), a negative R_H is also predicted, although the Hall mobility for this case would be small. A negative R_H would also arise if lone-pair p-orbitals on different atoms were arranged such that they are directed towards a common point or aligned roughly parallel.

Thus far we have developed a theory for the Hall effect, and other transport properties, in terms of Mott's model of the density of states of an amorphous semiconductor, incorporating mobility edges. We must mention at this juncture that an alternative theory has been developed in terms of the motion of 'small polarons', and this will be discussed more fully in a later section.

5.4.4 Ionic conduction

Although in many non-metallic amorphous solids current is carried by electrons, in others containing a relatively large proportion of ions, particularly alkalis, *ionic conduction* may make a significant, even dominant, contribution to the total conductivity. Since many glasses, based particularly around the silicate compositions, often do contain a large amount of alkali ions, the likelihood of ionic conduction is obviously quite high, and we will therefore discuss this important form of conduction in what follows. A section on ionic conduction falls somewhat uneasily in a chapter dealing with electrons, but it is of interest to compare directly ionic and electronic conduction in non-crystalline systems, and the discovery of very high conductivities in certain glasses (so-called 'superionic' behaviour) offers the promise of technological applications and deserves discussion.

Ionic conduction is well known in crystalline solids (for a review see Salamon 1979, Chandra 1981), and both cations and anions can carry current in certain circumstances. Materials are called *superionic* if the conductivity is comparable with that of liquid electrolytes ($\sim 10^{-1}$–10^{-4} Ω^{-1} cm^{-1}); the mechanism giving rise to the (super)ionic conduction may be either of the 'point defect' or the 'molten sublattice' type. In the former, transport is through thermally generated Frenkel or Schottky defect pairs and conduction is therefore thermally activated; in the latter the number of ions of a particular type is less than the number of available sites in their sublattice and therefore a large number of ions are able to conduct with a lower activation energy than for the defect process. A simple expression for the ionic conductivity may be derived by assuming that the ionic motion occurs by field-enhanced thermally activated hopping between equivalent sites in the lattice. The action of the applied electric field E is to lower the energy of a site by an amount equal to qEa, where q is the charge on the ion and a is the intersite spacing. Therefore the probability of ionic hopping in the direction of the field increases, and the net number of ions per volume hopping in the direction of the field can be shown to be:

$$\phi = \frac{nv_{ion}qaE}{k_B T} \exp[-W/k_B T]$$

where n is the number of available ions per unit volume, v_{ion} is the vibrational frequency for an atom constrained in the potential well at a site, W is the barrier height between sites, and it has been assumed that $qaE \ll k_B T$. The current density is then just $j = qa\phi$ and the conductivity is given by $\sigma = j/E$, i.e.:

$$\sigma_{ion} = \frac{na^2 q^2 v_{ion}}{k_B T} \exp[-W/k_B T] \qquad [5.89]$$

Thus a plot of $\ln \sigma T$ v.s. $1/T$ should be linear if ionic conduction is dominant. Equation 5.89 has been derived using a very simple model, and further complications involving ion–ion interactions, etc., should be employed to give a more realistic theory. Nevertheless, the form of [5.89] is borne out by experiment. One final point concerning superionic conduction in crystalline solids is worthy of comment. It is noticeable that many materials exhibiting high ionic conductivities have pronounced interconnecting 'tunnels' or channels in the crystal structure which facilitate ion transport.

Many ion-containing glasses have been found to exhibit ionic conductivities behaving according to [5.89]. The activation energies in these cases must now be some form of average value since site equivalence no longer exists as for crystalline solids. Intriguingly, superionic behaviour has also been discovered in certain glassy systems, even though presumably there cannot exist the sort of interconnecting channels in the structure that occur in superionic crystals. Examples of these highly conducting ionic glasses are the mixed silver halide–silver oxosalt system $AgI-Ag_2XO_4$ (where $X = As, Cr$) which have room temperature conductivities in the range $10^{-2}-10^{-1} \Omega^{-1}$ cm^{-1} (Ingram and Vincent 1977), lithium-based niobate and tantalate and other glasses prepared by rapid-quenching techniques similar to those used for metallic glasses (Glass *et al.* 1978), and another lithium system, $LiCl-Li_2O-B_2O_3$ (Levasseur *et al.* 1977). These glassy materials have several advantages over the corresponding crystalline materials: the ionic conductivity is isotropic and not confined to certain crystallographic directions; the materials can often be prepared readily in thin-film form facilitating technological applications; and importantly the magnitude of the conductivity for most superionic glasses is *greater* than that of the devitrified polycrystalline product of the same composition.

The final topic we wish to discuss in this section is the so-called 'mixed-alkali effect'. Naively, it might be thought that as the relative proportion of a binary mixture of alkali oxide modifiers, added to a glass-former, is varied, the physical

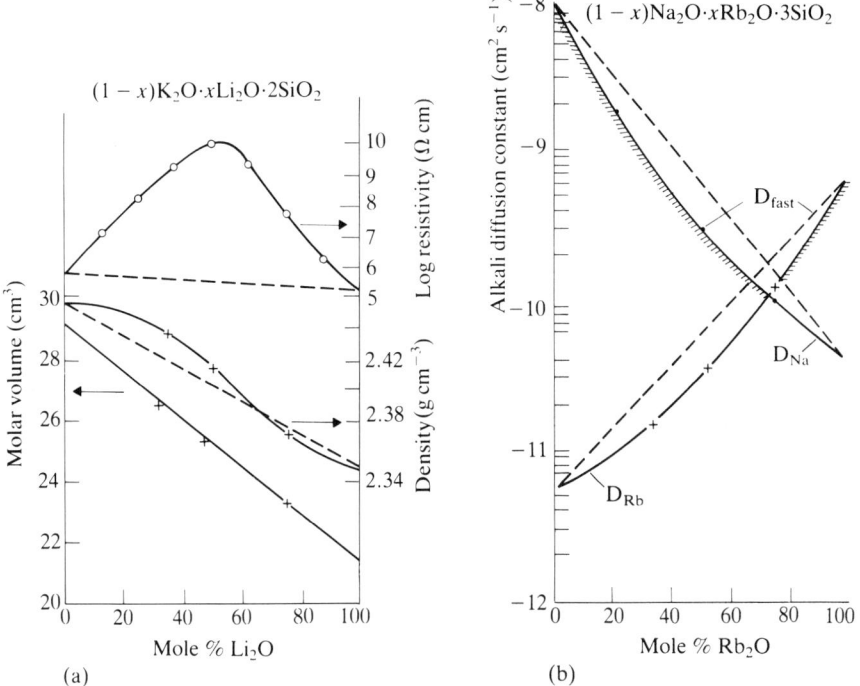

Fig. 5.33 The 'mixed-alkali' effect in silicate glasses.
(a) Electrical resistivity (150 °C) compared with molar volume (20 °C) and density (20 °C) as a function Li:K concentration (Day 1976).
(b) Plots of the alkali diffusion constants (400 °C) as a function of Na:Rb concentration. The hatched line corresponds to the dominant diffusion coefficient at a given concentration.

properties should change in a *linear* fashion from one extreme composition to the other. This is indeed the case for the molar volume and approximately true for the density; however, properties which are connected with alkali ion motion, such as conductivity, viscosity and dielectric loss, vary in an extremely non-linear fashion (Isard 1969, Day 1976). The mixed-alkali effect, which is this departure from additivity, is strikingly shown by the ionic resistivity in Fig. 5.33(a) for the K_2O–Li_2O–SiO_2 system. It can be seen that while the density deviates from linearity by $\leqslant 10\%$, the ionic resistivity shows a positive deviation from simple additivity by five orders of magnitude and exhibits a maximum at some intermediate composition. The extremal behaviour of the resistivity is caused in the main part by an equivalent maximum in the conduction activation energy, somewhat ameliorated by a minimum in the pre-exponential factor (ρ_0) at the same composition. The fact that for example the density does not pass through a similar extremum with the addition of a second alkali is a strong indication that the mixed-alkali effect is not due to gross structural changes, at least in the glass-forming network. Instead, it is the ionic mobility which is directly affected by the addition of a second alkali, and this is reflected in the behaviour of the diffusion constant. It is found invariably that the addition of a second alkali *lowers* the diffusion constant of the original alkali ion. This is demonstrated in Fig. 5.33(b) for the system Na_2O–Rb_2O–SiO_2, and the consequence of this behaviour is that the overall diffusion coefficient shows a minimum at some intermediate composition. It is therefore to be expected that transport properties should also exhibit an extremum, i.e. a minimum in the conductivity (or a maximum in the resistivity) but not necessarily at the same composition. Furthermore, the magnitude of the departure from additivity increases as the size (or mass) difference (or ratio) increases, i.e. the mixed-alkali effect is larger in the Cs_2O–Li_2O–SiO_2 system than in the Cs_2O–Rb_2O–SiO_2 system. Lastly, it is interesting to observe that the self-diffusion constant of an alkali ion whose concentration is in excess is always considerably higher than the diffusion constant of the impurity alkali (cf. Fig. 5.33(b)); this is in marked contrast to the case of metals where interstitial impurities always, and substitutional impurities often, have diffusion coefficients larger than that for self-diffusion.

Many theories have been proposed for this intriguing effect, but only those which address directly the problem of ionic diffusion and mobility, rather than that of the glass structure, would seem to offer any promise of success. A possible model for this effect has been proposed by Hendrickson and Bray (1972). They assert that the increase in activation energy for ionic mobility (and which subsequently occurs in expressions for conductivity, viscosity and the diffusion constant) arises from a dipole–dipole interaction. It is assumed that every alkali ion is bound to a negatively charged non-bridging oxygen atom, thereby forming dipole pairs. The interaction between these dipoles acts as a perturbation on the ion-pair oscillator, producing an energy splitting whose magnitude is approximately twice the interaction energy, and resulting in an extra additive term in the activation energy for mobility if the lower state is preferentially populated. This approach, which is based on perturbations to oscillator energies, therefore predicts that ions with different masses should have different activation energies, and hence a 'mixed isotope effect' should occur; the authors claim to observe such an effect in Li^6- and Li^7-containing borate glasses in a study of ionic diffusion by NMR. In fitting mobility and conductivity data exhibiting the mixed-alkali effect, satisfactory agreement was achieved only by assuming multiple ion–ion interations within a rather arbitrary interaction volume.

5.5 Small polarons

A charge carrier in a given state will always distort its surroundings, and the distortion will be more pronounced the more the carrier is localized. The distortion will tend to lower the energy of the carrier, and in extreme cases will lead to the carrier being self-trapped by the distortion its presence causes, and a 'polaron' is formed. The name comes from the early conjecture that self-trapping of an excess carrier might occur in a crystalline ionic (or polar) lattice; this circumstance is shown schematically in Fig. 5.34(a), and we refer to it as a 'dielectric polaron'. Polaron formation in its widest sense of course is not restricted solely to polar materials, but can occur also in covalent materials, in which the distortion is confined to neighbouring atoms which form some sort of bond when the carrier is trapped. Examples of this are V_k centres in alkali halide crystals, in which a hole trapped at a Cl^- ion causes a neighbouring Cl^- to be attracted (Fig. 5.34(b)); a similar

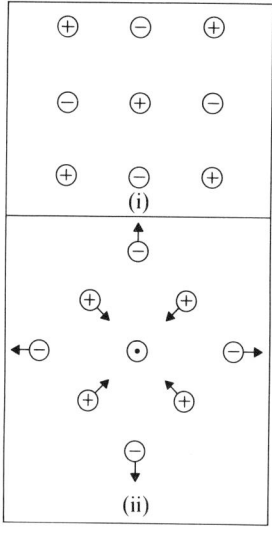 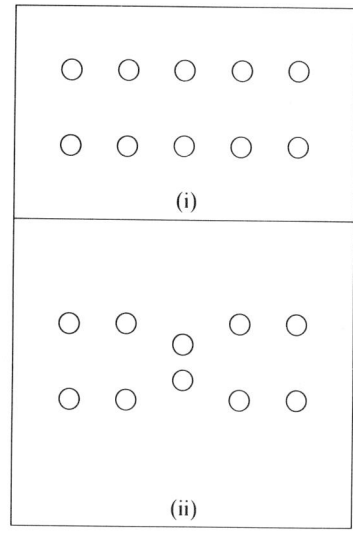

Fig. 5.34 Schematic illustration of polaron formation.
(a) 'Dielectric' polaron, in which an excess electron added to the ionic lattice shown in (i) causes the ionic readjustment shown in (ii);
(b) 'Molecular' polaron, in which an excess hole added to an array of rare gas atoms (i) causes the atomic rearrangement shown in (ii).

circumstance is for holes trapped in solid rare gases, and dangling bonds in chalcogenide glasses (see sect. 6.2.2) – Mott and Davis (1979) refer to these as 'molecular polarons'. The spatial extent of the distortion may be considerable, or it may be localized in the vicinity of the carrier, giving rise to polarons termed, not surprisingly, large and small, respectively. In both cases, the polaron has a larger effective mass than the free carrier since the induced distortion is carried with it as the carrier moves through the lattice. Emin has argued that the transport properties of many non-crystalline semiconductors can be understood on the basis of the motion of small polarons, without the need to introduce a mobility edge (for reviews see Emin 1973 and 1976 and Austin and Mott 1969); we will examine this proposal in what follows.

In the *adiabatic* regime, in which, because the electronic overlap (transfer) integral, J, is large, the vibrational motion of the atoms is so slow compared with that of an electron that the electron adjusts to the instantaneous position of the atoms (i.e. the Born–Oppenheimer approximation is obeyed), we may treat the system in the following manner. The energy of the system may be written in terms of a 'configurational coordinate', q, which may be taken to be the *displacement* of an atom in a molecular polaron of the V_k type (Fig. 5.34(b)), or the *dilatation* within the volume of a trap in band-tail states driven by the deformation potential (in which the potential energy of a carrier is proportional to the volume strain), or a measure of the polarization of the medium outside the trap for polarons of the dielectric variety (Fig. 5.34(a)). In these cases, and for a one-dimensional system, the lattice energy is of the form Aq^2, and if we assume that the energy of the electron (or hole) varies *linearly*

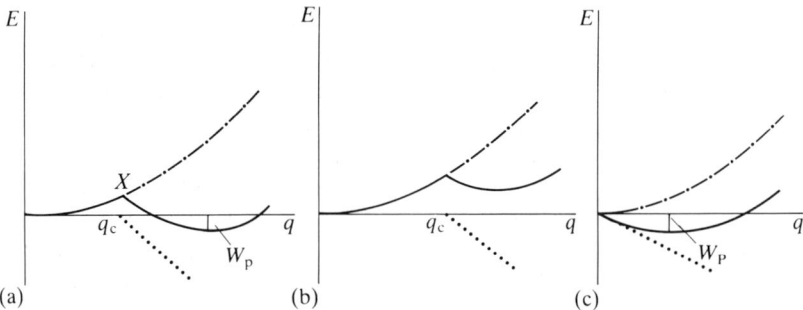

Fig. 5.35 The energy of a charge carrier in a deformable three-dimensional lattice. Solid curves represent the total energy as a function of the configurational coordinate q made up of a sum of two terms, electronic energy (dotted curve) and elastic energy (dash-dot curve).
Formation of a V_k-type polaron, (a) stable and (b) unstable. (c) Same quantities for a carrier trapped at a defect (after Mott and Davis 1979).

with q for small q as $-Bq$, the energy of the system is then $Aq^2 - Bq$. This is a minimum at $q = q_0$, where $q_0 = B/2A$, and thus the energy of the whole system is lowered by an amount termed the 'polaron energy', $-W_p$ (see Fig. 5.35(c)):

$$W_p = B^2/4A \qquad [5.90]$$

For a tail state, B is the deformation potential E_1, and $A = \tfrac{1}{2}\kappa_T a^3$, where κ_T is the bulk modulus and a^3 is of the order of the volume of the trap. Thus, in this case:

$$W_p = \frac{E_1^2}{2\kappa_T a^3} \qquad [5.91]$$

These considerations apply only to the one-dimensional 'molecular crystal model' (Fig. 5.34(b)) treated originally by Holstein (1959), or to a one-dimensional continuum model. In three dimensions, there is no stable bound state with a finite radius of distortion for a continuum model. This situation is analagous to the case of a quantum (square) well which only has a bound state if the depth and width are greater than certain critical values. Exactly the same is true here: a stable polaron (with $W_p < 0$) is formed only if q is greater than a critical value q_c (see Fig. 5.35(a)). However, if q_c is large, no stable polaron can form (Fig. 5.35(b)). On the other hand, it is believed (see e.g. Mott and Davis 1979) that for a strongly localized *defect*, $q_c = 0$, and energy will always be gained by distortion and a self-trapped carrier will be stable.

We now consider how a small polaron may move through a lattice, and to do so we focus first on a polaron in the molecular crystal model moving through a periodic lattice, for which there are two distinct modes of transport. The first arises when the carrier and its accompanying lattice distortion move between sites *without* emission or absorption of phonons; such a process is called 'diagonal'. The eigenstates of such a deformable lattice can be expressed by Bloch-type wavefunctions, as for a rigid lattice, but the bandwidth is $12Je^{-S}$ for a simple cubic lattice in the tight-binding scheme, where J is the electronic overlap integral and e^{-S} is the overlap integral between the two wavefunctions which describe the vibrational state of the system when the carrier resides on one or other of two neighbouring sites. This is reduced from the rigid-lattice bandwidth, $12J$, by the factor e^{-S} which can be considerable ($\sim 10^{-4}$–10^{-5}), and so the small-polaron bandwidth is very narrow (or alternatively the effective mass is high). The small polarons in this limit move as (heavy) free carriers, suffering occasional scattering by atomic vibrations. However, if the vibrational energies (or the random fluctuations in the site energies for a disordered solid) are comparable to the small-polaron bandwidth, 'non-diagonal' processes become dominant, in which phonons are exchanged when the carrier moves to a neighbouring site; such a process is termed 'hopping'. Thus, in crystals, small-polaron band motion will only occur at low temperatures ($k_B T \lesssim 10^{-4}$ eV), and it is debatable whether in amorphous semiconductors it should ever occur since the fluctuations in site energies in the band tails almost certainly must be much greater than $\sim 10^{-4}$ eV.

For a hopping event to occur, there must be an atomic distortion which momentarily brings the electronic energy level of an occupied site into coincidence with a neighbouring unoccupied site (Fig. 5.36), and this appreciable thermal distortion energy is the activation energy for the hopping process. In the simplest

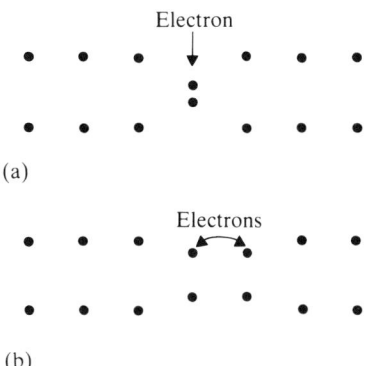

Fig. 5.36 Schematic illustration of a molecular polaron (a) and the excited state (b), i.e. the momentary coincidence of two polaron wells, which must be formed for an electron hop to take place.

case, when the energy difference between electrons on neighbouring sites is zero ($W_D = 0$) and if the centres are not so close that the distortion regions overlap, then it can be shown (see problem 5.2) that the hopping energy is half the polaron energy:

$$W_H = \tfrac{1}{2} W_p \qquad [5.92]$$

The mobility of hopping polarons μ_P may be obtained from the Einstein relation and the jump probability P. The latter consists of the product of two terms: P_1, the probability of a coincidence event,

$$P_1 = v_0 \exp(-W_H / k_B T) \qquad [5.93]$$

where v_0 is a typical phonon frequency; P_2, the probability of charge transfer (Holstein 1959), in the *non-adiabatic* regime (in which during each fluctuation the chance of an electron transfer is small) is

$$P_2 = \frac{1}{\hbar v_0}\left(\frac{\pi}{W_H k_B T}\right)^{1/2} J^2 \qquad [5.94]$$

where J, the electronic overlap integral, can be written as $J = J_0 \exp(-\alpha r)$, whereas in the *adiabatic* regime (in which during the time that an activated state persists, the electron makes several transitions back and forth), the probability of charge transfer is large, i.e. $P_2 = 1$. If we assume non-adiabatic motion, the mobility is finally:

$$\mu_p = \frac{ea^2}{k_B T \hbar}\left(\frac{\pi}{W_H k_B T}\right)^{1/2} J^2 \exp(-W_H/k_B T) \qquad [5.95]$$

The thermally activated nature of the carrier mobility is an essential feature of small-polaron theory (cf. the predicted $1/T$ behaviour for free carriers excited beyond the mobility edge in the absence of polaron formation, [5.62–5.64]). In the high-temperature regime in which hopping transport is dominant, it is expected that the d.c. conductivity should vary as:

$$\sigma = \sigma_0 \exp\left[-\frac{(E + W_H)}{k_B T}\right] \qquad [5.96]$$

where E is the activation energy required to produce an equilibrium number of carriers, and the pre-exponential factor is given by:

$$\sigma_0 = \frac{Ne^2 a^2 \nu P}{k_B T} \qquad [5.97]$$

where the polaron hopping activation energy from the term P_1 ([5.93]) has been made explicit in [5.96]. Note that although the mobility in the small-polaron picture is very small, the carrier density, $N \simeq 10^{22}$ cm^{-3}, is much higher than in non-polaronic theories to account for the experimentally observed pre-exponential factors.

The Hall effect for small-polaron transport has also been calculated. Band motion will result in the usual formula for the Hall coefficient R_H being obeyed ([5.83]) if $k_B T$ is small compared with the polaron bandwidth. For the case of small-polaron hopping transport, the three (or more) site interference mechanism must be employed, and for three-site interference in the non-adiabatic regime (Friedman and Holstein 1963):

$$\mu_H = \frac{ea^2}{\hbar} J \left[\frac{\pi}{12 k_B T W_H}\right]^{1/2} \exp(-W_H/3k_B T) \qquad [5.98]$$

Note that the activation energy for the Hall mobility is predicted to be *one-third* that of the conductivity mobility μ_p (cf. [5.95]), and since generally $J \gg k_B T$, therefore $\mu_H > \mu_p$. The site geometry, however, is crucial to the exact temperature dependence that is predicted, since $\mu_H \propto T^{-3/2} \exp(-W_H/3k_B T)$ for cubic geometry (see Emin 1973). The decrease in Hall mobility with temperature at low temperatures observed in many amorphous semiconductors (see e.g. Fig. 5.28(d)) has been interpreted by

Emin in terms of small-polaron transport (hole-like for the case of chalcogenide glasses). Consequently, μ_H is thermally activated according to [5.98], rather than decreasing with temperature due to the two-channel process favoured by Nagels. The various considerations pertinent to the sign of R_H discussed in section 5.3.3 are also applicable to small polarons (and indeed were first formulated for that case).

Another feature of electron transport which small-polaron theory is capable of explaining, without recourse to the concept of mobility edges, is the difference between the activation energy for conduction and that for thermopower often observed in amorphous semiconductors. In the mobility edge model, this is accounted for by assuming transport occurs by hopping in the band-tail states (cf. [5.71] and [5.80]). No such assumption is required in the small-polaron model, since even in the hopping regime, the polaron hopping energy W_H does *not* enter into the expression for the thermopower (see Emin 1973, 1976), which is found to be of the form (cf. [5.96]):

$$S = -\frac{k_B}{|e|}\left(\frac{E}{k_B T} + A\right) \qquad [5.99]$$

This equation is valid if the initial and final states interact equally with lattice vibrations (i.e. there is *no* vibrational energy transferred with a hop); the 'heat of transport' term, A, depends on the detailed density of states involved in the hopping, tending to zero if the bandwidth is much less than $k_B T$, but is generally expected to be of the order 1–10.

There is still considerable controversy as to whether electron transport in chalcogenide glasses is best described in terms of the mobility edge picture or in terms of small-polaron motion (Mott and Davis 1979). However, there are some glasses in which small-polaron formation indisputably occurs; these contain transition metal ions in two charge states, e.g. V^{4+}/V^{5+}, Fe^{2+}/Fe^{3+}, Cu^+/Cu^{2+}, in glassy systems such as V_2O_5–P_2O_5 and FeO–P_2O_5. Conduction is by the motion of electrons between, say, V^{4+} and V^{5+} ions and which, since lattice distortion is important, form small polarons.

Mott has suggested that for this process the d.c. electrical conductivity should be given by (Mott and Davis 1979):

$$\sigma = c(1-c)\frac{e^2 v_{el}}{R k_B T} \exp\left[-2\alpha R - W/k_B T\right] \qquad [5.100]$$

where c is the concentration of one type of ion, R is the interionic separation and α^{-1} is the radius of the localized wavefunction. The activation energy is given by:

$$W = W_H + \tfrac{1}{2}W_D \qquad [5.101]$$

for the case when there is a disorder energy W_D between sites. Mixed vanadate–phosphate glasses appear to obey [5.100] reasonably well, and exhibit an activation energy for conduction which decreases with decreasing temperature (see Mott and Davis 1979). This behaviour can be interpreted either according to conventional small-polaron theory which dictates that the activation energy should decrease to zero (band conduction) at low temperatures, or else in terms of 'variable-range hopping' of the small polarons (see sect. 6.3.3) which also predicts a

Electrons

continuously decreasing activation energy with decreasing temperature (Greaves 1973). The thermopower of vanadate–phosphate glasses obeys the Heikes–Ure (1961) formula:

$$S = \frac{k_B}{e} \ln\left(\frac{c}{1-c}\right) \qquad [5.102]$$

This equation is only valid if $W_D \ll k_B T$, i.e. if all the sites have the same energy.

5.6 Optical properties

We conclude this chapter with an account of the optical properties of amorphous semiconductors and insulators. In a sense, we have touched on this subject several times already in this book, albeit for different frequency regimes. Thus, IR absorption, at one end of the frequency spectrum, was discussed in the context of lattice vibrations (Ch. 4), and at the other end of the scale, UV and X-ray absorption were considered in the section dealing with experimental probes of the gross features of the electronic structure (sect. 5.2.2). We shall therefore not discuss these in detail again, but will rather concentrate first on the optical processes that occur for photons having energies comparable to that of the band gap. For most amorphous semiconductors, this means photons in the visible region of the spectrum since band gaps often lie in the range 1–3 eV, although obviously for insulators the energies are considerably larger (e.g. ~ 10 eV for SiO_2). The form of the energy dependence of the optical absorption typically observed is shown schematically in Fig. 5.37. We will divide this part of the section into two, dealing respectively with optical transitions at energies above and below the mobility gap. The reason for this division is that different microscopic mechanisms are responsible in each case, as will become apparent. The last part of this section deals with intraionic luminescence and absorption.

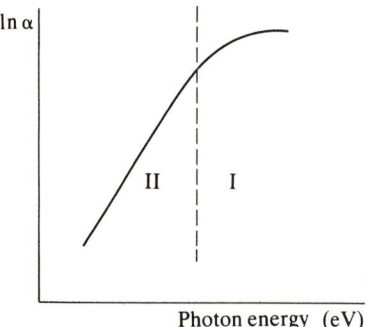

Fig. 5.37 Schematic illustration of optical absorption in amorphous semiconductors showing the interband region (I) and the Urbach edge region (II).

5.6.1 Interband absorption

Interband absorption in general has been dealt with in section 5.2.2; we are concerned here with the low-energy portion of that absorption occurring at photon energies corresponding to the band-gap energy and for which the absorption coefficient lies in the range $10^4 < \alpha < 10^6$ cm^{-1}. To calculate the expected absorption we may make use of the formalism already developed in terms of the random phase model [5.21]. It is normally assumed that the densities of states just beyond the mobility edges can be expressed in power-law form:

$$g_v(-E) \propto E^p \quad [5.103(a)]$$

$$g_c(E) \propto (E-E_0)^q \quad [5.103(b)]$$

where, as in [5.21], energies are measured from the valence band mobility edge and E_0 marks the mobility edge in the conduction band (i.e. it is the magnitude of the mobility gap). Inserting [5.103] into [5.21] gives:

$$\omega^2 \varepsilon_2(\omega) \propto (\hbar\omega - E_0)^{p+q+1} \quad [5.104]$$

If the form of the densities of states at both band edges are *parabolic*, i.e. $p=q=\frac{1}{2}$, then the photon energy dependence of the absorption becomes:

$$\omega\alpha(\omega) \propto \omega^2 \varepsilon_2(\omega) \propto (\hbar\omega - E_0)^2 \quad [5.105]$$

This form for the absorption is observed experimentally for many amorphous semiconductors, and some examples are shown in Fig. 5.38, for which a considerable part of the absorption plots linearly as $(\omega\alpha(\omega))^{1/2}$ v.s. $\hbar\omega$. Note that the extrapolation of such a plot gives a value for E_0. However, it is somewhat doubtful whether in reality E_0 does represent a true value for the optical gap, although it is often used as a

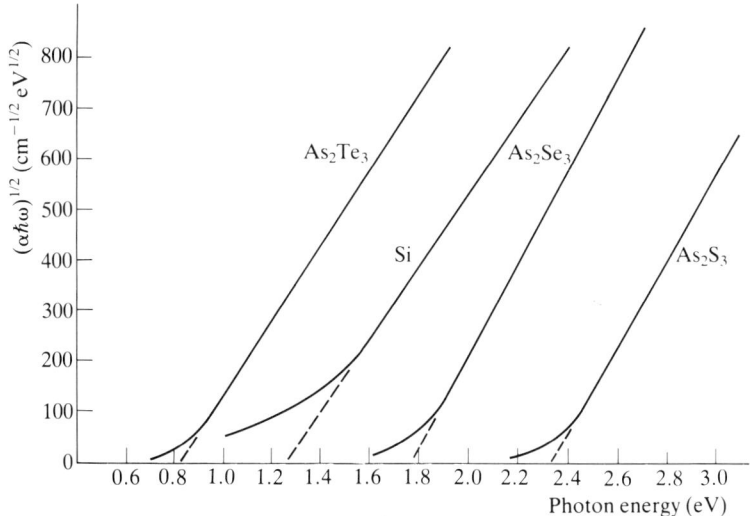

Fig. 5.38 Examples of optical absorption edges of materials whose spectral dependence is given by $\alpha\hbar\omega = C(\hbar\omega - E_0)^2$ showing the extrapolation to give values for the optical gap E_0 (after Mott and Davis 1979).

convenient marker of the gap in monitoring, for example, changes with temperature or pressure. Part of the uncertainty arises from the nature of the random phase approximation used in deriving [5.105]. In this, both extended–extended transitions *and* localized–extended transitions are possible, with the same matrix element, and hence transitions between, say, states at the valence band mobility edge and the localized conduction band tail states having a density of states $g_c \propto (E - E_0)$ will also give a quadratic energy dependence of $\omega\alpha(\omega)$ as in [5.105], only here E_0 represents the energy separation between E_v and E_A (Fig. 5.12). However, the magnitude of the absorption due to the latter process is likely to be considerably smaller than that for extended–extended transitions due to the low density of states in the tail state region.

The quadratic energy dependence of the absorption embodied in [5.105] is not universally observed: a-Se exhibits a linear dependence, and multicomponent chalcogenide alloys (e.g. Ge–As–Te–Si) which exhibit electronic switching behaviour show a cubic energy dependence. Still other materials, e.g. a-Ge, exhibit absorption edges which cannot be fitted to any simple power law, a reflection, presumably, of the fact that the densities of states cannot be described by a power law either. In this case, it is not possible to obtain a value representative of the optical gap from plots such as in Fig. 5.38; instead, an alternative option is to take the energy at which the absorption coefficient reaches some value (usually 10^4 cm^{-1}), and the optical gap is then conventionally denoted E_{04}.

5.6.2 The absorption edge

If the absorption mechanism detailed above were the only one operative, we would expect a very sharp 'absorption edge' marking the onset of optical transitions, for absorption coefficients less than $\sim 10^4$ cm^{-1}. Instead, what is found in practice for the great majority of amorphous semiconductors is an edge which is considerably less sharp and consequently tails well into the gap region, as shown schematically in Fig. 5.37. Such an edge is almost invariably found to obey accurately an exponential dependence on photon energy:

$$\alpha = \alpha_0 \exp\left[-\Gamma(E_e - \hbar\omega)\right] \qquad [5.106]$$

where E_e is an energy comparable to E_0, and Γ is a constant (at room temperature) having values in the range $10-25$ eV^{-1}. So widespread is this behaviour that it seems to be a general feature of amorphous semiconducting materials (except perhaps a-Si and Ge). Similar behaviour is also observed in some crystalline solids, notably the alkali halides, although in these cases Γ is temperature dependent near 300 K, behaving as $\Gamma \sim 0.8/k_B T$, but temperature independent below this temperature. These absorption edges are termed 'Urbach edges' after their discoverer, and the usage extends to the amorphous case also. Several examples of such edges are shown in Fig. 5.39.

The temperature variation of the Urbach edge of amorphous materials is somewhat complicated, but resembles the behaviour observed in crystalline solids, albeit in a different temperature regime. For temperatures between ~ 77 K and T_g, the glass-transition temperature, the slope Γ is temperature independent although the edge does shift in a parallel fashion to lower energies with decreasing temperature, presumably reflecting the temperature shift of the optical gap. Below

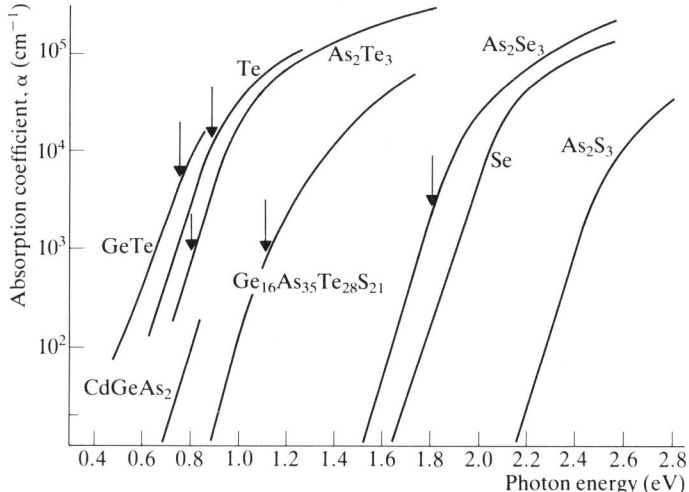

Fig. 5.39 Optical absorption edges for a variety of amorphous semiconductors at room temperature showing Urbach behaviour. The arrows mark the value of $2\Delta E_\sigma$ for those materials for which the electrical conductivity is activated with energy ΔE_σ (after Mott and Davis 1979).

~ 77 K, the edge becomes practically temperature independent, while above $\sim T_g$, the $1/T$ temperature dependence observed in crystalline materials is seen in amorphous solids too.

The most obvious explanation for the Urbach edge is that the densities of states at the band edges increase exponentially with energy as $g(E) \propto \exp(\Gamma E/2)$. However, the relative constancy of Γ between materials makes this argument seem unlikely. In fact, the Urbach behaviour even in crystalline solids is not well understood, and so the level of understanding of the phenomenon in amorphous materials is on an even less secure footing. Several possible mechanisms have been proposed of which the following seems to be the most plausible.

Dow and Redfield (1970) suggested that the exponential edge arises from an electric-field broadening of an exciton line. Normally, the shape of an exciton line is gaussian, but in the presence of a uniform electric field Dow and Redfield showed in a numerical study that the edge becomes accurately exponential over appreciable ranges in energy. The basic mechanism is one of field ionization in which the electron can tunnel out of its quasi-bound state in the coulomb well caused by the hole, into the deep and wide potential outside (resulting from the local microfield) and which admits new states lying in the previously forbidden gap. If the outer, microfield-induced potential trough is sufficiently wide and deep, the stationary-state electron wavefunction will have most of its amplitude there, with an *exponentially* small amplitude near the hole. Since the optical absorption is proportional to the square modulus of this amplitude, the exponential nature of the Urbach edge becomes clear. The internal electric fields that are responsible for this behaviour could arise from a variety of sources: static spatial fluctuations in potential (e.g. accompanying bond-angle distortions), density variations, charged defect centres or longitudinal optic phonons (possibly 'frozen-in' at T_g and having a gaussian distribution characteristic of T_g). In fact, any internal electric field of magnitude $\sim 10^6$ V cm^{-1}

which has a gaussian distribution will give, after suitable averaging, an exponential form for the absorption edge similar to those observed, and so the almost universal Urbach behaviour is somewhat less surprising, since it is rather insensitive to the precise details of the microfield distribution.

The position of the Urbach edge in a given amorphous material may be shifted by a variety of externally applied perturbations, which may be temperature, pressure or electric field. The first we have already considered. The application of pressure causes a parallel shift of the Urbach edge to lower energies in the few amorphous semiconductors which have been studied in this way. The sign of this shift may be related to the similar shift in the position of the optical gap (determined by E_{04} or from a plot of $(\alpha \hbar \omega)^{1/2}$ v.s. $\hbar \omega$) observed in chalcogenide glasses. These materials contain a large proportion of low coordinated chalcogen atoms, around which, by definition, there exists a considerable volume free of bonding charge – the so-called 'bond-free solid angle' (Kastner 1973). The application of pressure is then supposed to cause atoms to move together preferentially in the non-bonding directions (thereby affecting non-bonding, lone-pair electron interactions), rather than to compress the primary covalent bonds directly. This has the effect of simply broadening the width of the bands (J) without affecting the position of their centroid, and thus the band gap decreases:

$$\left(\frac{\partial E_0}{\partial P}\right)_T = -\left(\frac{\partial J}{\partial P}\right)_T$$

Precisely the opposite behaviour is observed for tetrahedral amorphous semiconductors (Ge, Si, GaAs, etc.), where the application of pressure causes an *increase* in the optical gap. In these materials, the bond-free solid angle is considerably less, and in this case pressure is transmitted directly to the bonds. Using the simple Penn model for the gap, in which E_g is some 'average' energy separation between valence and conduction bands, E_g depends on the bond length r_1 approximately as $E_g \propto r_1^{-2.5}$ and thus bond compression produces an increase in E_g. Thus the pressure dependence now comprises two terms, one involving a shift in band centres and the other involving a band broadening:

$$\left(\frac{\partial E_0}{\partial P}\right)_T = \left(\frac{\partial E_g}{\partial P}\right)_T - \left(\frac{\partial J}{\partial P}\right)_T$$

Presumably, in tetrahedral materials, the first term is greater than the second to account for the observed pressure dependence.

The application of an external electric field to a sample also changes the optical absorption in the region of the Urbach edge of amorphous semiconductors, and this forms the basis of the technique known as 'electroabsorption'. The changes in the absorption are small, necessitating the use of modulation techniques to render them easily observable. The Dow–Redfield model for the Urbach edge naturally predicts that the edge should be field sensitive, but the experimental data obtained so far (mainly on chalcogenide glasses) do not seem to offer a coherent picture, either among themselves, or with a simple application of the Dow–Redfield theory. It is often found that the Urbach edge shifts to lower energies and becomes steeper with increasing electric field, but this can only be reconciled with the field-broadened exciton mechanism if part of the absorption is insensitive to an external field, possibly reflecting the existence of local anisotropy in the amorphous solid resulting in a non-uniform distribution of microfields (Sussmann *et al.* 1975).

5.6.3 Intraionic absorption and luminescence

In this section we deal briefly with the intraionic spectra resulting from transitions which occur within the electron levels of transition metals or rare earths dissolved purposely or inadvertently in glassy matrices. The optical absorption properties of such ions, particularly transition metals, are immensely important since they are the colouring agents in oxide glasses, and have been employed in artistic and functional applications of 'glass' ever since it was first manufactured. The scientific interest in such absorption processes stems not so much from a desire to understand these processes (which are in fact well understood from studies in the crystalline state), but rather in their use as a probe for the local structure, since the absorption spectra are affected by the local configuration in subtle ways.

The interest in the intraionic luminescent properties of such ions comes about because of the considerable potential that rare earth (e.g. Nd) doped glasses have as high-power lasing materials for possible use in thermonuclear fusion programmes. In this case, the scientific effort has been directed into understanding the luminescent process itself, in order to optimize the lasing action.

We discuss first the absorption spectra of transition metal (T) and rare earth (RE) doped glasses. A comprehensive survey of the early experimental data on this subject is contained in the classic monograph by Weyl (1959) and more recent developments are given by Bamford (1977) and Wong and Angell (1976). If we consider firstly the 3d-transition series for simplicity, the 3d levels which are degenerate in the atomic state become split in the condensed state by the crystal fields exerted by neighbouring, negatively charged anions (or 'ligands'). Whether the energy of a given d-orbital is raised or lowered with respect to the free-atom level depends on the configuration of the ligands around the transition metal ion; generally either octahedral or tetrahedral coordination by the ligands is preferred, although distortions from the symmetrical configurations are not uncommon in glassy systems. If the transition metal ion has an incomplete d-shell, transitions between the crystal-field split d-levels are possible and are responsible for the intraionic absorption we wish to discuss. However, all d–d transitions are formally forbidden by the so-called Laporte rule, which is simply the quantum-mechanical selection rule, $\Delta l = \pm 1$ for electronic transitions. However, the transitions are weakly allowed if some mixing of the 3d-transition metal orbitals with metal 4p- or ligand p-orbitals can take place. This can arise in several ways: orbital mixing cannot occur in ideal centrosymmetric (e.g. octahedral) complexes under static conditions, but thermal vibrational disorder removes the centre of symmetry as does the topological disorder inherent in a glass; furthermore, tetrahedral ligand coordination allows much more d–p mixing, with the result that typically the absorption intensity due to ions in a tetrahedral configuration is much more intense (by a factor of $\simeq 10-10^2$) than that due to the same ions in an octahedral environment. Thus, this effect alone can give some structural information if the absorption spectra of an ion in the *same charge state* in different network-forming matrices are compared.

The particular charge state of a transition metal ion, as well as the nature and coordination of the surrounding ligand atoms, are obviously factors which determine the energy of the resulting absorption band. In order to understand these factors more fully, it is necessary to describe briefly the ligand-field theory of the d-level splitting in various transition metal complexes; extensive treatments of this topic are given in standard inorganic chemistry texts (e.g. Cotton and Wilkinson 1980) and so only a brief outline will be given here. The simplest version of ligand field theory (crystal field theory) treats the ligand ions as point charges placed at the centroid positions of the ions, considers only electrostatic interactions and neglects any covalent interactions between metal atom and ligand. This simple theory gives a clearer physical picture of the nature of the d-level splittings than is afforded by the more quantitatively accurate molecular orbital theory. Consider firstly a d^1 configuration (i.e. Ti^{3+}) in an *octahedral* environment. The d_{z^2} and $d_{x^2-y^2}$ orbitals which project along the axes will interact strongly with the two negative ligand ions at the polar positions, and hence the coulomb repulsion for these two (equivalent) states will be high; the three remaining orbitals, d_{xy}, d_{yz} and d_{zx} are radially directed between the x, y and z axes and therefore do not interact appreciably with either the two polar or the four equatorial ligand ions. Hence the coulomb repulsion is less than for the other two d-orbitals, and the d_{xy}, etc., orbitals are said to be 'stabilized'. The splitting between the two sets of levels for the octahedral case is termed Δ_0 and is illustrated in Fig. 5.40, where it can be seen that the (d_{z^2}, $d_{x^2-y^2}$) orbitals (given the group symbol e_g) are pushed up in energy more than the (d_{xy}, d_{yz}, d_{zx}) orbitals (given the symbol t_{2g}) are pushed down; this preservation of the 'centre of gravity' of the

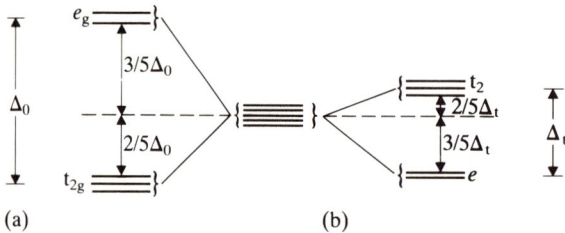

Fig. 5.40 Energy level diagram showing the splitting of a set of d-orbitals by (a) octahedral and (b) tetrahedral crystal fields (Cotton and Wilkinson 1980).

levels occurs whenever the interactions are purely electrostatic in origin. It should be noted at this stage that a d^9 configuration ion (e.g. Cu^{2+}) can be treated in the same manner as the one-electron case, except that now there is a single *hole* in an otherwise filled d-band; this has the effect of *reversing* the orbital ordering shown in Fig. 5.40. If we now consider a d^1 ion in a *tetrahedral* environment, the ligand ions now lie at four of the corners of a cube with its principal axes lying parallel to the coordinate axis system. Now it is the (d_{xy}, d_{yz}, d_{zx}) orbitals which suffer more coulomb repulsion than the (d_{z^2}, $d_{x^2-y^2}$) set and which are therefore stabilized, in contrast to the case of octahedral symmetry (see Fig. 5.40); note too that the energy splittings themselves are inverted with respect to the octahedral case and that the overall tetrahedral splitting Δ_t is smaller than Δ_o i.e.:

$$\Delta_t = \tfrac{4}{9}\Delta_o \qquad [5.107]$$

(The g subscript – *gerade* – is omitted for the tetrahedral configuration because it refers to orbitals which are centro-symmetric, which they are not in the tetrahedral case.)

For the case of more than one d-electron (or hole), the situation is more complicated since now the coupling of the electronic orbital, angular and spin momenta must be considered. In the Russell–Saunders (L–S) approximation (in which coupling between L and S is neglected), the states of the many d-electron *free* ion, labelled S, P, D. F and G, in turn can be split by ligand fields, and each of the crystal field states has the same spin multiplicity as the free ion states from which they derive. An example of the type of splitting expected is illustrated in Fig. 5.41 for the case of a d^2 configuration. Although the number of possible optical transitions for such a level diagram appears to be very large, in fact the additional selection rule which only allows transitions between states having the same spin multiplicity means that in this case only the $^3T_1(F) \rightarrow {}^3T_2(F)$, $^3T_1(F) \rightarrow {}^3T_1(P)$ and $^3T_1(F) \rightarrow {}^3A_2(F)$ transitions will be spin allowed. For the case of a d^n ion in a *tetrahedral* environment the same level inversion occurs as for the d^1 case; hence the $^3A_2(F)$ state becomes the ground state of a d^2 ion in a tetrahedral crystal field (or of a d^8 ion in an octahedral site). In the simple crystal field model, the overall stabilization

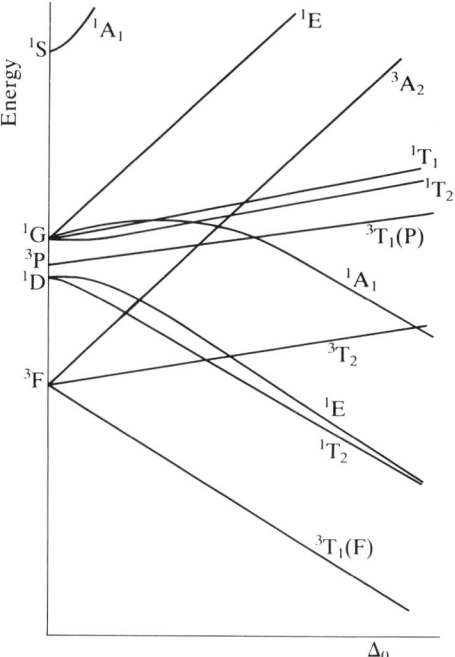

Fig. 5.41 Effect of the ligand field strength on the energies of the Russell–Saunders free ion states for a d^2-ion in an octahedral field (Cotton and Wilkinson 1980).

Electrons

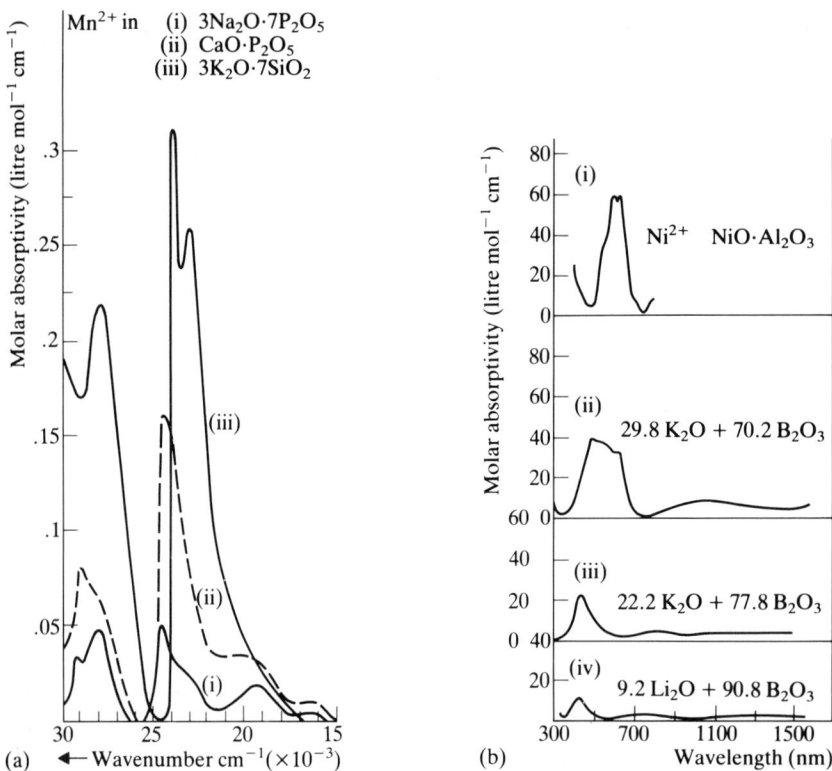

Fig. 5.42 Intraionic absorption in transition metal ions in glassy matrices.
(a) Mn^{2+} in phosphate and silicate glasses. Octahedral sites favoured in (i) and tetrahedral in (ii).
(b) Ni^{2+} in borate glasses and crystalline $NiO \cdot Al_2O_3$ (i) Regular tetrahedral site; (ii) tetragonally distorted tetrahedral site; (iii) tetragonally distorted octahedral site; (iv) regular octahedral site (Wong and Angell 1976).

energy E_s of a d^n ion in say an octahedral field can simply be calculated by adding a stabilization energy of $\frac{2}{5}\Delta_0$ for each electron in a t_{2g} state and subtracting a destabilization energy of $\frac{3}{5}\Delta_0$ for each electron in an e_g state (cf. Fig. 5.40). Note that for the d^5 configuration (i.e. Mn^{2+}) if it is in the 'high-spin' state, i.e. with an electron in each of the t_{2g} and e_g levels and therefore with none paired, $E_s = 0$ and no transitions at all are allowed since they are all spin forbidden. Thus very little absorption is expected from this ion in an octahedral site, as indeed is observed experimentally (Fig. 5.42(a)); Mn^{2+} in a (distorted) tetrahedral configuration is much more absorptive because of the enhanced d–p orbital mixing now involved. Thus Mn^{2+} is an excellent ion for probing the local site geometry in various glassy matrices, even though its overall absorptivity is very low. In this regard, Ni^{2+} and Co^{2+} are even better candidates since their absorptivities are typically three orders of magnitude greater than Mn^{2+} and they have characteristically different absorption bands when occupying either a tetrahedrally or an octahedrally coordinated site. However, because the d^8 configuration (i.e. Ni^{2+}) is overwhelmingly stabilized in the octahedral configuration ($E_s = \frac{6}{5}\Delta_0$), it is not often found in tetrahedral sites, a problem not shared by Co^{2+} (d^7), although it in turn suffers from the disadvantage that the absorption bands for octahedral and tetrahedral configurations tend to overlap in energy. This site-selective characteristic of Ni^{2+} is illustrated in Fig. 5.42(b), where tetrahedral symmetry is imposed on the Ni^{2+} by the crystal lattice in crystalline $NiO \cdot Al_2O_3$, and the absorption band observed is due to the $^3T_1(F) \rightarrow {}^3T_1(P)$ transition (the splitting is due to spin–orbit coupling effects), whereas in the glasses, the Ni^{2+} tends to favour an octahedral site and a much weaker absorption band appears at a higher energy (due to $^3A_2(F) \rightarrow {}^3T_1(P)$ transitions).

We turn now to luminescent processes. Luminescence is a general term describing the radiative recombination of carriers (electrons and holes) excited by a variety of methods: light (photoluminescence), high-energy electrons (cathodoluminescence), or an electric field (electroluminescence). Electron–hole excitations may also lower their energy by *non*-radiative recombination, emitting phonons rather than photons, and often the two processes are competitive, occurring in parallel, although generally one channel or the other is the more probable. The term luminescence also embraces two phenomena which have been differentiated in the past, namely 'fluorescence' and 'phosphorescence'. The term fluorescence is generally reserved for those radiative transitions for which the emitted photon energy is different (generally lower) than the energy of excitation; the energy difference arises because of appreciable non-radiative energy loss suffered by the excited carriers (e.g. via thermalization down in energy through a continuum of states) until radiative recombination occurs. Phosphorescence, on the other hand, refers to the radiative re-emission of absorbed energy at a time delay after (optical) excitation; such behaviour can occur if a carrier is promoted to an excited state from which the transition back to the ground state is forbidden (e.g. by symmetry), and must therefore transfer to a state from which the transition is allowed before radiative recombination can take place. Luminescent processes are also discussed in section 6.3.1 in connection with structural defects, although here we are concerned with intraionic processes involving transitions between crystal-field split d or f levels.

An example of such a fluorescent ion in the 3d series which we have discussed at length already in this section is Cu^+. This typically absorbs in the UV at $\simeq 40\,000$ cm^{-1} ($\simeq 5$ eV) and luminesces in the visible region at $\simeq 20\,000$ cm^{-1} ($\simeq 2.5$ eV), and the emission is believed to be due to the transition from the $3d^{10}$ excited state to the $3d^9 4s^1$ ground state, which is Laporte forbidden. The difference in excitation and luminescence energies is ascribed to a 'Stokes shift', i.e. the same two levels are involved in both absorption and emission processes but because of a strong coupling between the d-electrons of the ion and the surrounding lattice (through the ligands), there is a change in configuration when the ion is electronically excited and hence the minima in the vibronic potential wells for each of the two levels do not coincide spatially; as a consequence, the excitation energy is greater than the luminescent energy (see Fig. 6.15).

However, lanthanides (rare earths) are perhaps more important than transition metals as luminescent species because of the potential that complex silicate glasses doped with rare earths have as high-power laser materials. In the lanthanides, the intraionic electronic transitions of interest are f–f transitions, since the 4f levels are also split by the local crystal fields experienced by an ion. However, unlike the d-electrons in transition metals, the 4f-orbitals in the lanthanides have appreciable electron density close to the nucleus and so the outer 5s and 5p electrons shield them to a certain extent from surrounding fields; thus the f-level splittings are much less dependent on the environment (such as coordination and type of ligand) than is the case for 3d-splittings in transition metal ions. For this reason the site symmetries of lanthanide ions are less readily obtained from absorption spectra. Luminescence can also result from f–f transitions, and it is these which make lasing action in 4f-ion doped glasses (and crystals) possible.

Lasing action can occur if a 'population inversion' is achieved, i.e. a state in which there are more species in an optically pumped excited state than exist in the ground state. In order to achieve a population inversion by optical pumping, a two-level system (i.e. ground state and one excited state) is insufficient, since in this case the maximum ratio of population of the excited state to the ground state is 1:1, i.e. saturation is all that can be achieved. What is required is at least one additional, intermediate excited state (2) to which electrons can relax (non-radiatively) after initial optical excitation to a higher-lying excited state (3), and from which they can subsequently decay radiatively to the ground state (1) (although an alternative recombination channel is by radiative decay from (3) directly to (1)); a population inversion between levels (1) and (2) can thus easily be obtained in this way. An example of a lasing material which makes use of this level configuration is the Er^{3+} glass system and the ruby laser (Cr^{3+} in crystalline Al_2O_3). Certain other systems (e.g. Nd^{3+}:glass) form a four-level system in which the extra level (1'), close in energy to the ground state, is the one to which radiative decay occurs; this has the advantage that all ions which are pumped to level (3) contribute to the population inversion since level (1') is not populated in this case. Nd^{3+} lases at a variety of mean wavelengths in glassy matrices (0.92–1.37 μm), including the line at 1.06 μm which is caused by the $^4F_{3/2} \rightarrow {}^4I_{11/2}$ transition, and which is seen in the Nd^{3+}–(K, Ba silicate glass system) as well as in the Nd^{3+}–YAG (crystalline yttrium aluminium garnet) laser.

Selective amplification in a given direction of the isotropically emitted spontaneous radiation is achieved by mounting the laser rod between two mirrors, which constrain the light emitted parallel to the rod to be repeatedly passed back and forth through the rod. Light amplification occurs by definition when the light intensity at a given point is increased after a round trip through the rod. The condition for this can be written as (Snitzer 1973):

$$R_1 R_2 \exp[(\beta N - \alpha)2L] > 1 \qquad [5.108]$$

where R_1 and R_2 are the reflectivities of the end mirrors, L is the length of the rod, α is the optical absorption coefficient, $N (= N_2 - N_1 > 0)$ is the population inversion. β is the gain coefficient per ion which can be expressed as:

$$\beta = \frac{A\lambda^4}{8\pi c n^2 \Delta\lambda} \qquad [5.109]$$

where n is the refractive index, A is the Einstein coefficient for spontaneous emission and $\Delta\lambda$ is the line width for the luminescent transition of wavelength λ. For a lasing ion in a glassy matrix, the line width will be much greater of course than for the same ion in a crystalline host because of the wide distribution of site symmetries inherent in a glass. Hence, β will be smaller, which means that higher pumping intensities are required to reach the laser threshold, thereby effectively ruling out continuous mode operation for glass lasers. However, this apparent disadvantage can be turned to advantage in the Q-switched, mode-locked operation of a laser. In this, the reflectivity, R_2, of the exit mirror is kept at a low value during the initial period of pumping, so allowing a very large population inversion to build up; if R_2 is now increased (Q-switched), this large amount of stored energy is released suddenly. One method of achieving this in practice is to introduce an absorbing medium (e.g. a dye) which limits the transmittance from the lasing rod to the exit mirror until the light intensity is sufficiently high that the absorbing transition of the dye is saturated and it becomes transparent. This 'mode-locking' causes the Q-switched pulses (of duration $\simeq 10^{-8}$ s) to break up into much shorter pulses of duration $\simeq 10^{-12}$ s; the importance of the relatively broad-band laser emission of lanthanide ion glasses lies in the fact that the broader the emission frequency spectrum, the shorter the pulse width. Thus, these materials hold the promise of forming the basis of very high power, very short pulse lasers.

Problems

5.1 Show that in the Weaire–Thorpe model, the valence band limits are $(V_2 - V_1)$ and $(V_2 + 3V_1)$, and the conduction band limits are $(-V_2 + 3V_1)$ and $(-V_2 + V_1)$.

5.2 Show that the polaron hopping energy (the energy required to produce two polaron wells having the same electronic energy, cf. Fig. 5.36) for the case when there is a disorder energy, i.e. a difference in ground state energies, W_D, between two polaron wells, is:

$$W = W_H + \tfrac{1}{2}W_D + W_D^2/16W_H \qquad [5.1P]$$

where $W_H = \tfrac{1}{2}W_p$. For the case where $W_H \gg W_D$, [5.1P] reduces to [5.101].

5.3 Show that the Kubo formula ([5.47]) yields the standard Drude formula ([5.48]) for the frequency-dependent conductivity of an electron gas, given that the electron wavefunctions are written as

$$|\alpha\rangle = \sum_k a_k^\alpha |k\rangle \qquad [5.2P]$$

where the a_k^α are independent random gaussian variables whose variance is given by:

$$\langle |a_k^\alpha|^2 \rangle = \left(\frac{\pi}{lk^2\Omega}\right) \frac{1}{[(k-k_\alpha)^2 + l^2/4]} \qquad [5.3P]$$

Bibliography

M. H. Brodsky (ed.) (1979) *Amorphous Semiconductors* (*Topics in Applied Physics*, vol. 36) (Springer-Verlag).

J. C. Garland and D.B. Tanner (1978), *Electrical Transport and Optical Properties of Inhomogeneous Media* (AIP).

S. S. Mitra (ed.) (1976), *Physics of Structurally Disordered Solids* (Plenum).
N. F. Mott and E. A. Davis (1979), *Electronic Processes in Non-Crystalline Materials* (OUP).
J. Wong and C. A. Angell (1976), *Glass: Structure by Spectroscopy* (Dekker).
F. Yonezawa (ed.) (1981), *Fundamental Physics of Amorphous Semiconductors (Solid-State Sciences*, vol. 25) (Springer-Verlag).
J. M. Ziman (1979), *Models of Disorder* (CUP).

6 Defects

6.1	**Introduction**
6.2	**Types of defect**
6.2.1	Dangling bonds (positive U)
6.2.2	Dangling bonds (negative U)
	The polaronic model
	The charged dangling bond model
6.2.3	Vacancies
6.2.4	Density defects
6.2.5	Dislocations and disclinations
6.3	**Defect-controlled properties**
6.3.1	Photoluminescence
6.3.2	Electron spin resonance
6.3.3	Direct current hopping conductivity
6.3.4	Alternating current conductivity
6.3.5	Drift mobility
	Problems
	Bibliography

6.1 Introduction

Thus far in this book, we have dealt in the main with 'ideal' amorphous solids, considering their structure and their electronic and vibrational properties. However, real amorphous materials cannot be prepared to meet this exacting standard, and the introduction of structural defects or imperfections is often an unavoidable consequence of the drastic methods of preparation needed to avoid the initiation of crystallization in producing an amorphous solid; in addition, structural defects can be considered to be present in thermal equilibrium in certain liquids and these will be frozen-in on vitrification of a melt.

The notion of a defect in the structure of an amorphous solid might at first sight seem somewhat strange. Surely, the reader may ask, if the structure of the solid is random, is it not rather a contradiction in terms to talk about a defect in an amorphous matrix? Or phrased in a different way, might not the structure of a non-crystalline solid be thought of as being *completely* defective, as in the so-called 'dislocation model' of amorphous structures (in which a crystal is considered to contain so many intersecting dislocations that an aperiodic structure results (Ninomiya 1977))? These points of view are not, however, particularly helpful; instead we assert that the concept of a defect in an amorphous material *is* a valid one, although of course any given defect can only be defined with reference to some non-defective state. For crystalline materials, this is obviously the perfect single crystal lattice. For non-crystalline solids, we may take 'ideal' amorphous structures represented for example by a continuous random network for a covalent material or a dense random packing of spheres for a metal, as discussed extensively in Chapters 3 and 7.

The importance of defects, and the reason why we devote a chapter to them, lies in the fact that many properties of amorphous materials can be defect controlled, as indeed is also the case for crystalline solids. Among these number magnetic properties, opto-electronic behaviour, vibrational properties and mechanical characteristics. In many cases we shall see that the behaviour resulting from the presence of defects can completely dominate that due to the intrinsic material. Furthermore for certain materials, e.g. the chalcogenide glasses, the ideal amorphous state is impossible to achieve experimentally since structural defects are present even in thermal equilibrium in the melt and are consequently frozen-in on vitrification.

The first part of this chapter is therefore devoted to a description of the different types of defect that are believed to occur in amorphous solids, with examples of their occurrence in a variety of materials. Some of these kinds of defect are analogous to those found in crystals and are familiar from conventional materials science (see e.g. Henderson 1972, Stoneham 1975); others are unique to the amorphous phase. The last part of the chapter deals with a discussion of the various properties that are exhibited by such defects, stressing in particular the differences in behaviour from that expected for intrinsic (defect-free) materials.

On a first reading, the reader may prefer to peruse sections 6.2.1 and 6.2.2 dealing respectively with the two types of dangling bond found in covalent amorphous semiconductors or insulators, perhaps also looking at the two sections describing volume defects, that on voids and vacancies (sect. 6.2.3) and that on density (free-volume) defects (sect. 6.2.4). Following these introductory sections, those describing experimental probes for structural defects, namely photoluminescence (sect. 6.3.1) and electron spin resonance (sect. 6.3.2), may be of interest, and finally the sections dealing with the effect of defects on electronic properties, such as d.c. conductivity (sect. 6.3.3) and drift mobility (sect. 6.3.5) could be read. Comprehensive reviews of structural defects and their influence on physical properties have been given by Davis (1979), Mott (1980) and Robertson (1982).

6.2 Types of defect

6.2.1 Dangling bond (positive U)

This is perhaps conceptually the most straightforward form of point defect: it is simply a broken or unsatisfied bond in a covalent solid. (A dangling bond has no meaning in a solid formed from non-directional bonds, such as in a metal, ionic salt

Fig. 6.1 Isolated dangling bond in a tetrahedrally coordinated continuous random network simulating the structure of a-Si or a-Ge.

or a rare gas.) Although the concept is a simple one, in practice complications often occur, such as atomic reconstructions. A representation of a dangling bond in a CRN is shown in Fig. 6.1. A simple dangling bond normally contains *one* electron and is electrically neutral; however it is amphoteric (being able to donate or accept an electron), and under certain circumstances the electronic occupancy can change, varying concomitantly the charge of the centre. An important point to note is that an *isolated* dangling bond is a perfectly feasible entity in an amorphous solid, but *not* in a crystalline solid where crystallographic constraints preclude it. A moment's thought will lead to the realization that if an atom is removed from, say, a tetrahedrally coordinated diamond cubic lattice, *four* dangling bonds remain

Defects

pointing into the vacancy (see sect. 6.2.2), and this is the least number that can result from the introduction of vacancies. Equally, dislocations in crystals introduce many dangling bonds along the dislocation line. Thus, the possibility of the existence of single isolated dangling bonds is a consequence of the presence of structural randomness and for this reason we discuss them separately from vacancies.

The energy levels for electron states associated with an isolated dangling bond may be discussed in terms of a simple molecular orbital picture. If we consider the case of a tetrahedral semiconducting material, e.g. a-Si, having an atomic electronic configuration s^2p^2, the atomic levels hybridize to form four sp^3 molecular hybrids, each of which may admit a bonding or antibonding orbital; solid-state interactions then broaden the molecular levels into bands separated by a band gap (Fig. 6.2(a)). A dangling bond, or non-bonding orbital, containing a single electron will therefore have an energy level lying at the zero energy for the sp^3 hybrids, viz. near the *middle of the gap* (if atomic relaxations are neglected). Thus, structural defects such as dangling bonds are expected to introduce electron states deep into the gap (which is otherwise empty in the ideal case except for band tailing – see Fig. 5.12(a)); the precise position of the energy levels, however, will depend crucially on factors such as structural relaxation around the defect or the electronic character of the states at the top of the valence band and the bottom of the conduction band from which the eigenfunction of the defect state derives. Note the analogy between the behaviour of electron states and vibrational modes associated with local defects already

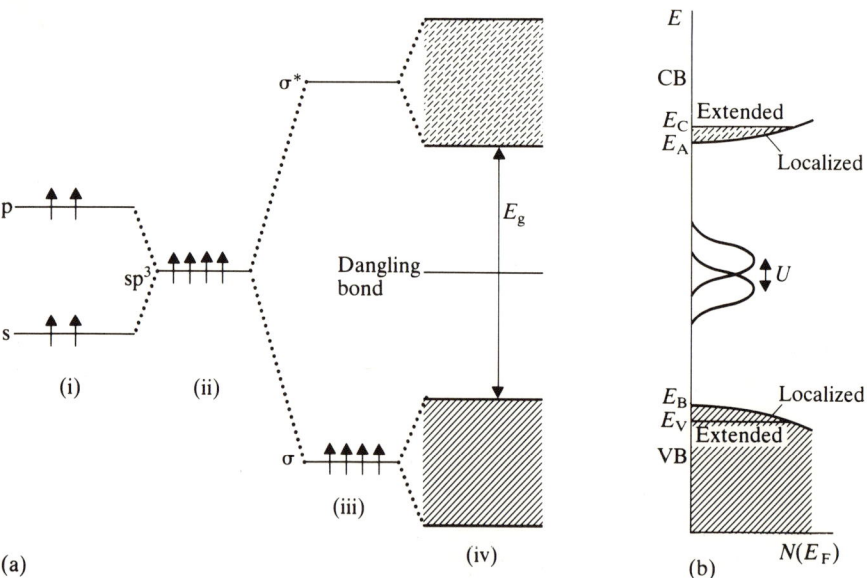

Fig. 6.2 (a) Schematic representation of the origin of valence and conduction band states for a tetrahedrally bonded semiconductor (i) atomic s- and p-states; (ii) sp^3-hybrid states; (iii) bonding (σ) and antibonding (σ^*) states; (iv) broadening of and σ- and σ^*-states into valence band (VB) and conduction band (CB).
(b) Density of states for such a band scheme, showing the localized band-tail states. A dangling-bond band is also near mid-gap, together with the band for double occupancy higher by U in energy (for a positive-U system).

mentioned in section 4.3.2; like the vibrational excitations, we expect the electron states in the mobility gap arising from dangling bond defects to be highly localized.

The density of states in the gap for an amorphous semiconductor containing isolated dangling bond defects might be as shown schematically in Fig. 6.2(b). The dangling bond level is broadened by disorder (e.g. by bond-angle fluctuations, etc.) into a band; the lower mid-gap band in Fig. 6.2(b) corresponds to the neutral dangling bond containing a single electron (spin) and is 'donor-like' (i.e. neutral when occupied), and the upper band corresponds to a different charge state of the *same* defect, namely when an extra electron is placed in it, and is 'acceptor-like' (i.e. neutral when empty). Under normal circumstances, the net energy cost for this addition is positive; it costs an extra energy U (the Hubbard or correlation energy):

$$U = \langle e^2/4\pi\varepsilon_0\varepsilon r_{12} \rangle \tag{6.1}$$

where r_{12} represents the appropriate separation for two electrons at the same site, and a configurational average has been taken in evaluating U.

The density of states depicted in Fig. 6.2(b) was first proposed by Davis and Mott (see Mott and Davis 1979), and can account for the following features. The material is expected to be optically transparent to photons of energy less than that corresponding to the mobility gap, a prediction not shared by the Cohen–Fritzsche–Ovshinsky (1969) model (see sect. 5.3.2 and Fig. 5.13(b)) in which tail states derived from the valence and conduction bands extend well into the gap and overlap near mid-gap – these states would be neutral when full and empty, respectively. A further consequence of the Davis–Mott model is that, like the CFO model, it predicts that the Fermi level (E_F) is 'pinned' near mid-gap by the presence of the defect states; if the two defect bands are separated in energy, the Fermi level will lie in the gap between them, or if, as in Fig. 6.2(b) they overlap, E_F will lie in the centre of the region of overlap. Although in this situation, we have E_F lying in a band and one might therefore at first sight think that the material should be metallic, it must be remembered that the material is in fact a 'Fermi glass', i.e. E_F lies in a band of *localized* states, and hence the material will remain a semiconductor. The evidence for a pinned Fermi level has been cited in section 5.4.1; for example plots of $\ln \sigma$ v.s. $1/T$ (for extended-state conduction) stay linear down to very low temperatures (although an additional conduction mechanism due to electron hopping between defect levels may intercede at low temperatures – see sect. 6.3.3), and amorphous semiconductors in general cannot readily be doped, i.e. the inclusion of impurities has little effect on electrical properties. Finally, since the dangling bonds in this model suffer a *positive* Hubbard U, most of them will be neutral, possessing a single electron spin. Therefore these centres will be paramagnetic, exhibiting electron spin resonance signals and Pauli paramagnetism (sect. 6.3.1).

The materials which come closest to realizing the Davis–Mott model for states in the gap are probably evaporated or sputtered a-Si or Ge films. These amorphous thin films exhibit a large ESR signal ($\sim 10^{19}$–10^{20} spins cm^{-3}) in their as-prepared state, and a concomitant high density of states in the gap, as measured by techniques such as the 'field effect' (for an illustration of its application and use see Fig. 6.3). No structure in the density of states, such as that shown schematically in Fig. 6.2(b), has come to light using this or other techniques for *un*hydrogenated a-Si or Ge; indeed the density of states appears to be so high that little variation in the field effect can be

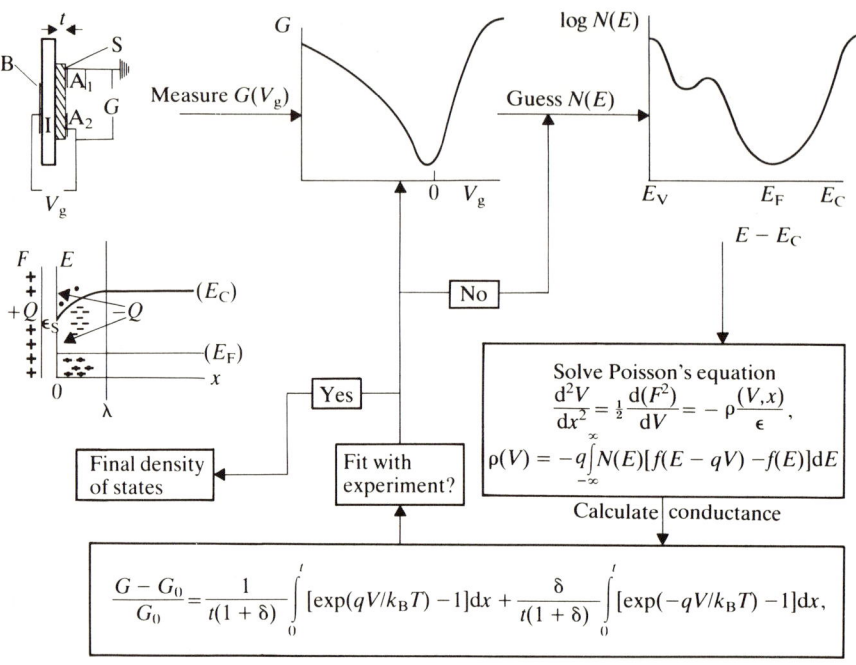

Fig. 6.3 Block diagram showing the principles of the determination of the density of states using field-effect measurements. The field-effect transistor configuration is shown, where S is the sample film (thickness $t \simeq 1$ μm); I, the insulator; A_1, A_2, gap (drain) electrodes (gap $\simeq 100$ μm); B, field electrode; V_g, gate voltage; G, surface conductance. The electron energy profile through the sample film when a positive potential is applied to **B** is also shown: the space charge $-Q$ in the film is composed of excess electrons in extended and localized states (W. E. Spear and P. G. Le Comber (1972) *J. Non-Cryst. Sol.* **8–10**, 727). In Poisson's equation, ρ is the charge density, F is the field, $f(E)$ is the Fermi–Dirac function. In the expression for the surface conductance, δ is the ratio of hole to electron current, and G_0 is the conductance, both under flat-band conditions (Powell 1981).

detected, although in the energy region which can be explored, the density of states appears to be uniform. The very high densities of gap states in these materials naturally explain the observation that they cannot be doped: any impurity which is introduced will donate or accept an electron to or from the defect gap states, but the density of states is so high and the number of impurity atoms so relatively small, that the Fermi level is not shifted significantly and the material is not doped.

This is *not* the case for *hydrogenated* a-Si:H alloys, prepared for example by the glow-discharge decomposition of silane, which have such a low residual density of gap states that the substitutional introduction of P, As or B can easily dope the materials n- or p-type, changing the electrical conductivity by many orders of magnitude (see Fig. 5.24) (Spear 1977). Presumably, the presence of the hydrogen acts in the main part to satisfy any dangling bonds, thereby removing the defect states in the gap, replacing them with Si—H bonding states deep in the valence band, and rendering the material dopable. This simple picture is not sufficient to account completely for the role of the hydrogen, however, since it is found that some 10 times more H (~ 1–10 at. %) is introduced by the glow-discharge process into electronic grade a-Si:H than is needed to satisfy all the dangling bonds evinced by

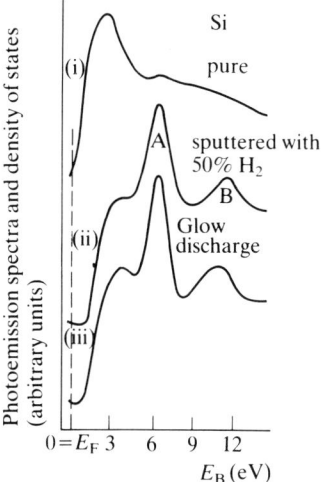

Fig. 6.4 Ultra-violet photoemission spectroscopy spectra (taken with HeII radiation) of the valence band density of states for (i) sputtered a-Si; (ii) sputtered a-Si:H; (iii) glow-discharge deposited a-Si:H (Von Roedern *et al.* 1979). Note the erosion of the valence band edge (arising from Si 3p-like states for a-Si) upon addition of hydrogen, and the growth of H-related peaks near 6 eV (A) and near 11 eV (B). These are ascribed to Si 3p/H 1s bonding states and Si 3s/H 1s bonding states, respectively.

ESR in evaporated a-Si; moreover the introduction of H progressively erodes the top of the valence band as well as removing states from the gap, as shown by photoemission studies (Fig. 6.4) and the finding that the optical gap increases on hydrogenation by ~ 0.4 eV for ~ 15 at.% H (Moustakas 1979). The role of hydrogen in the deposition process and the subsequent electrical activity of this potentially useful electronic material is still a subject of some considerable controversy and active study.

Although the isolated dangling bond is a possible entity in a CRN, more often dangling bonds occur at voids or other internal surfaces (a circumstance discussed in sects. 6.2.2 and 6.3.2), or else reconstruct so as to lower the total energy, thereby changing the electron occupancy and hence the charge state of the defects. Such atomic reconstructions at defects are not unique to the amorphous state, but also occur in crystalline materials. As an example we may consider the case of crystalline silicon which, under different circumstances, appears to exhibit the two forms of electron–lattice coupling that drive such reconstructions, namely covalent-like or ionic-like. It is a well-known feature of crystalline semiconductor surfaces (corresponding to predominant lattice planes) that they reconstruct and so change the periodicity at the surface. For instance, the *unrelaxed* (111) surface of Si has one half-filled (neutral) dangling bond per site and thus should be a paramagnetic two-dimensional metal (with one electron in each orbital and the Fermi level lying in the middle of the band of states so formed). Experimentally, it is found that this is not so; the surface atoms reconstruct so that a 2×1 superlattice is formed and the surface becomes a diamagnetic semiconductor. One model for this reconstruction is based on an ionic-like electron–phonon coupling mechanism; alternate rows of atoms are raised or lowered, changing their hybridizations towards s^2p^3 or sp^2 respectively,

Defects

and consequently also changing the electron occupancies of the dangling bonds (Fig. 6.5(a)) (Applebaum and Hamann 1976). The raised atoms (s^2p^3) acquire a net negative charge, and we may refer to the configuration to which they tend as T_3^-, where T represents a tetrahedral type atom, and the subscript and superscript refer respectively to the atomic coordination and charge state; correspondingly, the lowered surface atoms (sp^2) tend towards a T_3^+ configuration. The ionic character of the electron–lattice coupling leading to this sort of reconstruction is thus clearly seen. Recent angle-resolved photoemission results have cast some doubt on the ionic buckling model, and another form of reconstruction involving π-bonding (Fig. 6.5) has been proposed (Pandey 1981). An illustration of the other type of

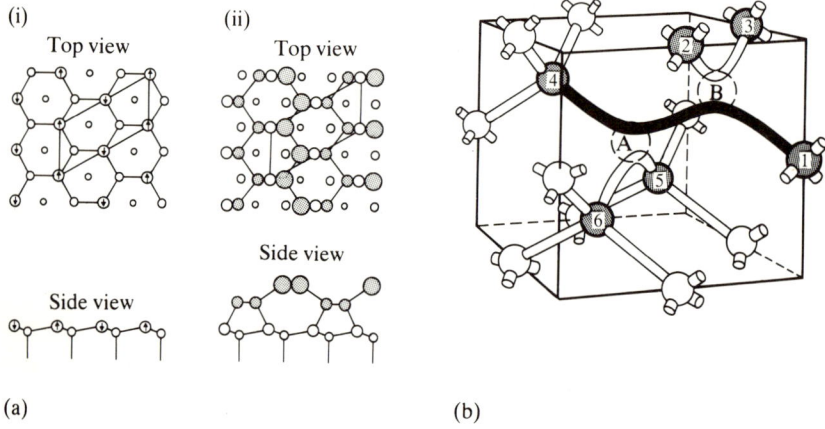

Fig. 6.5 Different forms of atomic reconstruction in crystalline silicon:
(a) 2 × 1 reconstruction of the (111) surface according to (i) the 'ionic' buckling model of Haneman (1961), and (ii) the π-bonded chain model of Pandey (1981);
(b) divacancy showing the extended bond reconstructions (A and B mark the sites of the original vacancies) (Spear 1974).

electron–lattice coupling, namely covalent-like, is afforded by the divacancy in crystalline Si. The removal of two neighbouring atoms from the lattice results in six dangling bonds pointing into the resulting vacancy. These dangling bonds can then form weak 'bent' bonds between neighbouring hybrid orbitals and even weaker bonds between more distant orbitals, if the electron occupation of the orbitals is such as to allow the bond formation (Fig. 6.5(b)).

6.2.2 Dangling bond (negative U)

A feature of the Si (111) surface reconstruction, and also of the distortion which occurs in *mono*vacancies in crystalline silicon (to be discussed in sect. 6.2.3), is that the defects exhibit a so-called 'negative effective Hubbard or correlation energy', U_{eff} (Anderson 1975); i.e. spin pairing of electrons occurs at the defect centres, and the net process is exothermic because of the stronger (ionic or covalent) electron–phonon coupling which is a feature of such systems, despite the endothermic cost in energy to place two electrons on the same site ([6.1]).

The polaronic model

The concept of electron pairing mediated by a negative U_{eff} mechanism was first proposed by Anderson (1975) to apply to certain covalent amorphous materials, particularly chalcogenide glasses. The behaviour of these materials proved mysterious for some time, since although certain experimental techniques (e.g. photoluminescence, sect. 6.3.1 and a.c. conductivity, sect. 6.3.4) demonstrated that there existed states in the gap due to defects, as in the case of a-Ge or a-Si, no ESR signal could be detected, in marked contrast to the large signals observed in the tetrahedrally bonded amorphous semiconductors. This apparently contradictory behaviour is understandable if the electron–phonon interaction is sufficiently strong in these materials to produce a negative U_{eff} and hence render spin pairing energetically favourable.

Anderson's theory is in fact couched in the most general terms and we will discuss this first before turning our attention to the applications that have been made of the theory to specific point defect configurations. The model considers a random lattice of sites connected by a *spectrum* of 'bonds', each populated by a pair of up and down spin electrons, and which span the range of electron pair configurations from conventional covalent bonds, through 'weak' bonds, to non-bonding lone pairs; the electron pairing mechanism is a strong electron–phonon (polaron) coupling leading to a negative U_{eff}. Note that the concept of a well-defined *structure* of an amorphous semiconductor does not apply in this model, nor is there a distinction between defects and the bulk; large variations in bond lengths, bond angles and coordination number are allowed, the only constraint being that a given electron pair forms a 'bond' having the configuration with the lowest energy. Qualitative features of the model are the following:

(a) The electron system is diamagnetic (i.e. spin paired) thereby explaining the absence of an ESR signal in e.g. chalcogenide glasses.

(b) There is a continuous distribution of electron energies, reflecting the spectrum of bonds allowed, and in particular in the simplest case, the density of occupied *two*-electron states extends up to the Fermi level, with unoccupied states at higher energies; i.e. there is *no* gap in the two-electron density of states (Fig. 6.6(a)). However, these two-electron states are not experimentally accessible. The only states which are experimentally accessible are one-electron states for which there is a gap of magnitude U in the density of states (Figs 6.6(b) and (c)) since thermal excitations which add a single electron to a given state (and concomitantly leave a hole in the parent state) involve pair-breaking processes which cost the Hubbard energy U ([6.1]).

(c) The Fermi level is pinned in the middle of the one-electron gap, i.e. at the top of the occupied two-electron density of states, since any electrons injected into the material will pair up, populating two-electron states just above the Fermi level, rather than shifting E_F substantially.

Note the similarities between the density of states in the Anderson model and the Cohen–Fritzsche–Ovshinsky (CFO) model; both have electron states which extend throughout the gap region, thereby producing a 'pseudo-gap' rather than a

Defects

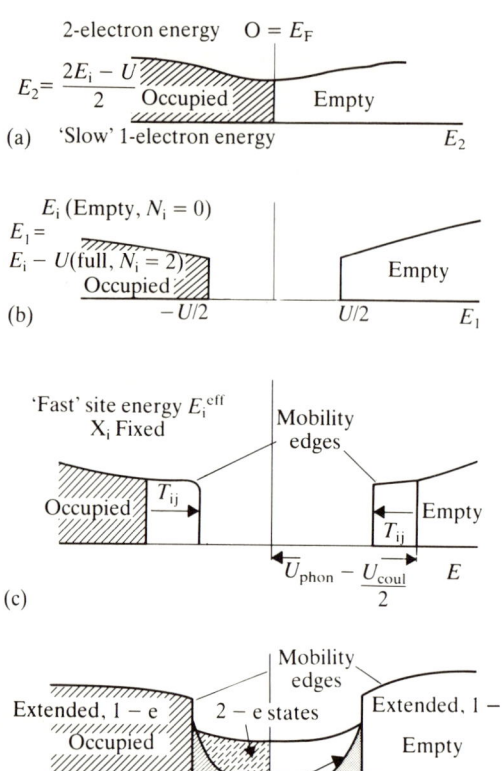

Fig. 6.6 Schematic densities of states for an amorphous semiconductor suffering a negative effective Hubbard energy (Anderson P. W. (1976) *J. de Phys.* **C4**, 339). (N.b. $U = U_{phon} - U_{coul.}$)
(a) Two-electron density of states. Note that the occupied levels extend to the Fermi level.
(b) One-electron density of states (for 'slow' states, i.e. excluding hopping) showing the gap of magnitude U between occupied and unoccupied states.
(c) The effect on the one-electron density of states of including hopping, i.e. 'fast' states.
(d) Overall density of states with the inclusion of randomness, i.e. band tails.

true gap. The difference between the two is that whereas in the CFO model these are *one*-electron states, in the Anderson model they are *two*-electron states although there *is* a true gap in the one-electron density of states.

Let us now consider the Anderson model in more detail. The Hamiltonian can be written as a sum of three parts, an electronic term, a contribution pertaining to the lattice (or phonons), and a term describing the electron–phonon interaction. The electronic term can be written as:

$$\mathcal{H}_e = \sum_{i,\sigma} E_i \hat{n}_{i\sigma} + U \sum_i \hat{n}_{i\sigma} \hat{n}_{i-\sigma} + \sum_{i,j,\sigma} T_{ij} c_{i\sigma}^\dagger c_{j\sigma} \qquad [6.2]$$

where the summations are over the sites i (or j), and over the spin states of the electrons σ. The parameter E_i is the energy of an electron localized on site i, U is the Hubbard electron repulsion energy ([6.1]) which occurs for double occupancy at a site, T_{ij} represents the hopping interaction with neighbouring sites which is neglected henceforth, and $\hat{n}_{i\sigma}$ ($=c_{i\sigma}^\dagger c_{i\sigma}$) is the occupancy of site i by an electron with spin σ, where $c_{i\sigma}^\dagger(c_{i\sigma})$ is the creation (annihilation) operator. The lattice or phonon term can be written as:

$$\mathcal{H}_1 = \hbar\omega \sum_i b_i^\dagger b_i \qquad [6.3]$$

where ω is the phonon frequency, taken to be constant in the Einstein approximation, and $b_i^\dagger$, b_i are phonon creation and annihilation operators. The electron–phonon coupling term can be, in the simplest form, taken to be linear:

$$\mathcal{H}_{e-l} = \lambda \sum_{i,\sigma} \hat{n}_{i\sigma}(b^\dagger + b) \qquad [6.4]$$

where λ is the electron–phonon coupling constant. This cross term can be eliminated by introducing a displaced phonon operator, following Licciardello et al. (1981):

$$d_i = b_i + \frac{\lambda \hat{n}_i}{\hbar\omega} \qquad [6.5]$$

where $\hat{n}_i = (\hat{n}_{i\sigma} + \hat{n}_{i-\sigma})$. If this is substituted into [6.3] and [6.4], the total Hamiltonian reduces to a sum of pure electron and phonon terms $\mathcal{H} = \tilde{\mathcal{H}}_e + \tilde{\mathcal{H}}_1$, i.e.:

$$\mathcal{H} = \sum_{i,\sigma} E_i \hat{n}_{i\sigma} + U \sum_i \hat{n}_{i\sigma} \hat{n}_{i-\sigma} - \frac{\lambda^2}{\hbar\omega} \sum_i \hat{n}_i \hat{n}_i + \hbar\omega \sum_i d_i^\dagger d_i \qquad [6.6]$$

where the first three terms represent $\tilde{\mathcal{H}}_e$ and the last $\tilde{\mathcal{H}}_1$. This can be rearranged to give:

$$\mathcal{H} = \sum_{i,\sigma}\left(E_i - \frac{\lambda^2}{\hbar\omega}\hat{n}_{i\sigma}\right)\hat{n}_{i\sigma} + \left(U - \frac{2\lambda^2}{\hbar\omega}\right)\sum_i \hat{n}_{i\sigma}\hat{n}_{i-\sigma} + \hbar\omega \sum_i d_i^\dagger d_i \qquad [6.7]$$

Hence the *effective* Hubbard correlation energy for two spin-paired electrons at the same site is given by:

$$U_{eff} = U - \frac{2\lambda^2}{\hbar\omega} \qquad [6.8]$$

which is negative if $(2\lambda^2/\hbar\omega) > U$, i.e. for strong electron–phonon coupling.

In general, the equilibrium eigenfunction can be written as $|n_i, v_i\rangle$, where n_i and v_i are the number of electrons and phonons respectively, localized at site i. In the low frequency limit, $v_i = 0$, and the energy eigenvalues of the Hamiltonian ([6.6]) for n electrons occupying a site are then given by:

$$E_n = nE + U\delta_{2,n} - \frac{\lambda^2 n^2}{\hbar\omega} \qquad [6.9]$$

where δ is the Kronecker delta. Alternatively, we may write the energy eigenvalues of the Hamiltonian, expressed as [6.7], explicitly:

$$\begin{aligned}n &= 0 & E_0 &= 0 \\ n &= 1 & E_1 &= E_{eff} \\ n &= 2 & E_2 &= 2E_{eff} - U_{eff}\end{aligned} \qquad [6.10]$$

where we have written

$$E_{\text{eff}} = E - \frac{\lambda^2}{\hbar\omega} \qquad [6.11]$$

Thus, we see from [6.10] that in those materials which have a negative U_{eff}, it is energetically favourable for electrons to pair up rather than remain unpaired, such that at certain sites there are two electrons and at others none, i.e. $E_2 + E_0 < 2E_1$.

The electron–lattice coupling may arise from two distinct mechanisms, as we mentioned earlier. The electron pairing may take the form of a covalent bond between two dangling bond orbitals, in which case an antisymmetric phonon mode couples with the electrons by modulating the two-centre covalent interaction which Harrison (1980) refers to as V_2; in this case we may write $\mathscr{H}_{e-1} = \sum_{i,\sigma} V_2^i(q) n_{i\sigma}$ (equivalent to [6.4]), where q is a 'configurational coordinate', e.g. a bond displacement. Alternatively, a lattice mode can modulate the energy of an orbital, thereby causing some orbitals to be doubly occupied (and negatively charged) and others unoccupied (and positively charged). In this ionic mechanism, the orbital hybridization depends on q via the bond angle, and $\mathscr{H}_{e-1} = \sum_{i,\sigma} V_3^i(q) n_{i\sigma}$ where V_3 is the ionic energy (Harrison 1980). Ngai et al. (1978) have further suggested that tunnelling states (sect. 4.5.4), called by them 'local rearrangement modes', may also couple to electrons in a similar manner to phonons and like them mediate a negative U_{eff} interaction.

The optical and thermal electronic excitations associated with the Anderson model are complicated, and the interested reader is referred to the articles by Anderson (1979) and Licciardello et al. (1981). It should be pointed out, however, that the Anderson model remains essentially a conceptual theory, and it has not been used to account in detail for experimental results for amorphous systems, e.g. chalcogenide glasses. Furthermore, it is not universally accepted that the generality of the Anderson model in predicting a very broad spectrum of 'bonds' in a negative U_{eff} material is a realistic description of *solid* amorphous semiconductors, although it probably is a reasonable approximation for *liquid* semiconducting systems. Instead, it is generally contended that a solid amorphous structure with well-defined bond lengths and reasonably well-defined bond angles is a better description; disorder in this case is mostly concentrated at point (coordination) defects, e.g. dangling bonds, and it is at these that electrons can suffer a negative U_{eff} via the Anderson polaronic mechanism. This conservation of bond length is probable in a solid because the bond energy is a highly non-linear function of bond length, and it is only in liquids, where there is considerable thermal energy available, that appreciable bond-length distortions are likely to be energetically favourable.

The charged dangling bond model

The importance of discrete dangling bond defects at which electrons experience a negative effective correlation energy was first stressed by Street and Mott (1975) and Mott et al. (1975a) and later by Kastner et al. (1976). These models were originally applied to the case of chalcogenide glasses, but various other authors have extended the idea of spin-paired electrons at dangling bond sites to other amorphous semiconductor systems as well. These bonding models have the virtues of being

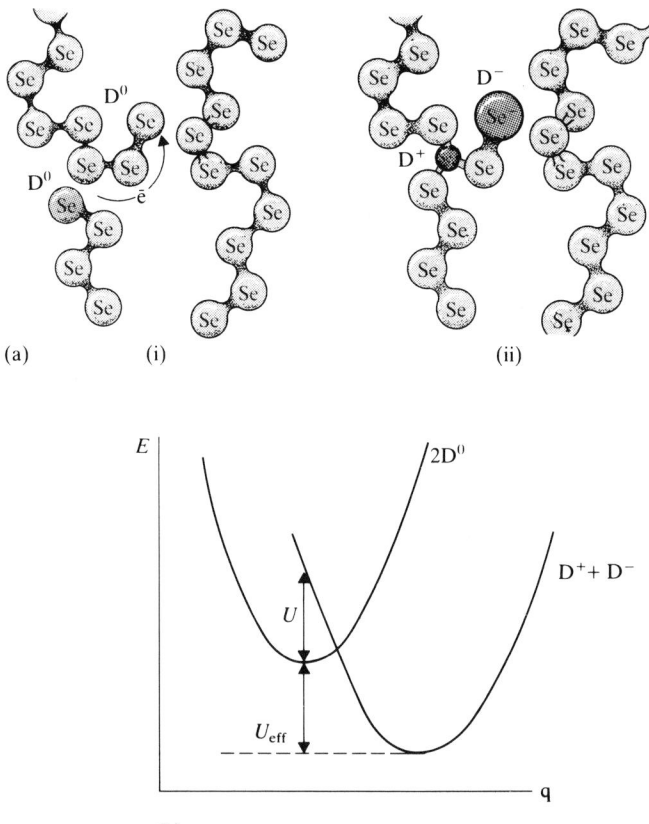

Fig. 6.7 Formation of charged defects (valence alternation pairs) in chalcogenide glasses.
(a) Illustration of the formation of threefold coordinated D^+ (C_3^+) and singly coordinated D^- (C_1^-) defect centres by the exchange of an electron between two D^0 (C_1^0) centres (after D. Adler (1977) *Scientific American*, **236**, 47).
(b) Configuration-coordinate diagram for the formation of a D^+–D^- pair. Exchange of an electron between two D^0 centres to give a D^+–D^- pair at the *same* configuration costs the Hubbard energy U. The D^+–D^- centres subsequently relax to a different configuration and the overall energy is lowered by the effective correlation energy U_{eff}.

conceptually simple and of having been compared in detail with a variety of experimental data.

The essential features of the 'charged dangling bond' model may be understood by consideration of the monatomic system amorphous Se, illustrated in Fig. 6.7. The structure of amorphous Se, which is twofold coordinated, is believed to consist mainly of chains (with perhaps a few rings). Any chain end will be the site of a dangling bond orbital, which in the simplest case will contain an unpaired electron and be electrically neutral relative to the bulk; Mott *et al.* referred to this dangling bond defect as D^0, where the superscript indicates the charge state. They postulated that, following Anderson (1975), electrons residing at D^0 centres should experience a negative U_{eff} and electron pairing should be energetically favourable as a result of atomic rearrangements. Electron–lattice coupling is strong in amorphous chalcogenides because of the low atomic coordination resulting in a high degree of network flexibility, but more importantly because there exist non-bonding, lone-pair pπ orbitals at the chalcogen atoms which may be utilized in forming additional bonds, and which are energetically accessible since they are high-lying, forming the top of the valence band (Kastner 1972). The transfer of an electron from one D^0 centre to another produces one site which has the original dangling bond orbital containing two spin-paired electrons and which is consequently negatively charged (D^-), and the other which has an empty orbital which is then free to form a dative bond with the lone pair of a fully connected neighbouring atom; the defect now becomes threefold coordinated and positively charged (D^+) (see Fig. 6.7(a)). Note that in the latter case the site of the defect moves from the originally dangling bond to the atom providing the lone-pair electrons for the dative bond; this feature has important ramifications for systems containing more than one type of atom (see later). The repulsive Hubbard energy U involved in placing an extra electron on the same site to form a D^- centre is postulated to be outweighed by the energy gained in forming the extra bond at the D^+ site, rendering the reaction

$$2D^0 \rightarrow D^+ + D^- \qquad [6.12]$$

exothermic. This process can also be illustrated in terms of a configurational coordinate diagram (Fig. 6.7(b)). The appropriate configurational coordinate, q, in this case may be taken to be the sum of the distances between two D^0 centres and their respective nearest neighbour, but non-directly bonded, atoms, $q_1 + q_2$. Transfer of an electron between D^0 centres *without* a change in configuration costs the Hubbard energy U, but the lowest energy state for the two resulting oppositely charged centres lies at a *different* configurational coordinate because of the atomic relaxation (bonding) that takes place at the D^+ centre; the energy for this equilibrium configuration is lower by U_{eff} than that for the two D^0 centres.

Kastner *et al.* (1976) considered the same process of spin pairing at defects in amorphous chalcogenides referring to it as 'valence alternation', and introduced the useful notation C_3^+ (for D^+) and C_1^- (for D^-), where C stands for a chalcogen atom and the subscript and superscript refer, respectively, to the coordination and charge state of a defect site. Their description of the diamagnetic ground state of a chalcogenide glass in terms of charged defects is identical to that of Mott *et al.*, but the two treatments differ in their description of the neutral paramagnetic centres; Kastner *et al.* (1976) believed that the symmetrical C_3^0 centre was of lowest energy (see Fig. 6.8), whereas Mott *et al.* (1975a) considered the centre to lie in configuration somewhere between C_1^0 and C_3^0. Realistic calculations for a-Se have since indicated that the C_1^0 centre is much more stable than C_3^0, principally because of π-bonding interactions between the dangling bond orbital on the C_1^0 and a lone pair on its neighbouring C_2^0 atom (Vanderbilt and Joannopoulos 1980). Furthermore, the two approaches differ in the precise mechanism giving a negative U_{eff}; we have seen that Mott *et al.* proposed that this was due to the lattice relaxation that accompanies the additional bonding taking place at the $D^+(C_3^+)$ centre, very much in the spirit of Anderson (1975), whereas Kastner *et al.*, believing the C_3^0 to be the lowest energy neutral defect, assumed that [6.12] is exothermic if the energy Δ (by which antibonding states are pushed up exceeds that by which bonding states are depressed) is sufficiently large, i.e. $2\Delta > U$ (see Fig. 6.8). Note that the energies

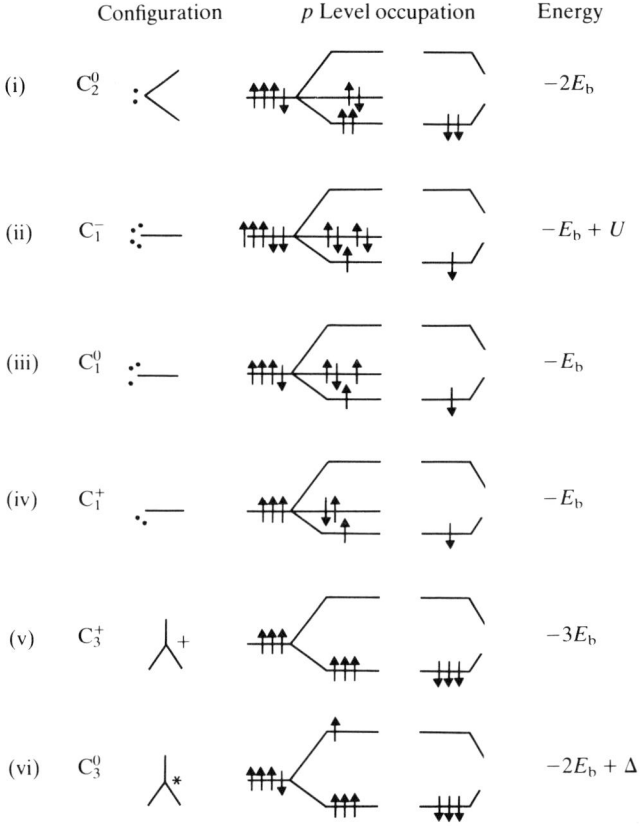

Fig. 6.8 p-Level occupations and energies for chalcogen defect configurations. E_b = bond energy, U = true correlation energy in a p–π orbital and Δ = antibonding repulsion. The electrons in the right-hand series of orbitals belong to the other atoms participating in the covalent bond (Kastner et al. 1976).

involved in the latter case are entirely electronic and any relaxation (polaronic) energies are neglected.

Kastner et al. also pointed out that since the creation of a valence alternation pair (VAP), C_3^+, C_1^-, from two normally bonded C_2^0 atoms

$$2C_2^0 \rightarrow C_3^+ + C_1^- \qquad [6.13]$$

involves essentially just a bond-flip conserving the total number of bonds, the creation energy U is low (~ 0.5 eV), much smaller than the energy to break a bond ($E_b \sim 2.5$ eV) (although Vanderbilt and Joannopoulos (1980) have argued that the creation of a dangling bond (C_1^0) is itself very low (~ 0.5 eV) because: (a) the repulsive energy of inter-core interactions is reduced on breaking a bond; and (b) π-bonding interactions further lower the energy – see above). The consequence of this is that a relatively high density of charged defects will be present in thermal equilibrium in a chalcogenide melt, which will be frozen-in on quenching to form a glass at a concentration corresponding to equilibrium at T_g. We may estimate the concentration of *randomly* distributed D^+, D^- centres as:

$$n_R \simeq n_0 \exp(-U/2k_B T_g) \qquad [6.14]$$

where n_0 is the density of chalcogen atoms. For glassy Se $n_0 \sim 10^{22}$ cm^{-3} and $T_g \sim 300$ K; this gives $n_R \sim 10^{18}$ cm^{-3}, a value in accord with estimates from photoluminescence and photo-induced ESR experiments (see later). Since the conjugate defects are equally and oppositely charged, it might be expected that they would tend to pair up in the melt under the action of their mutual coulombic attraction, thereby lowering the creation energy even further; in this case, the number of 'intimate valence alternation pairs' – IVAPs (Kastner et al. 1976) would be higher than n_R.

Note that VAP formation through overcoordination in chalcogenide glasses is facilitated by the existence of high-lying non-bonding electron states, which do not occur in other systems. In elemental group V ('pnictide') amorphous materials, the only non-bonding electrons are the s-states lying deep in the valence band (sect. 5.2.1), which can only be utilized if sp^3 hybridization occurs, thereby allowing the formation of overcoordinated P_4^+ centres, as well as the conjugate p-bonded P_2^- centres (P stands for pnictide). Because of the hybridization energy required, the energy balance for VAP formation is much finer than for the chalcogenide case; indeed it has been argued (Elliott and Davis 1979) on the basis of simple chemical bonding calculations as shown in Fig. 6.9 that a negative U_{eff} in e.g. a-As should only arise at those atomic sites which are considerably distorted, away from the

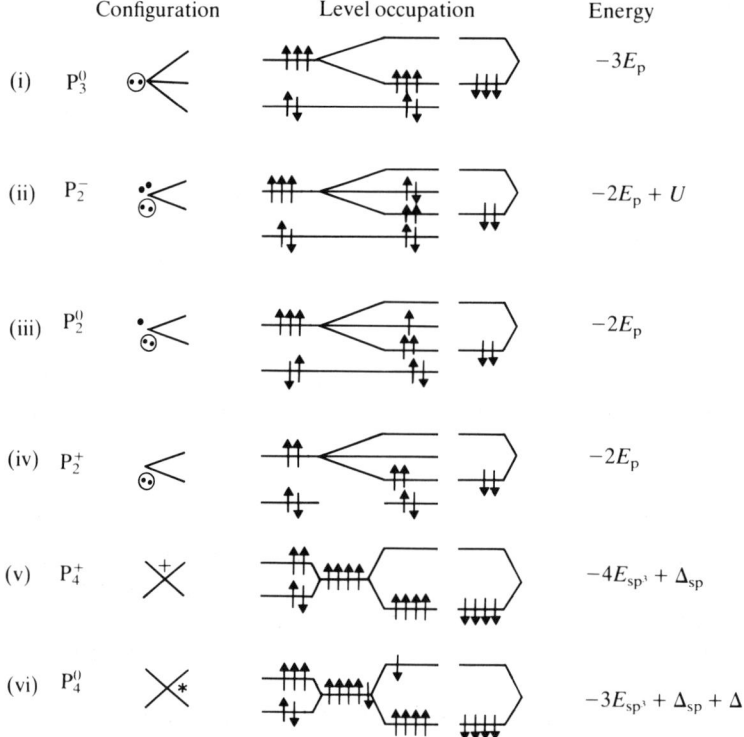

Fig. 6.9 Level occupation and energies for pnictide defect configurations, with E_p = p bond energy, E_{sp^3} = sp^3 bond energy, Δ_{sp} = s–p hybridization energy, Δ = antibonding repulsion energy and U = true correlation energy of a p lone-pair (π) orbital (after Elliott and Davis 1979).

equilibrium bond angle of 98° and nearer the tetrahedral angle of 109°, such that the hybridization energy has already been provided in part by the lattice. Otherwise, the p-bonded P_2^0 centres are energetically stable, and a-As is therefore a 'mixed' positive–negative U_{eff} system. Another such mixed system is possibly a-Si, for which since all atoms are sp^3 hybridized, overcoordination using s- or p-like non-bonding electrons cannot take place. Instead, it has been proposed that the ionic form of the Anderson negative U_{eff} polaronic mechanism discussed earlier is operative, with the hybridization tending to p-like at the doubly occupied sites T_3^-, and towards sp^2-like at the unoccupied sites T_3^+ (T standing for tetrahedral), but that a negative U_{eff} only obtains when the oppositely changed centres are sufficiently close that the attractive coulombic energy is gained in addition; otherwise the dangling bonds experience a positive U_{eff} and the paramagnetic T_3^0 centres are energetically stable (Elliott 1979a).

Thus far we have considered exclusively monatomic systems, for which there are only three distinct charge states of a dangling bond, D^+, D^0, and D^-. For alloys the situation becomes more complicated since each atom type, i, can be the site of a dangling bond, each capable of existing in three charge states D_i^+, D_i^0 and D_i^- (Street and Lucovsky 1979). For simplicity we will concentrate on binary alloys since among these number the important amorphous semiconducting systems, the arsenic chalcogenides (e.g. As_2Se_3) and the germanium chalcogenides (e.g. $GeSe_2$). Following Robertson (1982), we introduce the labels c and a for the more electropositive (cationic) and electronegative (anionic) species in the alloy, and observe that now *four* reactions similar in form to that in [6.12] become possible. Two are amphoteric reactions of the same type of defects, namely:

$$2D_c^0 \rightarrow D_c^+ + D_c^- \quad : \quad U_{\text{eff}}^c \qquad [6.15]$$

$$2D_a^0 \rightarrow D_a^+ + D_a^- \quad : \quad U_{\text{eff}}^a \qquad [6.16]$$

where U_{eff}^a and U_{eff}^c are the (true) intradefect effective correlation energies. In addition there are two interdefect reactions:

$$D_c^0 + D_a^0 \rightarrow D_c^+ + D_a^- \quad : \quad \tilde{U}_{\text{eff}}^{ca} \qquad [6.17]$$

$$D_c^0 + D_a^0 \rightarrow D_a^+ + D_c^- \quad : \quad \tilde{U}_{\text{eff}}^{ac} \qquad [6.18]$$

where we have distinguished the intersite correlation energy from the intrasite energy by the tilda label. They are related by differences in the orbital energies:

$$\tilde{U}_{\text{eff}}^{ca} = (\varepsilon_c - \varepsilon_a) + U_{\text{eff}}^a \qquad [6.19]$$

$$\tilde{U}_{\text{eff}}^{ac} = (\varepsilon_a - \varepsilon_c) + U_{\text{eff}}^c \qquad [6.20]$$

If U_{eff}^c and U_{eff}^a are negative, the Fermi level will be pinned (as will be shown later), whereas the sign of $\tilde{U}_{\text{eff}}^{ca}$ and $\tilde{U}_{\text{eff}}^{ac}$ determine the charge of the defects of each species, as well as removing unpaired spins from the gap. For instance if $\tilde{U}_{\text{eff}}^{ca} < \tilde{U}_{\text{eff}}^{ac}$ and $\tilde{U}_{\text{eff}}^{ca} < 0$, the stable defects will be D_c^+ and D_a^- (as long as there are equal concentrations of both types of atom).

The situation for alloy systems is further complicated with respect to monatomic materials since we must also specify the atomic nature of the nearest neighbours of a given dangling bond defect, henceforth given in parentheses. Robertson (1980) has shown that for a-As_2Se_3, $U_{\text{eff}}^c < 0$ and $U_{\text{eff}}^a < 0$ and that the relevant series of defects are $D_c: C_3^+(P_3)$, P_2^0, P_2^- and $D_a: C_3^+(P_2C)$, C_1^0, C_1^-, and since $\tilde{U}_{\text{eff}}^{\text{ca}} < \tilde{U}_{\text{eff}}^{\text{ac}} < 0$, D_c^+ and D_a^- are the stable configurations in the ground state. Note that P_4^+ is not favoured energetically because of the s–p hybridization energy required. For the case of a-GeSe$_2$, since overcoordination at the T sites cannot occur without the use of d-orbitals, the defects are $D_c: C_3^+(T_3)$, T_3^0, T_3^- and $D_a: C_3^+(T_2C)$, C_1^0, C_1^-, although U_{eff}^a and $\tilde{U}_{\text{eff}}^{\text{ca}}$ are still negative, there is evidence from trap-limited conduction (see sect. 6.3.5) that $U_{\text{eff}}^c > 0$. The reason for the difference in the sign of U_{eff}^c for a-As$_2$Se$_3$ and a-GeSe$_2$ may be ascertained by a consideration of the respective ionicities. The extra bond at D_c^+ is dative in character, and since electrons are being donated from anion to cation against a 'potential gradient', U_{eff}^c decreases with increasing ionicity (Robertson 1982). It must be admitted that conclusions as to the nature of the lowest energy defects which are based on simple chemical bonding arguments, as indicated in Figs 6.8 and 6.9, are unreliable since they ignore intercore repulsions; however self-consistent calculations of defect energies are very difficult and have to incorporate many approximations to be tractable (see Vanderbilt and Joannopoulos (1980a,b, 1981) for recent discussions). Nevertheless, the little work that has been done in this area has thrown up interesting features, e.g. that π-interactions are important in Se, but not in As$_2$Se$_3$, etc., and that the P_2 centre is more likely to be *positively* charged than negatively (i.e. P_2^-) because its level is calculated to lie close to the conduction band.

The existence of VAP defects in a-SiO$_2$ has been suggested by Greaves (1978, 1979) and Lucovsky (1979a,b, 1980), since the oxygen atoms are also chalcogens with a lone-pair valence band maximum. Both models agree that the negatively charged centre is a non-bridging oxygen ($D_a^- = C_1^-$), but disagree over the nature of the D_c^+ centre, Greaves suggesting T_3^+ and Lucovsky $C_3^+(T_3)$. We will discuss these valence alternation centres in more detail in connection with the radiation-induced, so-called E' cation centre observed in both vitreous and crystalline silica, together with a rival vacancy model, in section 6.2.3. (The letter E stands for electron centre and the single prime represents the occupancy.) Silica, in both amorphous and crystalline forms, exhibits a plethora of radiation-induced defects (see e.g. Wong and Angell 1976, Griscom 1978). In addition to the E' centre in which an electron is trapped at a Si atom, radiation-induced paramagnetic centres are also observed in a-SiO$_2$ at which a hole is trapped at a charged non-bridging oxygen (C_1^-). These oxygen hole centres may be of two types: in 'wet' silicas containing a high proportion of —OH impurities, the defect is believed to be a $C_1^0(T)$ in the vicinity of a neutral hydroxyl network terminator $\geqslant$Si—OH; in high purity 'dry' silicas with a very low OH content, the defect is believed to be a dangling bond on a peroxyl bridge, i.e. $C_1^0(C)$ (Stapelbroek and Griscom 1978, Robertson 1982). In addition to the intrinsic defects mentioned above, there are also many radiation-induced defects associated with impurities, such as Al or B; since these impurity-associated defects give rise to optical absorption bands in the visible, they are often termed 'colour centres' (see Wong and Angell 1976 and Griscom 1978).

We conclude this introductory section on the charged dangling bond model with a discussion concerning the ability of these centres to pin the Fermi level in the gap. We have alluded to this property of negative U_{eff} states previously when discussing Anderson's polaronic mechanism, but the concept is easier to understand for the case of dangling bond defects which give rise to discrete levels (or narrow bands) in the gap of an amorphous semiconductor. Whether U_{eff} is positive or negative affects dramatically the movement of the Fermi level when electrons are injected into the system either by the application of an external electric field (as in the field-effect experiment, Fig. 6.3) or by the addition of dopant atoms. Adler and Yoffa (1976) approached this problem by considering a single defect capable of being unoccupied, singly or doubly occupied, and used the grand partition function to calculate the Fermi level position as a function of the number of electrons per defect, n ($0 < n < 2$):

$$E_{\text{F}} = E_{\text{eff}} - k_{\text{B}} T \ln \left\{ \frac{2|n-1|}{1-|n-1|} \right\} \quad 0 < n < 1 \quad [6.21(a)]$$

$$E_{\text{F}} = E_{\text{eff}} + U_{\text{eff}} + k_{\text{B}} T \ln \left\{ \frac{2|n-1|}{1-|n-1|} \right\} \quad 1 < n < 2 \quad [6.21(b)]$$

for the case where $U_{\text{eff}} > 0$, and

$$E_{\text{F}} = E_{\text{eff}} - \frac{|U_{\text{eff}}|}{2} - \frac{k_{\text{B}} T}{2} \ln \left\{ \frac{2}{n} - 1 \right\} \quad [6.22]$$

for the case where $U_{\text{eff}} < 0$. For *positive* U_{eff} materials there is a level at E_{eff} for singly occupied states and lying above it at a separation U_{eff} a level corresponding to doubly occupied centres (see Fig. 6.10(a)). As the electron concentration, n, increases from zero at $T=0$ K, the Fermi level rises slowly in the lower singly occupied level. Note that the density of upper levels is equal to the density of the *filled* lower levels (a state cannot be doubly occupied for a positive U_{eff} material until it is first singly occupied). As soon as $n=1$, the Fermi level jumps discontinuously across the gap of magnitude U_{eff} to the upper level, and subsequently rises gradually through this level

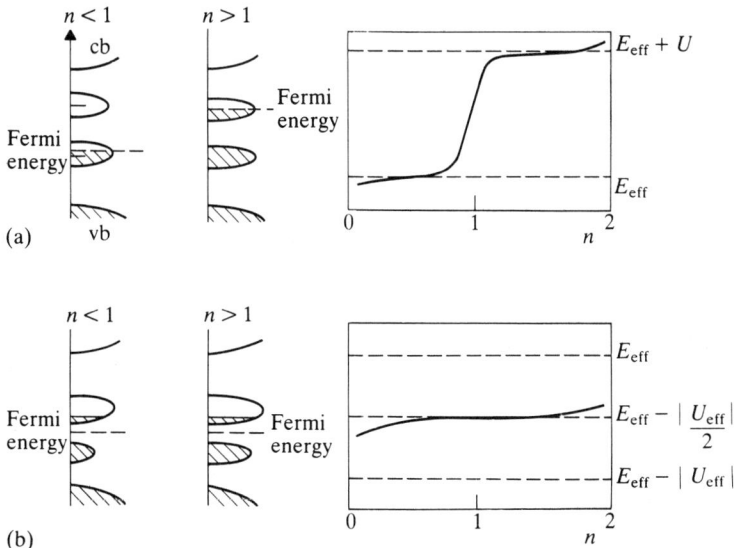

Fig. 6.10 Density of states and Fermi level as a function of electron occupation, n, for (a) $+U_{\text{eff}}$ defects; (b) $-U_{\text{eff}}$ defects (Adler and Yoffa 1976).

until $n=2$ and all states are doubly occupied. This behaviour can be seen by reference to [6.21] and is illustrated in Fig. 6.10(a). For *negative U_{eff}* defects, however, the levels corresponding to double occupancy lie *below* those for single occupancy, and the density of lower levels is equal to the number of electrons introduced and thus this band is always full (Fig. 6.10(b)). The injection of electrons into such a material therefore results in immediate double occupancy of levels without an intermediate $n=1$ stage. Thus, E_F remains pinned midway between the two levels at an energy $E_{eff} - U_{eff}/2$ for all values of n, except when n is very near 0 or 2, as given by [6.22] and shown in Fig. 6.10(b). Thus in a field-effect experiment, for example, the material *behaves* as if there were a high density of states deep in the gap, pinning the Fermi level there, whereas in reality there are *no* one-electron states at that energy (Frye and Adler 1981) – see Fig. 6.10(b). For a further discussion of the properties of negative U_{eff} charged dangling bond states see section 6.3.1, in particular Fig. 6.17.

6.2.3 Vacancies

Atomic vacancies are a widespread form of point defect in covalent, ionic and metallic crystalline materials. They are responsible for many of the electrical and optical properties of such materials, in addition to frequently controlling mass transport phenomena, such as diffusion, and hence many mechanical properties, e.g. creep (see e.g. Henderson 1972). Vacancies can be thermally generated, but more often are created by radiation damage processes, and they may occur either singly as monovacancies or as clusters, divacancies, etc. We discuss them here because after the single dangling bond, they are the most simple form of structural defect likely to be found in amorphous materials, and also they contain perforce a number of dangling bonds, which may reconstruct through the Anderson mechanism.

Recent studies on one of the most intensively studied vacancies in crystalline semiconductors, namely the monovacancy in Si, have indicated that this is a negative U_{eff} system in the Anderson (1975) sense. The monovacancy produced by electron irradiation at low temperatures has four sp^3 hybrid dangling bond orbitals pointing into the vacancy, and depending on the position of the Fermi level in the gap (determined by p- or n-type dopants) the vacancy may possess a variety of electron occupancies and hence charge states, the existence of which may be confirmed using ESR (for paramagnetic configurations) and diffusion (for diamagnetic configurations). In this way, the existence of V^{2-}, V^-, V^0, V^+ and V^{2+} charge states of the vacancy has been inferred.

Consider first the series of charge states V^{2+}, V^+ and V^0. It has been proposed (Baraff *et al.* 1980a,b) that the ESR-active V^+ charge state should always be metastable and as the Fermi level is moved through the gap by p-type doping, either the V^{2+} or the V^0 centre would be the more stable, the V^{2+} centre being more stable for E_F very close to the valence band edge (i.e. p^+-doping), and the V^0 centre in turn being more stable if E_F lies deeper than ~ 0.1 eV from the valence band edge. The reason for this negative U_{eff} behaviour, as we shall see in a moment, is the Jahn–Teller splitting of certain degenerate electron energy levels accompanying a lattice distortion. Note that this system is of 'pure' Anderson type: either charge state V^{2+} *or* V^0 is stable with respect to V^+ because of lattice distortion and consequent

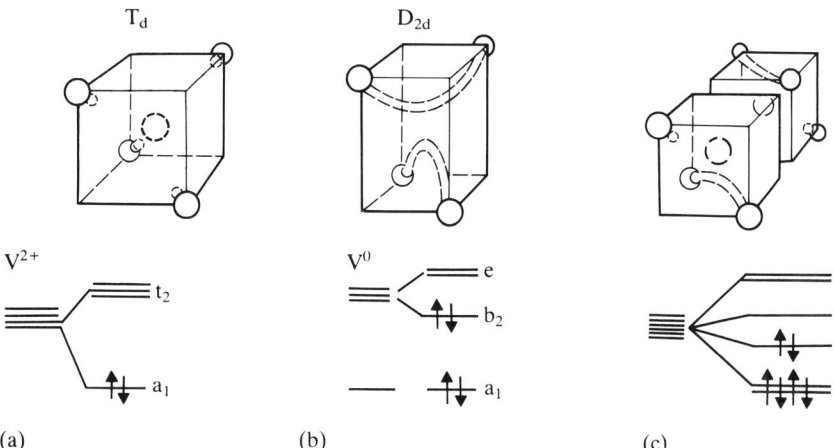

Fig. 6.11 Atomic configurations and level occupancies of a monovacancy in crystalline Si (Robertson 1982):
(a) undistorted V^{2+} centre (T_d symmetry);
(b) tetragonally distorted (D_{2d}) V^0 (and V^+) centre;
(c) neutral divacancy.

spin pairing, rather than the reaction $2V^+ \to V^0 + V^{2+}$ being exothermic, creating *pairs* of differently charged entities as in the charged dangling bond model.

The undistorted, 'ideal' monovacancy has four dangling bonds centrally directed into the vacancy, and has T_d symmetry; this is the configuration adopted by the V^{2+} centre, where the two electrons present occupy in a spin-paired fashion the lowest singly degenerate symmetric A_1 state which lies near the top of the valence band, leaving the triply degenerate T_2 state lying in the gap, unoccupied (see Fig. 6.11(a)). The T_2 level is susceptible to a Jahn–Teller distortion upon its occupation by any electrons, which in this case will be tetragonal with D_{2d} symmetry, and the level splits into a lower T_x state and a degenerate T_y and T_z state. The third electron in a V^+ centre will go into the T_x state, and the fourth in a V^0 centre will also go into the T_x level, further increasing the distortion and pairing the spins (Fig. 6.11(b)). Thus, the vacancy can be regarded as having distorted to give rise to a rebonding of the dangling bond orbitals. The Jahn–Teller distortion is such that as electrons are progressively introduced into the system by movement of the Fermi level through the gap, the V^{2+} centre transforms directly into the V^0 centre, without the production of the intermediate V^+ centre. Direct evidence for this behaviour has come from ESR studies (Watkins and Troxell 1980), and it has also been suggested that the V^0, V^- and V^{2-} series of charge states is similarly subject to a negative U_{eff}, accompanying a trigonal, rather than tetragonal, distortion (Baraff *et al.* 1980b). It is interesting to speculate that such Jahn–Teller driven reconstructions might also occur in a-Si and other similar materials, although it is perhaps doubtful whether vacancies, as such, exist in as-prepared material without being produced by irradiation, and dangling bonds are more likely to be the prevalent point defect in such tetrahedrally bonded materials. In the past, it has been suggested that divacancies (see Fig. 6.11(c)) are a ubiquitous defect in a-Si, and in particular a-Si:H, since the peaks in the gap-state density of states that had been observed by field effect in a-Si:H corresponded in energy with the energy levels found for the divacancy in c-Si (Spear 1974); however, it has recently been found that field-effect conductance measurements alone are incapable of detecting peaks in the gap-state density of states (Powell 1981), and so there is little direct evidence now for divacancies in these materials.

An example of a vacancy centre in an amorphous system which has been well documented is the so-called E' centre produced in vitreous SiO_2 by X-ray or γ-ray irradiation. This produces an optical absorption band at 5.8 eV, approximately in the middle of the gap (see e.g. Greaves 1978). Evidence from ESR indicates that this paramagnetic centre has a single electron localized in a single Si sp^3 orbital (see e.g. Griscom 1978, Friebele and Griscom 1979), which implies that the centre is associated with an oxygen vacancy. Two distinct microscopic models have been proposed for the E' centre. The first considers the paramagnetic centre to be a simple oxygen vacancy in which the two Si dangling bond sites relax *asymmetrically*, the T_3^+ site distorting to an almost planar sp^2 configuration, as also postulated for the case of the c-Si (111) surface and dangling bond defects in a-Si (see previously), and the T_3^0 centre remaining virtually undistorted (Fiegl et al. 1974). This configuration is shown in Fig. 6.12(a), where it can be seen immediately that the centre in its paramagnetic state is charged, although the nature of the precursor diamagnetic centre is uncertain; either it consists of two neighbouring T_3^+ centres (compensated elsewhere in the network by non-bridging oxygens C_1^-) and irradiation gives rise to *electron* trapping, or it consists of a pair of T_3^0 centres (presumably forming a weak bond between the dangling bond orbitals rendering the system diamagnetic) and irradiation gives rise to *hole* trapping. The other model that has been proposed for the E' centre in its excited paramagnetic state is an isolated T_3^0 dangling bond; however Greaves (1978, 1979) suggests that the diamagnetic ground state comprises a T_3^+ centre, whereas Lucovsky (1979a,b, 1980) proposes that it is a $C_3^+(T_3)$ defect (Figs. 6.12(b) and (c)). Detailed ESR experiments indicate that the Fiegl vacancy model is

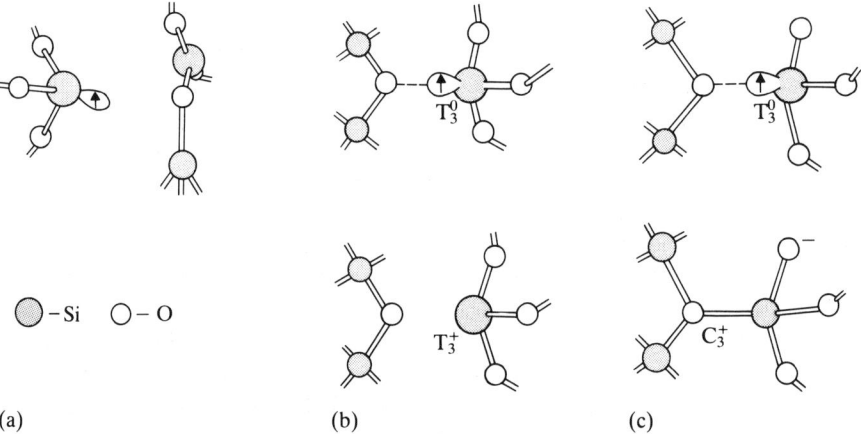

Fig. 6.12 Models for the E' centre in vitreous silica (after Robertson 1982):
(a) the Fiegl model (symmetrically relaxed O^- vacancy);
(b) the Greaves model T_3^0/T_3^+;
(c) the Lucovsky model $T_3^0/C_3^+(T_3)$.

preferable to the two charged defect variants, however, principally because of the hyperfine splitting that has been observed in v-SiO$_2$ enriched with ^{17}O and ^{29}Si; it was found that the spin on the Si atom interacts weakly with two other Si sites and three other O sites, which rules out the Greaves and Lucovsky models since in them there is an appreciable interaction with only one O atom. The same type of defect has been observed in irradiated crystalline SiO$_2$ (quartz), which has very similar optical absorption and ESR characteristics. However, there is a considerable amount of evidence that there are *two* distinct E' centres, the E$'_1$ and the E$'_2$, the former being essentially identical to the Fiegl model discussed above, and the E$'_2$ being similar but associated with an H atom (because of the ubiquitous OH impurity) and distinguished by the hyperfine interaction between the electron spin on the T$_3^0$ and the proton.

6.2.4 Density defects

The structural defects that we will discuss in this section cannot be categorized as point defects, like dangling bonds, mono- or divacancies, for example. These density defects range from voids (say < 100 Å), which may be regarded as an aggregate of a non-fixed number of vacancies, to fluctuations in density (or free volume), which may be regarded as vacancies which have become distributed throughout the material and which therefore cannot be regarded any more as well-defined point defects. We have mentioned in passing both of these forms of density defect earlier in this book.

It is rather a moot point whether one ought to regard these density variations as defects in the sense of our previous definition, namely departures from an 'ideal' amorphous structure, or whether they are an inevitable consequence of producing an amorphous solid for the material in question. In particular, are the voids – which are a universal feature of evaporated a-Si or a-Ge thin films – an inevitable consequence of the fact that these tetrahedrally bonded materials are 'over-constrained' (Phillips 1979 – see sect. 2.5.4), and do they therefore act as strain-relief centres for the excess strain otherwise incorporated into the random network if only perfect tetrahedral coordination is allowed? In addition, we have already seen that on the free-volume model (Cohen and Turnbull 1959 – see sect. 2.3.4) the inclusion of surplus free volume, principally in materials which have a significant degree of non-directional bonding (metals, ionic solids, polymers), is one of the factors that distinguish thermodynamically amorphous materials from their crystalline counterparts and is an inescapable feature of the amorphous state.

The presence of voids may be detected directly using electron microscopy (see Fig. 3.24(a)) for relatively large voids, and small-angle scattering (sect. 3.5.2) for smaller sized voids, and indirectly by a variety of techniques including density measurements (evaporated a-Ge and a-Si can be 3–15% less dense than their crystalline forms, whereas CRN models which are void free only have 1–3% density deficits). The presence of such voids influences to a great extent the physical properties of such films, and although annealing the samples to elevated temperatures does reduce the void concentration to a certain extent, they cannot be removed entirely before the crystallization temperature T_{cr} is reached. An illustration of the

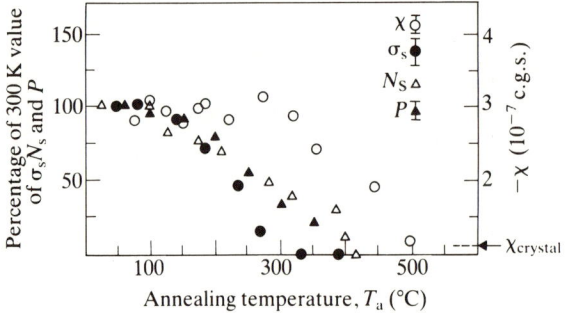

Fig. 6.13 Annealing properties of void-related properties of a-Ge. The properties are porosity (P), free spin density N_s, internal stress σ_s and diamagnetic susceptibility χ (Paesler *et al.* 1974).

widespread control that voids have on a variety of physical properties is shown in Fig. 6.13, where the unpaired electron spin density, the porosity and the internal stress of a-Ge films all decrease in a similar fashion on annealing, and they can be associated with a decrease in the void concentration; only the diamagnetic susceptibility retains its enhancement relative to the crystalline value until very close to T_{cr} indicating that it is associated instead with the bulk bonding states (Paesler *et al.* 1974). It was deduced that for freshly evaporated films, voids with an average radius of ~ 30 Å were present in a concentration of $\sim 10^{18}$ voids cm^{-3}. The presence of these voids has a profound influence on electronic properties too, because of the dangling bonds that are expected at the void surfaces, even allowing for the fact that many may reconstruct via one or other of the mechanisms mentioned in sections 6.2.2 and 6.2.3. Thus, a.c. conductivity and variable-range hopping d.c. conduction taking place through these states is observed, as is a large ESR spin signal ($\sim 10^{19-20}$ spins cm^{-3}), although isolated dangling bonds in the bulk (sect. 6.2.1) may make a contribution to these behaviours as well. The sensitivity of such evaporated films to changes in ambient gaseous conditions is also understandable, given the high degree of porosity characteristic of void aggregates and thus the ease with which gaseous impurities may interact with the electronically active electron states; this may explain the non-reproducibility of experimental results that bedevilled the field in the early days.

We turn now to the other form of density defect under consideration, namely those fluctuations in density which represent a distribution of free volume throughout the bulk. We have already dealt in section 2.3.4 with the free-volume concept and how it can account for mass-transport phenomena (e.g. viscosity) in the liquid phase, in particular in being able to account for the Vogel–Tammann–Fulcher relation for the viscosity ([2.6]). These free-volume defects also control mass-transport phenomena in glasses below T_g, such as diffusion, stress-induced flow (creep) and structural relaxations; Spaepen (1981) has reviewed these aspects for the case of amorphous metals.

There is much experimental evidence, e.g. from similarities in respective activation energies, that there is a close similarity, if not identity, between the defects responsible for diffusion and flow, but that these differ both structurally and in their concentration from those defects which control structural relaxation. The plausibility of an identity between diffusion and flow defects may be confirmed by

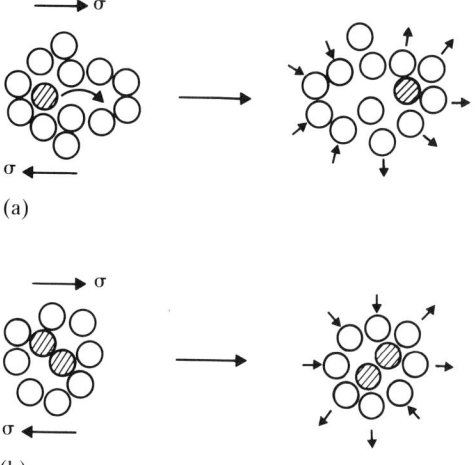

Fig. 6.14 Schematic illustration of the motion of free-volume defects which can give rise to diffusion or flow under the action of a shear stesss σ:
(a) single atom motion;
(b) two atom motion.

consideration of microscopic models for the defects. A diffusion defect must allow a change in nearest neighbours of a given atom, whereas a flow defect, upon the application of a shear stress, must allow a local shear transformation (which is then transferred elastically to the surface of the solid to produce a macroscopic strain). The free volume may be associated with a single atom (see Fig. 6.14(a)), which may hop from one site to another (accomplishing diffusive motion) with a concomitant collapse of the original cage of neighbouring atoms (producing a local shear strain). Alternatively, the free-volume defect may be associated with more than a single atom, and we show in Fig. 6.14(b) the motions of two atoms which can undergo a shear transformation and hence also contribute to diffusion and flow. It is to be noted that these diffusion and flow defects are rather localized, in contrast to relaxation defects which require long-range elastic stress fields to act as sinks in order that they may annihilate free volume to accomplish the irreversible process of structural relaxation (see Fig. 2.1).

We may discuss the influence that volume defects exert on the process of plastic flow in a little more detail, but at the outset we must distinguish between homogeneous and inhomogeneous flow: *homogeneous* flow occurs when each volume element of the sample contributes to the strain, and in amorphous metals this occurs for low stress levels at high temperatures; *inhomogeneous* flow, on the other hand, occurs for very high stress levels at low temperatures, and the strain is localized in a few shear bands. The plastic flow rate of a system subjected to a shear stress σ can be written for the flow defect model as (Spaepen 1981):

$$\dot{\gamma} = n_f \gamma_0 v_0 v_f \qquad [6.23]$$

where $\dot{\gamma}$ is the shear strain rate, n_f the concentration of flow defects, v_0 the volume of a defect, γ_0 the local shear strain resulting from the rearrangement of a defect and v_f the

Defects

net jump frequency of a flow defect under stress. (Equation 6.23 is in fact completely general, and can equally apply to the case of dislocation motion in crystals.)

In the absence of an applied stress, the defect will hop at an equal rate v'_f between one site and another. The application of an external stress, σ, will bias these jumps in a given direction by a stress-dependent factor $\beta(\sigma)$:

$$v_f = v'_f \beta(\sigma) \qquad [6.24]$$

where the unbiased jump rate can be written as:

$$v'_f = v_a \exp(-\Delta G/k_B T) \qquad [6.25]$$

and where v_a is the attempt frequency (of the order of a phonon frequency) and ΔG is the activation energy for motion, involving either strain energy and/or bond-breaking energy. The biasing factor $\beta(\sigma)$ may be written as the difference of two opposing fluxes, which gives:

$$\beta(\sigma) = \sinh\left(\frac{\sigma \gamma_0 v_0}{k_B T}\right) \qquad [6.26]$$

and if the stress is small, $\sigma < k_B T/\gamma_0 v_0$, [6.26] can be approximated as:

$$\beta(\sigma) \simeq \frac{\sigma \gamma_0 v_0}{k_B T} \qquad [6.27]$$

i.e. the flow is Newtonian. The shear viscosity, η, can therefore be defined as:

$$\eta \equiv \frac{\sigma}{\dot{\gamma}} = \frac{k_B T}{n(\gamma_0 v_0)^2 v'_f} \qquad [6.28]$$

Three temperature regimes for homogeneous flow may be distinguished (Spaepen 1981):

(a) $T > T_m$. In the fluid phase, the number of defects is of the order of the number of atoms, and single atom motion is dominant (cf. Fig. 6.14(a)) for which $v_0 \simeq \Omega$ (the atomic volume) and $\gamma_0 \simeq 1$. Furthermore, ΔG is small because of the minimal structural rearrangements that are necessary, and hence the overall temperature dependence of the viscosity is weak.

(b) $T \simeq T_g$. In this very viscous region ($\eta \simeq 10^{13}$ poise) the dominant parameter in [6.23] is the increase in the defect concentration with increasing temperature. This may be written as:

$$n = \frac{1}{\Omega} \exp\left(-\frac{cv^*}{v_f}\right) \qquad [6.29]$$

where v^* is the minimum hole size, v_f the free volume per atom and c is a geometric factor ($\simeq 1$). This equation, together with the expression for free volume derived in section 2.3.4, yields a Vogel–Tammann–Fulcher form for the viscosity. Since $v_0 \gamma_0 \simeq 1$, [6.27] is expected to apply still, and the flow should be Newtonian.

(c) $T < T_g$. Behaviour in this region is complicated, since unlike the case for $T > T_g$, the glassy solid is not in thermal equilibrium and structural relaxation to a higher viscosity state takes place. Thus, in order to perform true isoconfigurational creep measurements, the specimen must be annealed for a long time at a temperature higher than that of the test to ensure that structural relaxation does not take place during the time taken by the measurements. Since v_f continues to decrease with decreasing temperature because of the non-configurational part of the thermal expansion, the defect concentration continues to decrease below T_g ([6.29]), and the viscosity should have a non-Arrhenius temperature dependence. Such isoconfigurational experiments are difficult and experimental results in support of this model are wanting.

At very high stress levels, the strain is confined to the small volumes of shear bands which nucleate at edge cracks and propagate in a direction at 45° to the tensile axis. Fracture occurs along the shear band, and the fracture surfaces after failure display a 'vein' pattern characteristic of the Taylor instability developed when two solid surfaces separated by a fluid layer are pulled apart. These observations can be reconciled within the free-volume model by noting that dilatation at a crack tip will increase the free volume locally and hence dramatically lower the viscosity, rendering it liquid-like. This softening mechanism then allows flow on a 45° plane (corresponding to the direction of maximum shear stress), and the deformation can propagate through the material (Polk and Turnbull 1972).

6.2.5 Dislocations and disclinations

Dislocations are a ubiquitous form of defect in crystalline solids, controlling many of their mechanical properties. It has been suggested that dislocations, perhaps of a generalized form, also exist in amorphous solids. In particular, the similarity of the morphology of the fracture surfaces of glassy metals subjected to high-stress plastic flow to the slip bands resulting from plastic flow in single crystals suggests that dislocations may be responsible in the amorphous phase as well. However, we have seen in the previous section that this morphology can be accounted for by free-volume theory in which high stress softens the material locally on the shear plane by the introduction of extra free volume. It is in fact rather difficult to conceive of a dislocation in the usual sense in the absence of a periodic lattice. The only modelling study to date has been that of Chaudhari *et al.* (1979). By making a planar cut in a monatomic DRP model (Finney 1970) and displacing one side of the cut with respect to the other by a distance comparable to an atomic spacing, an edge or screw dislocation could be generated if the displacement were perpendicular or parallel, respectively, to the end line of the cut. Energy relaxation using Lennard–Jones potentials of the resulting defective clusters under zero stress (static) conditions indicated that the edge dislocation was unstable to reconstruction, whereas the screw dislocation was stable and retained its identity. It has not yet been established whether such contorted line defects in amorphous solids act in a similar manner to dislocations in crystals in facilitating plastic flow under stress.

Rivier (1979) has proposed that an additional type of line defect may exist in amorphous solids. This is termed a 'disclination' and is associated with rotations, in contrast with the case of dislocations already mentioned which are associated with translational symmetry. A disclination in an amorphous solid may be viewed as arising in the following way. Consider a tetrahedrally bonded material, e.g. Si, which in the (diamond cubic) crystalline form consists only of six-membered rings. The amorphous form, in contrast, contains a large variety of ring sizes, both odd and even, unless precautions are taken to construct only an even-ring CRN (Connell and Temkin, 1974). An odd-membered ring may be imagined to be produced by cutting a sixfold ring through the centre, adding or removing a wedge of material and reconnecting; the edges of the cut have been rotated to make space for the wedge and so the odd-membered ring surrounds the core of a disclination. Thus, if a sixfold ring is thought of as having zero local curvature, a fivefold ring has positive and a sevenfold ring has negative curvature. Rivier suggested that the odd-membered rings in a CRN can all be threaded through by one of these disclination lines, which avoid any even-membered rings and form closed loops in the body of the network or terminate at the surface.

It was further suggested that the elastic energy of the disclinations could be substantially reduced if fivefold and sevenfold rings are neighbours (i.e. share a common bond), in which case the disclination line is forced to adopt a sharp hairpin configuration in passing through the ring centres, and the net local curvature is therefore essentially zero. Thus it would appear that the strain energy of a CRN would be minimized in this respect if there were an equal number of five- and sevenfold rings. Steinhardt and Chaudhari (1981) have disagreed with this conclusion by pointing out that the ratio of five- to sevenfold rings in tetrahedral CRNs is in fact approximately 2:5, implying that the interaction between disclinations is screened. However, conventional strain energy relaxation of a CRN network does not alter the existing topology, and the answer to the question of whether the strain energy really is lowered in a CRN with an equal number of positive and negative curvature odd rings must await the construction of such a model. Steinhardt and Chaudhari do point out that the Steinhardt 201 atom and Polk–Boudreaux 519 atom models (with very similar ring statistics) do fit the RDF of a-Si well, and that any change in the ring statistics would worsen the fit. However, this is somewhat beside the point and does not prove or disprove the idea that the presence of disclinations might influence the properties of an amorphous network.

6.3 Defect-controlled properties

Thus far, we have described some of the more common forms of structural defect that can occur in amorphous solids. We turn now to a discussion of the physical properties which such defects govern, stressing in particular those aspects of the behaviour of amorphous solids which have not been addressed as yet in this book. In the main, we will consider those defect-controlled properties exhibited by amorphous semiconductors and insulators, leaving a discussion of the behaviour of amorphous metals until Chapter 7. Such properties are generally optical or electrical in nature, reflecting the unusual electronic characteristics of many defect configurations.

6.3.1 Photoluminescence

It must be stressed at the outset that the phenomenon of luminescence is *not* solely associated with the presence of structural defects; we have already seen (sect. 5.6.3) that luminescence can result from intraionic electronic transitions of transition metal (3d) or rare earth (4f) atoms in amorphous matrices (see e.g. Wong and Angell 1976). However, photoluminescence is also observed in many amorphous materials which do not contain such luminescent impurities, and therefore the phenomenon must also have a more general origin. Moreover, band-to-band photoluminescence is seldom if ever observed in amorphous materials; electron–hole pairs created optically by photons having energies comparable to the band gap do not generally recombine radiatively directly across the gap emitting a photon of energy also comparable to the band gap, but recombine non-radiatively or via other luminescence channels. These photoluminescence spectra are observed almost always at energies considerably lower than the energy of the exciting photons, and electron states in the gap are inevitably involved. These states can be band-tail states (see sect. 5.3.2), believed to be responsible for photoluminescence in a-Si:H, but are usually associated with structural defects (dangling bonds); moreover the deeper into the gap a level lies, the stronger will be the electron–phonon coupling and this will greatly influence the luminescence process. Extensive reviews on luminescence in amorphous semiconductors have been given by Street (1976), Fischer (1979) and Mott and Davis (1979).

Various mechanisms for radiative recombination which give rise to photons emitted at significantly lower energies than those of the exciting radiation are illustrated in Fig. 6.15. One possibility is that an electron excited from the valence band into the conduction band will recombine with defect states forming a band deep in the gap (Fig. 6.15(a)); another is that a similarly excited electron will thermalize down in energy through the deep-lying localized tail states (in the Cohen–Fritzsche–Ovshinsky model – sect. 5.3.2) until recombination occurs with a

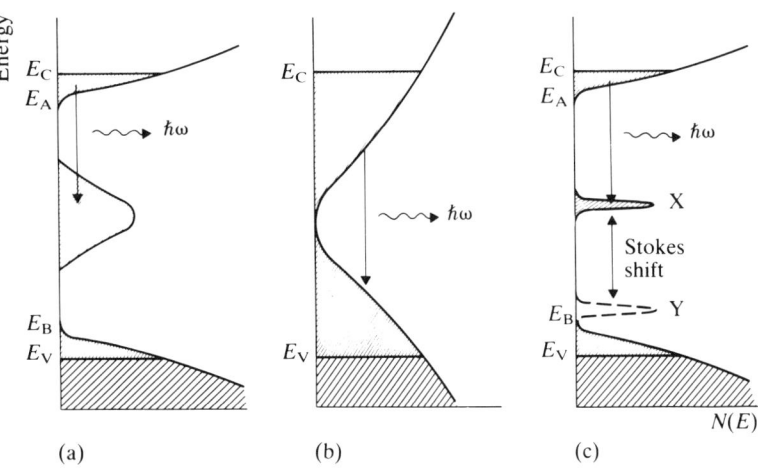

Fig. 6.15 Possible radiative recombination channels which give rise to a luminescent energy much less than the excitation energy: (a) and (b) assume no electron–phonon coupling, whereas (c) assumes strong electron–phonon coupling (Street 1976).

Defects

hole trapped at a comparably deep valence band tail state (Fig. 6.15(b)). Both these mechanisms assume negligible electron–lattice coupling. Alternatively, this coupling may be strong, and the energy of the level changes with its electron occupation, with the consequence that if an electron is excited to the conduction band from a defect gap state (at Y in Fig. 6.15(c)), relaxation at the centre takes place after excitation and radiative recombination at the *same* centre (X) will occur at a *different* energy; the difference in excitation and luminescence energies (in this case only) is termed the 'Stokes shift'. Thus all three mechanisms can give rise to luminescence at a considerably lower energy than that of the exciting light, but there are distinct differences between them. In particular, note that optical absorption can in principle take place at the *same* energy as the luminescence energy for mechanisms (a) and (b), but *not* for mechanism (c) (this is more clearly demonstrated in Fig. 6.16) and this is the test for a true Stokes shift.

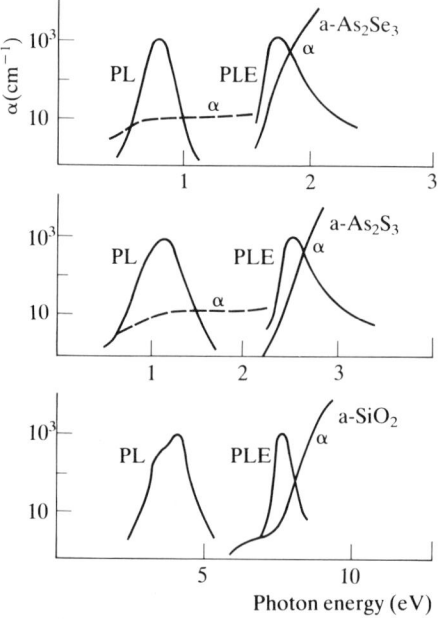

Fig. 6.16 Comparison of photoluminescence (PL), photoluminescence excitation spectra (PLE), absorption (α) and photo-induced absorption (dashed line) for three chalcogenide glasses, As_2Se_3, As_2S_3 and SiO_2. Note the similarity of PL and PLE spectra in all cases (after Robertson 1982).

One may distinguish between mechanisms (a), (b) and (c) by a consideration of photoluminescence and other optical behaviour, and as an example we may take the case of chalcogenide glasses. These are known to be substantially optically transparent for photon energies below the gap energy which rules out mechanism (b), since the optical absorption between the extensive tail states in the Cohen–Fritasche–Ovshinsky model is expected to be considerably higher than the residual Urbach tail. Furthermore, in view of this below-gap optical transparency, and the fact that photoluminescence occurs at approximately half the band-gap energy in these materials, obviously absorption and luminescence cannot occur at the same energy, ruling out mechanism (a) and leaving a Stokes-shift mechanism (c) as the only viable alternative. We will consider this case in a little more detail since the defects involved are believed to be the charged dangling bonds already discussed in section 6.2.2.

Experimental photoluminescence spectra for typical chalcogenide glasses (As_2Se_3, As_2S_3 and SiO_2) are shown in Fig. 6.16 where the photoluminescence (PL) spectrum is seen to be significantly Stokes-shifted from the excitation energy, shown in the figure as the photoluminescence excitation (PLE) spectrum (which is the energy dependence of the integrated PL intensity). Note that the peak in the PLE spectrum (i.e. the photon energy at which PL is most efficiently excited) occurs at relatively low values of the absorption coefficient ($\alpha \simeq 10^2$–10^3 cm^{-1}), well down the Urbach edge and consequently at a significantly lower energy than that corresponding to the band gap (for which $\alpha \simeq 10^5$ cm^{-1}). It is an interesting observation that *crystalline* As_2S_3 and As_2Se_3 exhibit PL spectra very similar to those of their amorphous counterparts; the Stokes shift and the width of the PL band are nearly identical. This marked similarity implies that the same luminescence mechanism is operative in both forms. However, this cannot be a universal conclusion, for the PL spectrum for c-Se shows no evidence for a Stokes shift; the small energy difference ($\simeq 0.2$ eV) that is observed between absorption and emission is ascribed to the fact that band-edge excitons are indirect and so phonons are involved in both absorption and PL processes (Fischer 1979).

The mechanism for PL in the charged dangling bond model is the following (Mott *et al.* 1975a, Street 1976). For simplicity we consider only electron recombination (the situation for hole recombination is similar). An electron may be photoexcited in one of two ways; either by direct excitation from a D^- centre (transition 1 in Fig. 6.17(a)) or else from a lone-pair tail state at the top of the valence band (transition 1' in Fig. 6.17(a)), in which case the hole produced is trapped by a D^- centre. In either case, the result is an electron trapped in a localized conduction band state, capable of recombining with a hole trapped at a D^- centre (transition 2 in Fig. 6.17(a)). The D^- with a trapped hole transforms into a D^0 centre, and therefore suffers a change in configuration. The overall luminescent process can thus be represented as:

$$D^- + h\nu_E \rightarrow D^0 + e \rightarrow D^- + h\nu_{PL} \qquad [6.30]$$

where $h\nu_E > h\nu_{PL}$. This process may be illustrated more graphically by means of a configuration–coordinate diagram (Fig. 6.17(b)), in which the potential curves shown are a sum of electronic and lattice components, and the distortion is clearly evinced by the shift in q of the potential minima. Photo-excitation occurs vertically in such a diagram at constant q, in accord with the Franck–Condon principle in which *optical* excitation of electrons takes place in a much shorter time than that in which the structure can relax (this is not so for *thermal* transitions as we shall see later). The excited D^0 centre will relax after the optical transition, losing energy non-radiatively by phonon emission until the potential minimum of the excited state is reached, whereupon the electron may recombine radiatively by means of another vertical transition at a Stokes-shifted energy (cf. Fig. 6.17(c)), or if lattice distortion in

Defects

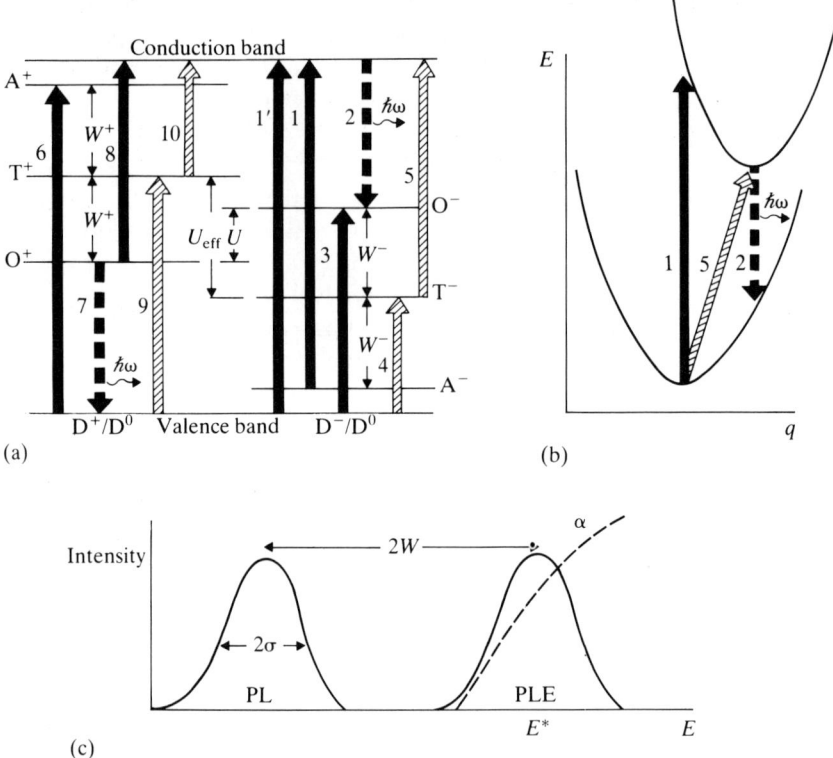

Fig. 6.17 Optical and thermal transitions between states in the gap arising from negative U_{eff} defects, e.g. in chalcogenide glasses.
(a) Levels in the gap and transitions between them. Levels labelled O and T are optical and thermal levels, respectively, and $A^+(A^-)$ is the level of $D^+(D^-)$ brought into the gap from the conduction band (valence band) by its charge. Solid arrows represent optical absorption transitions, dashed arrows represent optical emission (PL) transitions and hatched arrows represent thermal transitions. Define E_1 and E_2 by

$$D^+ + e \xrightarrow{(+E_1)} D^\circ, \quad D^\circ + e \xrightarrow{(+E_2)} D^-; \text{ then } U_{\text{eff}} = E_2 - E_1 < 0.$$

Transition D^-/D°
(1) $D^- \xrightarrow{(E_g - E_2 + W^-)} D^\circ + e_{cb}$ ⎫
(3) $D^\circ + e_{vb} \xrightarrow{(E_2 + W^-)} D^-$ ⎬ Optical

(4) $D^\circ + e_{vb} \xrightarrow{E_2} D^-$ ⎫
(5) $D^- \xrightarrow{(E_g - E_2)} D^\circ + e_{cb}$ ⎬ Thermal

Transition D^+/D°
(6) $D^+ + e_{vb} \xrightarrow{(E_1 + W^+)} D^\circ$ ⎫
(8) $D^\circ \xrightarrow{(E_g - E_1 + W^+)} D^+ + e_{cb}$ ⎬ Optical

(9) $D^+ + e_{vb} \xrightarrow{E_1} D^\circ$ ⎫
(10) $D^\circ \xrightarrow{(E_g - E_1)} D^+ + e_{cb}$ ⎬ Thermal

(b) Configuration-coordinate diagram for excitation of D^- to D°. Vertical arrows represent optical transitions; non-vertical arrow represents thermal transition.
(c) Schematic illustration of the photoluminescence (PL) and photoluminescence excitation (PLE) spectra in a chalcogenide glass. The PL is Stokes-shifted by energy $2W$. The absorption edge is also shown for comparison.

the system is strong and the two potentials in Fig. 6.71(b) are separated in q by a large amount, the system may transfer from the excited state to the ground-state configuration by passage over the small potential barrier separating them, and consequently lose energy non-radiatively by phonon emission.

The luminescence process can be expressed quantitatively by writing the potential in the configuration-coordinate diagram (Fig. 6.17(b)) as (Street 1976):

$$E^g(q) = Aq^2 \quad [6.31(a)]$$

$$E^e(q) = E^* + Aq^2 - \lambda q \quad [6.31(b)]$$

for the ground and excited state respectively. E^* is the energy separation of ground and excited states at $q=0$, and λ is the electron–lattice coupling constant if the coupling is assumed to be linear in q. The constant A is related to the appropriate vibrational frequency ω_0 by

$$A = \tfrac{1}{2}M\omega_0^2 \quad [6.32]$$

where M is the effective mass of the mode, and it is assumed for simplicity that ω_0 is the same in both ground and excited states.

The minimum of the potential curve for the excited state occurs at

$$q_0 = \frac{\lambda}{2A} \quad [6.33]$$

with energy

$$E_0 = E^* - W^- \quad [6.34]$$

where

$$W^- = \lambda^2/4A \quad [6.35]$$

The optical transition probability is given by the joint matrix elements for both electronic and vibrational wavefunctions, which in the adiabatic approximation can be treated separately. It is the vibrational matrix elements which determine the energy and line shape of a transition, and at low temperatures only the lowest vibrational state is populated, having the wavefunction $\psi(q) \propto \exp(-Aq^2/\hbar\omega_0)$. The absorption will be proportional to the matrix element, in turn proportional to $|\psi(q)|^2$. From [6.31(a)] and [6.31(b)] we have that:

$$h\nu = E^e - E^g = E^* - \lambda q \quad [6.36]$$

whereupon the absorption coefficient can be written as:

$$\alpha(h\nu) \propto \exp\left[-\frac{(h\nu - E^*)^2}{\sigma^2}\right] \quad [6.37]$$

where the half line width σ is given by:

$$\sigma = (2W\hbar\omega_0)^{1/2} \quad [6.38]$$

The emission intensity can also be calculated in a similar manner, giving:

$$S(h\nu) \propto \exp\left[-\frac{(h\nu - E^* + 2W^-)^2}{\sigma^2}\right] \quad [6.39]$$

Comparison of [6.37] and [6.39] shows that the luminescence band occurs at a lower energy than the absorption band, and that their difference in energy, the Stokes shift, is given by $2W$ (i.e. twice the polaron energy W). Note that in this model, at low temperatures, the position of the luminescence band (i.e. the Stokes shift) and the band width are related through the polaron energy W. At high temperatures, more vibrational levels are populated, and the band width becomes temperature dependent, i.e. $\sigma(T) \propto (\coth(\hbar\omega_0/2k_BT))^{1/2}$.

Optical (and thermal) transitions can be represented on an energy level diagram such as in Fig. 6.17(a), but matters are complicated since only *one*-electron transitions can be shown thus, and so D^-/D^0 and D^0/D^+ transitions must be shown separately. In Fig. 6.17(a) the level A^-, T^- and O^- refers respectively to the D^- level, and the thermal and optical levels of D^0; the equivalent positive quantities refer to the corresponding D^+/D^0 levels. The process that we have just considered, $D^- + h\nu_E \rightarrow D^0 + e$, can be represented on this diagram as an optical absorption transition from A^- to the conduction band edge (transition 1) whereupon the D^0 level relaxes to the optical state O^- and luminescence occurs by recombination to this level (transition 2), being Stokes-shifted by an energy $2W^-$. Note that the energy difference between levels O^- and O^+ is simply the true Hubbard correlation energy U ([6.1]) since it is the energy for the transfer of an electron between D^0 centres (i.e. $2D^0 \rightarrow D^+ + D^-$) at *constant* q, that is in the absence of relaxation.

Thermal transitions, on the other hand, require less energy than the corresponding optical transitions. This may be seen perhaps most clearly by reference to the configuration-coordinate diagram (Fig. 6.17(b)), where the non-vertical nature of these transitions is apparent. This arises because the lattice has time to relax during a thermal transition and so excitation is between the *minima* of the potential curves, at non-constant q, and therefore the thermal excitation energy is less by W^- (or equivalently W^+) than the corresponding optical excitation of a D^- (or D^+) to form a D^0. This process can be represented on the energy level diagram by the transition from the thermal level T^- to the conduction band (transition 5 in Fig. 6.17(a)). Note that the energy separation of the thermal levels T^- and T^+ is given by U_{eff}, since this now represents the energy of the reaction $2D^0 \rightarrow D^+ + D^-$ if lattice relaxation takes place.

Note that in this model for the photoluminescence in chalcogenide glasses due to Street (1976), radiative recombination only occurs if absorption takes place at or very near to a charged defect (D^- or D^+), in which case the photo-generated electron and hole are trapped in close proximity and consequently recombine with each other, i.e. so-called 'geminate' recombination. Otherwise, if a photon is absorbed in the glass, well away from a D^+ or D^- centre, it is presumed that the photo-created electron–hole pair (exciton) recombines non-radiatively, or else the exciton becomes self-trapped in the form of an intimate valence alternation pair (IVAP). This model for exciton recombination (Street 1977) is illustrated in Fig. 6.18, where the non-radiative path (III) and the self-trapping path (IV) are shown on a configuration-coordinate diagram, together with the atomic rearrangements that take place in forming the self-trapped exciton as an IVAP.

Note incidentally that in this model two possible absorption mechanisms exist. The first is a transition between the ground state and the uncoupled exciton (transition I, Fig. 6.18), and for which it is expected that the electric-field broadening

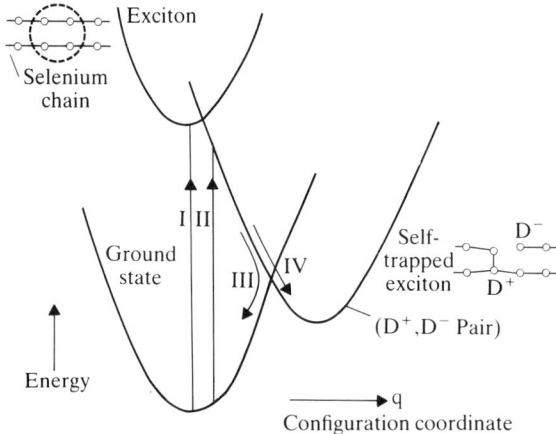

Fig. 6.18 Schematic illustration of the model for self-trapping of excitons in chalcogenide glasses (Street 1977). Optical excitation (I) to an exciton state can be followed by two non-radiative decay channels, either directly back to the ground state (III) or to the metastable self-trapped exciton state, which can be regarded as a D^+–D^- pair (IVAP). The absorption transition (II) illustrates the below-gap absorption which may contribute to the Urbach edge.

mechanism of Dow and Redfield (1972) would broaden the exciton absorption line into an exponential tail, observed experimentally as the Urbach edge (sect. 5.6.2). The second possible optical absorption mechanism is a transition directly to the self-trapped exciton state (transition II, Fig. 6.18), for which it has been calculated (Street 1977) that the absorption should be gaussian in form, cf. [6.37] (and not exponential as observed), although of comparable magnitude. One may speculate that this contribution represents the field-insensitive component of the absorption detected in electro-absorption measurements (sect. 5.6.2). A further consequence of this model is that it may be the cause of the photostructural changes exhibited by chalcogenide glasses exposed to band-gap radiation (see sect. 3.4.1); the self-trapped exciton state (IVAP) is metastable once formed, and reversion to the ground state (i.e. a normally bonded structure, C_2^0) can only occur by thermal activation over the potential barrier separating the metastable and ground states, and this is presumed to occur only if the material is annealed to the glass-transition temperature, T_g.

Evidence for the presence of charged centres (D^+, D^-) in the ground state of chalcogenide glasses comes from several aspects of photoluminescence, the shape of the excitation spectrum, and the temperature and time dependence of the luminescence. The Street–Mott model obviously takes as its basis the concept of a diamagnetic charged-defect ground state, and on this the peaked shape of the photoluminescence excitation (PLE) spectrum can be rationalized as follows: the decrease in PLE at low energies is understandable if the PL intensity is proportional to the number of photons absorbed (in the vicinity of charged defects), in turn proportional to the absorption coefficient, $\alpha(h\nu)$, which of course decreases with decreasing energy according to the Urbach rule. The decrease in PLE at *higher* energies comes about because now photons are not absorbed at or near D^+ or D^- centres (cf. transition 1, Fig. 6.17(a)), but instead at normally bonded sites (cf. Fig. 6.18), and in this case the electron–hole pairs recombine geminately in a non-radiative fashion, and at even higher photon energies the photo-carriers are created beyond the mobility edge with sufficient kinetic energy to enable them to diffuse away from each other, thereby precluding the possibility of geminate recombination. In this case, therefore, the decrease in PLE simply reflects the decreasing number of photons that are absorbed at radiative, charged-defect sites, and is not a consequence of a competing non-radiative *surface* recombination mechanism (as invoked for the case of a similar drop-off in PLE efficiency at high photon energies observed in *crystalline* semiconductors).

The PL intensity decreases with increasing temperature, and the temperature dependence obeys the empirical relationship (Street 1976):

$$I_{PL} = I_{PL}^0 \, e^{-T/T_0} \tag{6.40}$$

The quantum efficiency of the PL can be written as:

$$y = y_0 \left(\frac{p_r}{p_r + p_{nr}} \right) \tag{6.41}$$

where y_0 is the fraction of carriers initially captured by radiative centres, p_r is the radiative recombination rate and p_{nr} is the competing non-radiative rate, arising from the thermal escape of a carrier from a radiative centre. It is expected that p_r should be temperature independent, but obviously p_{nr} will be markedly temperature dependent, and experimentally it is found that its activation energy varies as T^2 (from [6.40]), ranging from $\sim 10^{-3}$ eV at low temperatures (~ 4 K) to $\sim 10^{-1}$ eV at higher temperatures ($\simeq 100$ K). The extremely small value of $\simeq 1$ meV for thermal escape can only be understood if the carrier escapes from a *neutral*, excited centre (i.e. the precursor state must be charged); if on the other hand the excited state were charged, the coulombic attraction between it and the carrier (electron or hole) would be substantial, giving rise to a much larger activation energy. Street (1976) has accounted for the temperature dependence of the PL intensity embodied in [6.40], by assuming that the loosely bound carrier (e.g. an electron) escapes from the radiative D^0 centre by tunnelling between potential fluctuations occurring at the band minima (see sect. 5.3.6).

Recently much effort has been devoted to the study of the time dependence of photoluminescence in amorphous semiconductors. The evolution of the PL intensity and peak position is monitored as a function of time in the range 10^{-8}–10^{-1} s following the absorption of a laser pulse (of width ~ 10 ns). The shifts in the PL peak with time delay after the exciting pulse for two amorphous semiconductors, a-As_2S_3 and a-Si:O:H, are shown in Fig. 6.19 (Street 1980). The initial (< 100 ns) rapid shift to lower energies of the PL peak position is ascribed in both cases to thermalization (i.e. non-radiative energy loss) of the photo-created carriers through tail states prior to radiative recombination. For time delays $\geqslant 10^{-4}$ s the PL peak position again shifts to lower energies in both cases, and this effect has been ascribed to tunnelling of the carriers leading to radiative recombination, resulting in a lifetime τ given by:

$$\tau = \tau_0 \exp(2R/R_0) \tag{6.42}$$

where R is the electron–hole separation, R_0 is the effective Bohr radius of the larger particle and $\tau_0 \simeq 10^{-8}$ s. Since the carriers are created having a broad distribution

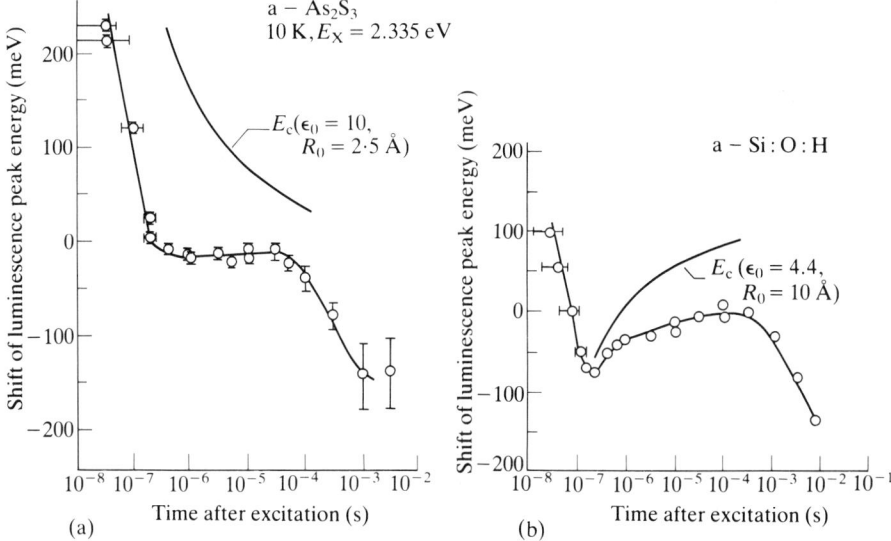

Fig. 6.19 Time-resolved photoluminescence spectra of two amorphous semiconductors. (a) a-As$_2$S$_3$; the upper curve is the coulomb interaction E_c assuming the recombination centre is a D$^+$–D$^-$ pair with Bohr radius 2.5 Å. (b) a-Si:O:H; the upper curve is again the appropriate coulomb interaction (Street 1980).

of separations, this gives rise to a time-dependent shift in PL peak position since different carrier separations are sampled at subsequent time delays t, corresponding to $t = \tau$.

However, for our present purpose of determining the nature of the charge state of the radiative centre, the important range of time delay is between those just discussed, namely $10^{-7} < t < 10^{-4}$ s. Note that for a-As$_2$S$_3$ the PL peak position is effectively constant in this region, whereas for a-Si:O:H it increases markedly in energy. This behaviour can be understood by consideration of the coulomb contribution to the PL energy arising from the interaction between the photo-created carriers. If the radiative centre is *neutral* before excitation, there is an attraction between electron and hole of magnitude:

$$E_c = e^2/4\pi\varepsilon\varepsilon_0 R \qquad [6.43]$$

and so the PL energy becomes $E_p = E_p^0 - E_c$, meaning that with increasing time delay (or equivalently by probing more widely separated centres, cf. [6.42]) E_p increases. This is obviously the case for a-Si:O:H (Fig. 6.19(a)), and Street has ascribed the PL process there to recombination between electrons and holes, trapped in conduction and valence band tail states, respectively (the addition of O to a-Si:H reduces the dielectric constant ε from 11.5 to 4.4, thereby increasing the size of the coulomb effect). Defects in this material, e.g. T$_3^0$ centres, are presumed to be non-radiative. If the radiative centre is *charged* before excitation (e.g. D$^-$ or D$^+$ in chalcogenides) no coulomb effect is expected since the carrier is bound to a neutral centre in the excited state (e.g. $e + $D^0), and therefore $E_p = E_p^0$. Thus, the absence of a significant shift in the PL peak for a-As$_2$S$_3$ (Fig. 6.19(b)) in the intermediate time delay domain lends considerable support to the isolated charged-defect model for PL in chalcogenides.

Finally, if recombination takes place between *donor–acceptor pairs* (or D^+–D^- pairs), having a distribution of separations, there is a coulombic interaction but of opposite sign, viz. $E_p = E_p^0 + E_c$, and therefore E_p should *decrease* with increasing time delay. The fact that this behaviour does not appear to be observed in chalcogenide glasses seems to rule out the IVAP involvement in the PL process (Kastner 1980, Higashi and Kastner 1979) that has been proposed to explain an apparent lack of electric or magnetic field dependence of the PL intensity, which would otherwise be expected if an electron were loosely bound to a neutral D^0 centre as in the isolated charged dangling bond model. Time-resolved PL studies are capable of furnishing much information concerning the recombination mechanism as well as the nature of the luminescent centres, and as this is still very much an active field of research, the precise details of the luminescence processes in amorphous semiconductors are probably not as yet finalized.

6.3.2 Electron spin resonance

Electron spin resonance is a valuable tool for the study of defects in amorphous solids, most particularly for those materials, free of paramagnetic ions (e.g. transition metals), which are covalently bonded, so conduction electron spin resonance is negligibly small (because of the few carriers thermally excited to the conduction band in such semiconductors), and for which the defects are therefore predominantly coordination defects, such as dangling bonds, and hence are capable of existing in a paramagnetic state if they contain an unpaired electron. A general description of ESR has already been given in section 3.2.4, and therefore will not be repeated again here.

Paramagnetic defects may arise in one of two ways: either the defects are present in a paramagnetic form as a result of the preparation process (e.g. in evaporated a-Si), or else they are created (say optically or thermally) either from preexisting diamagnetic defects or from the perfectly bonded diamagnetic host lattice. Obviously the strength of electron–electron interactions, and hence the sign of U_{eff}, is crucial to the existence or otherwise of paramagnetic centres. In view of this and for reasons of continuity we will concentrate first on chalcogenide glasses since for these materials photoluminescence and electron spin resonance are very closely related.

Electron spin resonance signals in pure chalcogenide glasses are generally observed only after optical excitation (see Bishop *et al.* 1977); in the ground state, these materials are invariably diamagnetic (Agarwal 1973). It was this observation of course that prompted Anderson (1975) to propose his model for spin pairing. All chalcogenide materials studied so far, e.g. Se, and As and Ge chalcogenide alloys, exhibit a resonance near $g = 2$, which is reasonably broad (~ 500 gauss wide) and which has been ascribed to an unpaired spin predominantly localized on a single chalcogen atom, viz. a C_1^0 centre. In addition, As-containing glasses, as well as a-As itself, exhibit an additional, much broader resonance (~ 1500 gauss wide), which has been ascribed to an electron trapped in an As orbital containing a significant amount of s-character (perhaps a P_4^0 centre); the width of the resonance arises from the unresolved structure of the hyperfine interaction with the As ($I = 3/2$) nucleus. In all cases, identification of the nature of the centre is made possible by simulation of the ESR spectrum by means of powder-pattern modelling (see sect. 3.2.4).

Identification of these ESR centres with their precursor charged dangling bond states is made possible by the close correlation between ESR and photoluminescence. Experimentally, it is found that if the PL intensity is monitored under continuous optical excitation using near-band-gap light (for which $\alpha \simeq 10^3$ cm^{-1}), the intensity *decreases* with continuing excitation: i.e. the luminescence 'fatigues'. Concomitant with this PL fatigue is a growth of the ESR signal. The explanation for this behaviour follows on from the discussion of the photoluminescence mechanism given in the last section. If it is assumed that the luminescent centre is a D^-, the hole of an optically generated electron–hole pair will be trapped, resulting in an electron loosely bound to a D^0 centre. In the model for the temperature dependence of PL due to Street (1976), it is the hopping away of the electron before recombination takes place that is responsible for the decrease in PL intensity with increasing temperature. It has been proposed (see e.g. Bishop *et al.* 1977) that this is also the mechanism by which the PL intensity fatigues under continuous excitation, and the end result of this process of course is a metastable D^0 centre, which being paramagnetic, will give rise to an ESR signal. Just as the PL signal has disappeared by ~ 150 K, so too does the ESR signal vanish on warming the sample to approximately the same temperature, presumably by the diffusion of excess carriers to the D^0 centres followed by non-radiative recombination. At the same time as the ESR signal grows at the expense of the PL intensity, an optical absorption band appears at energies near mid-gap. The presence of this photo-induced absorption can be understood by reference to Fig. 6.17(a) with the supposition that metastable D^0 states are created by the fatiguing process. It can be seen that the optical levels for D^0 lie near the middle of the gap, and so the two transitions involving this centre, namely $D^0 + e_{vb} \to D^-$ and $D^0 \to D^+ + e_{cb}$ (transitions 3 and 8 respectively), involve energies much less than that of the band gap. Note also that optical transitions of this sort involving the mid-gap absorption band *regenerate* luminescent centres (i.e. D^- or D^+), and so illumination of a fatigued sample with *half-band-gap* light should restore the luminescent intensity and concomitantly reduce ('bleach') the ESR signal, as indeed is observed (Bishop *et al.* 1977).

There is a certain degree of controversy surrounding the density of photo-induced spins observed in such experiments. There is some evidence that the spin density saturates (at $\sim 10^{18}$ cm^{-3}), indicating that pre-existing centres are being excited (viz. D^+ or $D^- \to D^0$), but other experiments show no sign of saturation even for very long illumination times, and the continuous increase in spin signal seems clear evidence for defect creation (perhaps by the mechanism illustrated in Fig. 6.18), and this behaviour has been indirectly linked with the photostructural changes observed in chalcogenide glasses. This matter therefore remains somewhat uncertain as indeed does the link between ESR and PL fatigue, since there is some evidence that the fatigue process involves some non-radiative channel other than that which produces D^0 centres. An additional interesting complication is that while very similar PL behaviour is observed in *crystalline* chalcogenide materials, they do *not* exhibit an ESR spectrum, photo-induced or otherwise.

Electron spin resonance has played a significant role in determining the nature of radiation-induced centres in other materials too, notably silica. Again two types of radiation-induced process can be visualized: irradiation by neutrons or charged particles such as protons or alpha particles results in a high degree of damage and defects are created through atomic displacement by knock-on processes; whereas irradiation with electromagnetic radiation or electrons can create (paramagnetic) defects by electronic excitation, e.g. by bond rupture, or can transform pre-existing diamagnetic defects into a paramagnetic state by the trapping of the photo-electrons or holes. We have discussed previously (sects 6.2.2 and 6.2.3) an example of a defect in silica, namely the E' centre (see Fig. 6.12). This has axial symmetry, and hence the g-tensor has just two values $g_\|$ and $g_\perp$; $\Delta g = g - g_s$, is small and negative with $|\Delta g_\perp| > |\Delta g_\||$.

Another example, very much related to the chalcogen ESR centres described above, are the so-called 'oxygen hole centres' (OHCs) observed in γ-irradiated silica-based glasses (Stapelbroek et al. 1979); these have also been discussed briefly in section 6.2.2. The g-tensor in this case is completely anisotropic, with $g_1 \simeq g_s$ (cf. [3.50]), $g_2 \simeq 2.01$, but a broad distribution of g_3-values (with $\langle g_3 \rangle \simeq 2.07$) found from detailed simulation of the line shape. The distribution in g_3-values is presumed to arise from gaussian distributions in the energy level splittings. The nature of such centres can be determined by consideration of the g-values, in conjunction with the observation of the hyperfine splittings characterized by the six-line spectrum observed in ^{17}O ($I = 5/2$) enriched v-SiO$_2$, which arises because of the interaction between electron and nuclear spins.

In general, the wavefunction of an unpaired spin can be written as:

$$|\psi\rangle = \alpha|ns\rangle + \beta|np\rangle + \delta_i|k_i\rangle \qquad [6.44]$$

where $|ns\rangle$ and $|np\rangle$ are atomic orbitals of the central atom, $|k_i\rangle$ are other orbitals of the central atom and ligand orbitals, and the coefficients α and β are given by:

$$\alpha^2 = \frac{A_{\text{iso}}}{A_{\text{ns}}} = \frac{(A_\| + 2A_\perp)}{3A_{\text{ns}}} \qquad [6.45]$$

$$\beta^2 = \frac{A_{\text{aniso}}}{A_{\text{np}}} = \frac{(A_\| - A_\perp)}{3A_{\text{np}}} \qquad [6.46]$$

Here A_{ns} and A_{np} are atomic hyperfine coupling constants, and $A_\|$ and $A_\perp$ are the hyperfine coupling constants obtained from the experimental spectrum for the centre in question (in the case of axial symmetry of the hyperfine tensor). Furthermore, the wavefunction coefficients are related via a normalization condition $\alpha^2 + \beta^2 + \gamma^2 = 1$, where $\gamma^2 = \sum_i \delta_i^2$. In addition, for the case of axial symmetry of the centre, such as exists for a dangling bond sp^3 orbital, there is a relation between the angle θ subtended at the central atom by the dangling bond orbital and a normally bonded orbital and the hyperfine coefficients:

$$\tan \theta = -[2(1 + \beta^2/\alpha^2)]^{1/2} \qquad [6.47]$$

Hence detailed analysis of the ESR spectrum yields values for the wavefunction coefficients α and β and therefore the nature and environment of the orbital in which the unpaired spin resides.

In this manner, the OHCs are found to be predominantly associated with oxygen p-orbitals, and therefore the centres arise when a hole is trapped either in a non-bridging (C$_1$) orbital, or else in a lone-pair orbital of a normally bonded, i.e. bridging, atom (C$_2^0$). However, there is strong evidence for the presence of *two* distinct oxygen hole centres in any sample of vitreous silica, a 'wet' OHC associated predominantly with high —OH-containing material, and a 'dry' OHC observed mainly in low —OH-containing silica (Stapelbroek et al. 1979). The detailed nature of the differences of such defects are not known, however; it is an obvious assumption that the wet OHC is associated with a proton (in an —OH group), but the absence of an observable proton hyperfine interaction dictates that the unpaired spin and the proton must be separated by a large distance ($\geqslant 3$ Å).

A related defect is the so-called Al centre which is believed to be an unpaired spin on a bridging oxygen bonded to a Si atom and a substitutional Al atom; the hyperfine interaction with ^{27}Al ($I = 5/2$) atoms gives rise to a six-line spectrum. This centre is of interest since it is also found in crystalline α-quartz, and the optical absorption associated with it is responsible for the colour of smoky quartz.

We turn our attention now to another class of amorphous semiconductors which have been much studied by ESR, namely the tetrahedrally bonded materials, a-Si and a-Ge (see Stuke 1977 for a review). In contrast with the photo-induced spins observed in chalcogenide glasses, evaporated a-Si and a-Ge thin films exhibit an ESR signal in the absence of optical excitation, and a much higher spin density is observed ($\simeq 10^{19}$–10^{20} cm^{-3}). A single resonance line is observed in these materials, with both the line width and the g-value being smaller in a-Si than for a-Ge ($g_{Si} \simeq 2.005$, $g_{Ge} \simeq 2.02$). Hydrogenated materials, on the other hand, have a much reduced spin density, $\leqslant 10^{16}$ cm^{-3} (and often less than the detectable limit for glow-discharge samples), although the ESR signal is considerably increased *during* continuous optical excitation (Street 1980, Street and Biegelsen 1980). While it is reasonably certain that the spin defects are T_3^0 centres (i.e. single dangling bonds), the location of such centres remains somewhat contentious. Originally it was thought that for evaporated materials the unpaired spins resided at void surfaces, in view of the correlation of the spin density with various void-related properties (e.g. porosity) upon annealing (see Fig. 6.13). The early observation of ESR signals on cleaved surfaces of c-Si supports this contention, but the later discovery that on perfect clean surfaces spin pairing occurs (see sect. 6.2.1) cast some doubt on the internal surface origin of spin centres in a-Si and Ge, leading to the conclusion that at least some of the spins are isolated dangling bonds in the bulk (see Fig. 6.1).

An interesting correlation exists between the ESR observed in evaporated a-Si or Ge and another defect-related property, namely d.c. hopping conductivity (see sect. 6.3.3). It is found that the ESR line width (defined as the peak-to-peak separation ΔH_{pp}), measured in terms of the magnetic sweep field, i.e. in gauss) can be deconvoluted into two components, a temperature-independent term, $\Delta H_{pp}(0)$ (caused mainly by g-value broadening through the spin–orbit coupling, but also in part by unresolved hyperfine structure) and a strongly temperature-dependent term, $\Delta H_{pp}(T)$. This latter term is approximately proportional to the hopping conductivity, $\sigma_h(T)$, which results from the hopping of electrons between localized defect states (see sect. 6.3.3). The line broadening then results from a decrease in lifetime caused by an enhancement of spin–lattice relaxation due to the electron motion.

Finally we consider a technique which links ESR and PL, and which has been extensively used in the study of crystalline semiconductors, but only very recently has been applied to the study of amorphous materials. This is 'optically detected magnetic resonance', ODMR – for an extensive review of the technique see Cavenett

(1981). In essence, the method consists of monitoring the photoluminescence intensity at the same time as microwave radiation excites resonance between the Zeeman levels of the centre, split by an external magnetic field: instead of the absorption of the microwaves being used as a monitor of resonance as in a conventional ESR experiment, the PL intensity is measured in its place. The resonance condition is found as usual by sweeping the magnetic field and the g-value for the resonant centre is determined from the position of the maximum or minimum of the PL intensity. The advantage of ODMR is that, since microwave absorption is due to electronic transitions to levels otherwise little populated in thermal equilibrium, an *increase* in the PL intensity at resonance implies that transitions from such a level are radiative, whereas the observation of a *decrease* in PL intensity means that a non-radiative channel is being explored. In this way, therefore, identification of a PL emission band with a given centre becomes possible.

As an example it has been found that the PL band observed in glow-discharge a-Si:H does not comprise a single emission band, but instead is composed of three distinct bands, centred at 1.4, 1.25 and 0.9 eV, whose relative intensities vary with different sample preparation conditions, but the 1.25 eV emission is dominant in good electronic quality, high quantum efficiency material. Optically detected magnetic resonance studies indicate two main features: a quenching resonance at $g = 2.005$ is observed when the high energy ($\simeq 1.5$ eV) part of the PL band is monitored, whereas an enhancing resonance at $g = 2.007$ is found when the low energy ($\simeq 0.9$ eV) part is probed. The former appears to be associated with a dangling bond state (T_3^0), which mainly offers a non-radiative channel but with a small probability for radiative recombination, giving rise to the 0.9 eV band. There is some dispute concerning the higher energy PL components, since it has variously been ascribed to distant-pair processes (see Cavenett 1981) or to geminate recombination of carriers trapped in tail states (Biegelsen *et al.* 1978). Finally, note that ODMR is just one example of spin-dependent recombination. The ability of microwaves to invert (flip) the spin state of a free electron or hole can dramatically influence the recombination rate, manifested also in photoconductivity experiments (see Solomon 1979).

6.3.3 Direct current hopping conductivity

We have previously considered the mechanism for d.c. electrical conductivity of amorphous semiconductors in the 'ideal' case, i.e. in the absence of structural defects when the gap is therefore free of states except perhaps for tail states (sect. 5.4.1). In real materials, which may contain a high concentration of defects and consequently have a high density of states in the gap, electron transport can take place via such defect states, and the magnitude of this defect-controlled conductivity may greatly exceed that due to conventional band conduction. Moreover, the temperature dependence of this form of conduction can have a distinctive form, easily distinguishable from the simple activated behaviour characteristic of band conduction. It is therefore of interest to discuss such behaviour in some detail.

We have seen previously in this chapter that structural (bonding) defects give rise to electron states lying within the gap of amorphous semiconductors and these will therefore be localized. However, one of the definitions of localization is that the configurationally averaged conductivity associated with the states at energy E is zero, $\langle \sigma_E \rangle = 0$; i.e. localized states by themselves cannot carry current. Thus, conduction involving localized states (of whatever origin) can only take place by means of transitions of electrons from full states to neighbouring empty states, usually with phonon assistance. It is this transfer of electrons between sites localized at different positions in space that is commonly termed 'hopping', although we prefer to reserve the usage for transitions which involve energy activation.

The mechanism of hopping conduction was first proposed to account for the electrical conduction that takes place in crystalline semiconductors (e.g. Si) which have been doped (say n-type) and compensated. 'Impurity conduction' occurs due to the hopping of electrons between filled localized donor states (which are randomly situated), and states which are empty because of the compensation by acceptors: the acceptors also create a coulomb potential which perturbs the donor energy levels, meaning that phonons are necessary for the carriers to overcome the energy differences introduced.

It is an obvious extension of the model to apply it to the case of conduction in amorphous semiconductors, in which all the states lying in the mobility gap are localized, rather than as in the case of impurity conduction where only the states lying in the impurity (e.g. donor) band are localized in the Anderson sense, if the disorder in potentials is sufficiently great. Much work has been done on this subject, for it is an interesting if formidable theoretical topic, but the level of the mathematics often goes far beyond the level aimed at in this book. We will therefore describe the various approaches that have been employed, stressing the physical aspects and leaving the interested reader to study the mathematical details in the references given.

The simplest approach is perhaps that given in the original application of the hopping mechanism to amorphous solids by Mott (1969) – see also Mott and Davis (1979). Consider two localized sites, one filled and at or slightly below the Fermi energy E_F, and the other empty and above E_F; their energy and spatial separations are Δ and R, respectively. The hopping transition rate γ will be determined by three factors: (i) the probability of the existence of a phonon of energy Δ, given by the Boltzmann factor $\exp(-\Delta/k_B T)$ – the presence of the phonon is necessary to conserve energy in the hopping process; (ii) the probability of electron transfer between the sites – if it is assumed that each state is exponentially localized, with the same localization length α^{-1}, $\psi \simeq \exp(-\alpha r)$, and if an electron transfer is by means of tunnelling, the probability is given by the overlap, i.e. $\exp(-2\alpha R)$; (iii) a rate term which can be regarded as an attempt frequency v_0, and which will depend on the strength of the electron–phonon coupling and the phonon density of states, but will be only weakly dependent on R or Δ. The hopping transition rate is then given by:

$$\gamma = v_0 \exp(-2\alpha R - \Delta/k_B T) \qquad [6.48]$$

(Note that for hops downward in energy the probability is just $v_0 \exp(-2\alpha R)$ by the principle of detailed balance.) The conductivity may be calculated by making use of the Einstein relation between mobility and diffusion coefficient $\mu = eD/k_B T$, where $D = \gamma R^2/6$ if the conduction process can be thought of as arising from a random, diffusive motion, and if the density of states at the Fermi level $N(E_F)$ is constant, so that $k_B T N(E_F)$ are the concentration of electrons contributing:

$$\sigma = \frac{e^2}{6} R^2 v_0 \exp(-2\alpha R - \Delta/k_B T) \qquad [6.49]$$

At low temperatures, where the number and energy of phonons are both small, hopping to near neighbours (thereby minimizing the $\exp(-2\alpha R)$ term) is unfavourable because of the large energy separations encountered on average. Instead, it is more favourable for the electron to tunnel to more distant sites, at the expense of the $\exp(-2\alpha R)$ term, but at a saving in the $\exp(-\Delta/k_B T)$ term since there is a higher probability that more distant sites will have smaller energy separations. Mott (1969) therefore considered that the factor $(2\alpha R + \Delta/k_B T)$ should be optimized, subject to the condition that there is at least one state at a given spatial and energy separation, i.e.

$$1 = \frac{4\pi}{3} N(E_F) \Delta R^3 \qquad [6.50]$$

If [6.50] is substituted into the factor $(2\alpha R + \Delta/k_B T)$, maximization of the exponent yields for the optimum hopping distance:

$$\bar{R} = \left[\frac{9}{8\pi\alpha N(E_F)k_B T}\right]^{1/4} \qquad [6.51]$$

Substitution of $\bar{R}$ into [6.50] then gives a value for the optimized energy separation $\bar{\Delta}$, and if these are both inserted into [6.49], this yields the final expression for the conductivity:

$$\sigma = \sigma_0 \exp[-A/T^{1/4}] \qquad [6.52]$$

where

$$\sigma_0 = \frac{e^2}{6} v_0 N(E_F) \left[\frac{9}{8\pi\alpha N(E_F)k_B T}\right]^{1/2} = \frac{e^2 v_0}{2(8\pi)^{1/2}} \left[\frac{N(E_F)}{\alpha k_B T}\right]^{1/2} \qquad [6.53]$$

and

$$A = 2.1[\alpha^3/k_B N(E_F)]^{1/4} \qquad [6.54]$$

Thus, we see that 'variable-range hopping' in a band of constant density of states is characterized by an $\exp(-AT^{-1/4})$ temperature dependence of the d.c. conductivity. This behaviour has been observed in several amorphous semiconductors at low temperatures, where the activated nature of band conduction (sect. 5.4.1) ensures that it is smaller than the contribution from variable-range hopping: Fig. 5.26(b) shows schematically the change in conduction mechanism on lowering the temperature. Variable-range hopping has been observed principally in materials which contain positive U defects (see sect. 6.2.1), e.g. a-Si or a-Ge. As an example we show in Fig. 6.20 data for films of a-Si evaporated in ultra-high vacuum, in which it can be seen that for these films of thickness > 500 Å, the $T^{-1/4}$ temperature dependence expected of variable-range hopping is well obeyed. Note that using [6.52] it is possible in principle to obtain independent estimates for $N(E_F)$ and α from experimental values of both the pre-exponential factor σ_0, i.e. the $T=0$ intercept ([6.53]) and the exponent A ([6.54]). However, almost universally it is found that the

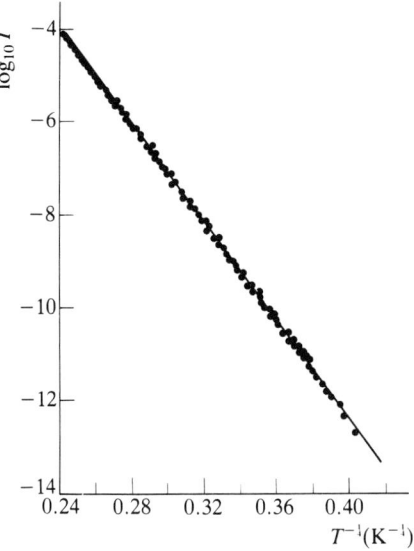

Fig. 6.20 Illustration of the $T^{-1/4}$ temperature dependence of the conductivity characteristic of variable-range hopping. The data shown are for ultra-high vacuum evaporated a-Si, and the current I which is plotted may be converted to conductivity by $\sigma = 1.39 \times 10 I \, \Omega^{-1} \, \text{cm}^{-1}$ (M. L. Knotek (1975) *Sol. St. Comm.* **17**, 1431).

pre-exponential term is anomalously high, yielding unphysically large values for $N(E_F)$ or v_0, depending on which is given. This behaviour is not fully understood, but may have to do with single-phonon theories that have been developed to obtain values for v_0. Nevertheless, the exponent A appears to be better defined, and much use has been made of it to extract values for $N(E_F)$. However, $N(E_F)$ cannot be obtained independently of α, but given assumed values for α^{-1} (usually $\sim 10 \, \text{Å}$), estimates in the range 10^{18}–$10^{19} \, \text{cm}^{-3} \, \text{eV}^{-1}$ are commonly obtained. It is of interest to note that if $N(E_F) = 10^{19} \, \text{cm}^{-3} \, \text{eV}^{-1}$ and $\alpha^{-1} = 10 \, \text{Å}$ then [6.51] gives a hopping distance of $\bar{R} = 80 \, \text{Å}$ at $T = 100 \, \text{K}$.

Many different theoretical approaches dealing with the hopping problem have been developed since Mott's early treatment. Apsley and Hughes (1974, 1975) considered the hops to take place in a four-dimensional space (one energy and three spatial coordinates), and using a different averaging technique to optimize the 'range', obtained an expression of the same form as [6.52] for the case of a constant density of states at low temperature and electric fields, but with some differences in the values of the pre-exponential and exponent factors. They also obtained an expression for the electric field dependence of the hopping conductivity (i.e. non-ohmic behaviour) which is observed at high fields for systems in the hopping regime. The effect of the applied field F is to modify the energy of a hop between two sites i and j, separated by a distance R_{ij}, by an amount $eFR_{ij} \cos \theta$, where θ is the angle between the directions $\boldsymbol{F}$ and $\boldsymbol{R}_{ij}$. At high fields, the quantity eFR_{ij} becomes greater than the average hopping energy, and an electron then only moves by means of downward hops with the emission of a phonon with each hop. In this case, *no* temperature dependence of the hopping conductivity is expected, and it takes the limiting form:

Defects

$$\sigma = \sigma_0^{HF} \exp(-B/F^{1/4}) \qquad [6.55]$$

(since the critical field is $F \simeq 3/(4\pi e N(E_F) R^4)$ and the conductivity is proportional to $\exp(-2\alpha R)$). In fact Apsley and Hughes (1975) obtained for the pre-exponential and exponent factors respectively:

$$\sigma_0^{HF} = \frac{Ne^2 v_0}{\alpha^2} \left[\frac{\alpha^4}{4\pi e N(E_F)}\right]^{1/4} \frac{1}{F^{1/4}} \qquad [6.56]$$

and

$$B = \left[\frac{64\alpha^4}{\pi e N(E_F)}\right]^{1/4} \qquad [6.57]$$

Thus by making measurements at low fields (in the $T^{-1/4}$ regime) and at high fields (in the $F^{-1/4}$ regime), independent estimates for α and $N(E_F)$ may be obtained from the exponents alone ([6.54] and [6.57]).

A more rigorous treatment of the hopping process is afforded by percolation theory (see sect. 5.3.6). The electron states having energies in the gap are randomly distributed in an amorphous solid, and the transition rate γ_{ij} between a pair of such sites has the form given in [6.48]. This situation can be represented as a series of conductances G_{ij} connecting any given pair of sites whose value is given by:

$$G_{ij} \propto \frac{e^2}{k_B T} \gamma_{ij}(R_{ij}, E_i, E_j)$$

However, because of the random distribution of site separations and energies, there is therefore an enormous spread in transition rates or equivalent conductances in such a system. The overall conductivity of the amorphous solid modelled as a resistor network can be written as $\sigma = \tilde{G}/\tilde{R}$, where $\tilde{G}$ is some characteristic value of the conductances in the network and $\tilde{R}$ is some characteristic length-scale of the system. The essential feature of the percolation approach is to assert that $\tilde{G}$ is given by G_c, the critical percolation conductance, defined as the largest conductance such that there is a connected path through the network for all $G_{ij} > G_c$. Thus it is neither the abundant small conductances, nor the relatively scarce large conductances which determine the conductivity, but the critical conductance which establishes a percolation path: resistors for which $G_{ij} \ll G_c$ are effectively shorted out and those for which $G_{ij} \gg G_c$ form isolated regions which cannot therefore contribute to current throughout the sample.

Thus, solution of the hopping problem hinges on the evaluation of the critical percolation conductance for the network in question. If we rewrite [6.48] for a pair of sites i and j, introducing the energies of each site explicitly, the condition $G_{ij} > G_c$ may be written as (Ambegaokar et al. 1971):

$$[2\alpha R_{ij} + (|E_i| + |E_j| + |E_i - E_j|)/2k_B T] < \ln\left(\frac{v_0}{\gamma_c}\right) \qquad [6.58]$$

where $\gamma_c = k_B T G_c / e^2$. Equation 6.58 may be re-expressed in dimensionless terms as:

$$\frac{R_{ij}}{R_{max}} + \frac{(|E_i| + |E_j| + |E_i - E_j|)}{2E_{max}} < 1 \qquad [6.59]$$

where

$$R_{max} = \frac{1}{2\alpha} \ln\left(\frac{v_0}{\gamma_c}\right) \quad [6.60(a)]$$

$$E_{max} = k_B T \ln\left(\frac{v_0}{\gamma_c}\right) \quad [6.60(b)]$$

Thus the inequalities $R_{ij} < R_{max}$ and $|E_i| \leq E_{max}$ must independently be satisfied for [6.58] (or [6.59]) to hold. The total number of sites per unit volume such that $|E_i| \leq E_{max}$ is:

$$n = 2N(E_F)E_{max} \quad [6.61]$$

assuming $N(E)$ to be constant in the region of E_F. The n sites are randomly distributed in space and the term (E_i/E_{max}) is randomly distributed between -1 and 1. The critical path is established if [6.59] is satisfied, but part at least of this network must span the sample. This requirement can be expressed in the form:

$$nR_{max}^3 = \eta_c \quad [6.62]$$

where η_c is a constant of order unity. Ambegaokar et al. (1971) estimate $\eta_c \simeq 4$ from a consideration of the percolation limit for 4-D (three spatial and one energy) hyperspheres. (Other authors, e.g. Pike and Seager (1974), use Monte Carlo techniques to obtain similar answers.) Combining [6.60], [6.61] and [6.62] gives:

$$\ln\left(\frac{v_0}{\gamma_c}\right) = \left(\frac{4\alpha^3 \eta_c}{k_B T N(E_F)}\right)^{1/4} \quad [6.63]$$

which gives an expression for the hopping conductivity of the Mott form $\sigma = \sigma_0 \exp[-A/T^{1/4}]$ where,

$$A = \left(\frac{4\alpha^3 \eta_c}{k_B N(E_F)}\right)^{1/4} \quad [6.64]$$

Thus far we have considered hopping in three dimensions exclusively, where if the density of states is constant, the conductivity has the Mott $T^{-1/4}$ form. Consider now, however, the case when the length of one dimension of the system is reduced below that of the hopping length, $\bar{R}$ ([6.51]). In this case, hopping motion is effectively confined to *two* dimensions and it can be shown (see problem 6.1) that a different temperature dependence results. In fact, the equation for variable-range hopping in this regime is:

$$\sigma = \sigma_0 \exp[-C/T^{1/3}] \quad [6.65]$$

where

$$C = \left(\frac{3\alpha^2}{dk_B N(E_F)}\right)^{1/3} \quad [6.66]$$

where d is the film thickness and the constant varies slightly between different treatments. Note carefully that it has been assumed that 2-D behaviour results because the hopping is forced to be two dimensional because the film thickness is less

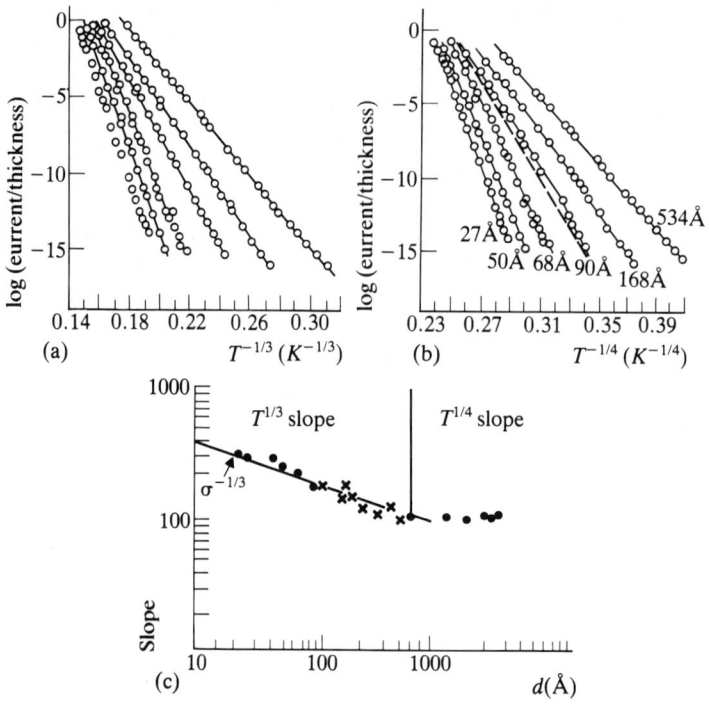

Fig. 6.21 Illustration of variable-range hopping in a-Ge and the transition from 3-D to 2-D behaviour. Data are plotted versus (a) $T^{-1/3}$ and (b) $T^{-1/4}$ (Knotek *et al.* 1974); the dashed lines are straight. The thickness dependence of the slope of such plots is shown in (c), clearly showing the transition from 2-D to 3-D hopping (Knotek 1974).

than the hopping distance, and in this case the density of states in [6.66] is the *bulk* value (cm^{-3} eV^{-1}). If, however, a truly two-dimensional random system is considered (as for example at the 'inversion layer' in MOSFETs – see Mott *et al.* 1975b), the 2-D density of states (cm^{-2} eV^{-1}) must be used, and there is therefore no thickness term d in [6.66]. Shown in Fig. 6.21 are experimental data for thin films of a-Ge of different thicknesses, deposited by evaporation in ultra-high vacuum (Knotek *et al.* 1974). Clear evidence is seen for a change in temperature dependence from $T^{-1/4}$ to $T^{-1/3}$ for films thinner than $\simeq 500$ Å, lending support to the notion of 2-D variable-range hopping. Further confirmation of this comes from the film thickness dependence of the slope C of $\ln \sigma$ v.s. $T^{-1/3}$ plots (Fig. 6.21(c)): the slope varies as $d^{-1/3}$ as expected from [6.66], but no variation of the slope A is found in the 3-D regime. It should be noted from [6.54] and [6.66] that α and $N(E_F)$ appear together in a different functional form in the exponents for 3-D and 2-D hopping, and so varying the thickness of the films and measuring the temperature dependence of σ in the two regimes offers another method of obtaining the quantities α and $N(E_F)$ independently, in addition to that using the field dependence mentioned earlier.

6.3.4 Alternating current conductivity

We have seen in the previous section that upon the application of a d.c. electric field, conduction can take place among localized gap states in amorphous semiconductors by a mechanism of phonon-assisted hopping, which is simply activated with a constant activation energy if hopping is between (nearest) neighbours at constant range ([6.49]), but with effectively a temperature-dependent activation energy (i.e. $T^{-1/4}$ behaviour) if the hopping is between neighbours at variable ranges ([6.52]). This d.c. behaviour could be thought of as the $\omega=0$ limit of a *frequency-dependent* (a.c.) conductivity, $\sigma(\omega)$, observed upon the application of an alternating (sinusoidal) field of frequency ω if the hopping mechanism is the *same* in both cases. We will see that for some situations this is a valid approach, but in others, notably for those materials which do *not* exhibit d.c. hopping conductivity but instead only band conduction, it is not valid and the a.c. conductivity which is observed must therefore result from a completely different mechanism. We see at once, therefore, that the phenomenon of a.c. conductivity is inherently more complicated than is the case for d.c. conductivity, where most authors now agree on the essential physics of the problem (except perhaps for the pre-exponential) but differ in their derivation of the constant in e.g. [6.54].

Alternating current conductivity is also interesting in that it has been observed in every amorphous semiconductor or insulator measured (in contrast to d.c. hopping conductivity), and also because in the frequency range ~ 10–10^6 Hz in almost every case the qualitative behaviour is the same, i.e.:

$$\sigma(\omega) = C\omega^s \qquad [6.67]$$

where the exponent s is near unity ($\leqslant 1$) and may be weakly temperature dependent, and C is independent of frequency and also only weakly dependent on temperature. Thus a.c. conduction is, like the thermal anomalies observed at low temperatures (sect. 4.5) unique to the amorphous phase, and furthermore it seems to be a general property shared by most if not all amorphous semiconductors and insulators. We will therefore at the outset establish those principles which mainly determine this behaviour, and leave until later a fuller discussion of the various mechanisms that have been proposed to account in detail for the phenomenon.

The concept of a frequency-dependent conductivity should not appear strange since it has appeared in another guise, albeit for a different frequency range, earlier in this book. Thus, a.c. conductivity is related to the optical absorption coefficient, $\alpha(\omega)$ (cf. [5.22]), since both involve energy loss within the solid, by:

$$\sigma(\omega) = \frac{n_0 \alpha(\omega)}{377} \qquad [6.68]$$

where n_0 is the refractive index, and the units of $\sigma(\omega)$ and $\alpha(\omega)$ are Ω^{-1} cm^{-1} and cm^{-1}, respectively. As an example of this relationship, we show in Fig. 6.22 data for a-As_2Se_3 plotted in this fashion for a frequency range spanning 13 decades. Although a plot of this sort is interesting, it does not imply of course that the same mechanism responsible for the loss (a.c. conductivity or optical absorption) is operative for all frequencies. Hence in this section dealing with defect-related properties, we consider only the linear region in the frequency range $\simeq 10$–10^6 Hz, since this is determined by electronic hopping between defects, whereas optical absorption obviously arises from completely different phonon or electronic mechanisms.

We will demonstrate first the origin of a linear frequency-dependent con-

Defects

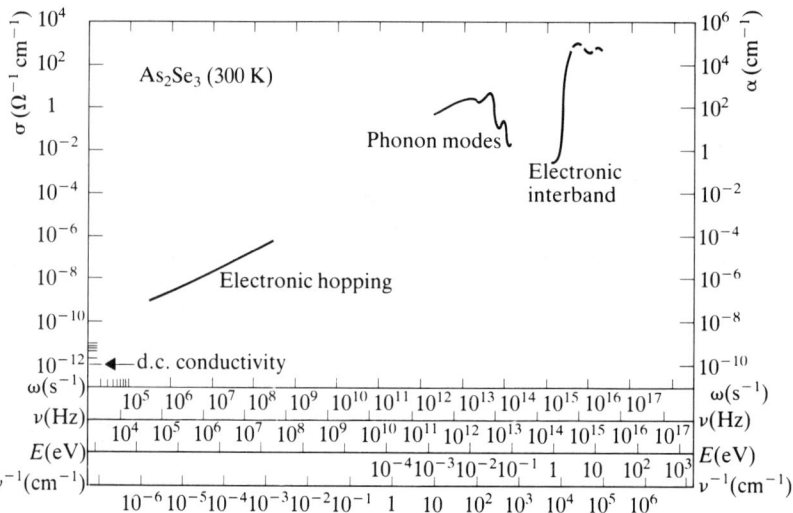

Fig. 6.22 Comparison of absorption (or equivalently conductivity) in different frequency regimes for a-As$_2$Se$_3$. The equivalence of various frequency and energy scales is also shown (Mott and Davis 1979).

ductivity. Consider the application of a harmonically varying electric field $E(t) = E_0 \sin \omega t$ to a sample, producing a polarization $P(t)$. These are related via Fourier transforms to the complex quantities $E(\omega)$ and $P(\omega)$ in the frequency domain, where:

$$P(\omega) = \varepsilon_0 \chi(\omega) E(\omega) \qquad [6.69]$$

where ε_0 is the free-space permittivity. The dielectric susceptibility is in general a complex function of frequency:

$$\chi(\omega) = \chi'(\omega) - i\chi''(\omega) \qquad [6.70]$$

The imaginary part, $\chi''(\omega)$, is termed the 'dielectric loss', since the current due to this is in phase with the applied field. The susceptibility is further related to the dielectric constant ([5.16]) by:

$$\varepsilon_1(\omega) = 1 + \chi'(\omega) \qquad [6.71(a)]$$

$$\varepsilon_2(\omega) = \chi''(\omega) \qquad [6.71(b)]$$

Finally the a.c. conductivity is related to the dielectric loss by:

$$\sigma(\omega) = \varepsilon_0 \omega \chi''(\omega) \qquad [6.72]$$

We now assume that the dielectric response of the amorphous solid to the applied alternating field is Debye-like, i.e. the polarization, which is caused, say, by the transition of a charge between two preferred sites, decays with first-order kinetics on removal of the exciting field, with a characteristic relaxation time, τ, viz. $dP/dt = -P/\tau$. This has the solution $P(t) = P_0 \exp(-t/\tau)$, and Fourier transforming $f(t) = (P_0/\tau) \exp(-t/\tau)$ yields finally the frequency dependence of the susceptibility:

$$\chi(\omega) = \frac{\chi(0)}{(1 + i\omega\tau)}, \text{ or:}$$

$$\chi'(\omega) = \frac{\chi(0)}{(1+\omega^2\tau^2)} \qquad [6.73(a)]$$

$$\chi''(\omega) = \frac{\omega\tau\chi(0)}{(1+\omega^2\tau^2)} \qquad [6.73(b)]$$

If τ is a constant, the dielectric loss $\chi''(\omega)$ will exhibit a peak at a frequency given by $\omega\tau = 1$. This behaviour is observed in many glassy hydrocarbon polymers, for instance (see e.g. Wong and Angell 1976), where rotational motion of polar side-groups between two preferred orientations is responsible for the effect. In this case, the a.c. conductivity varies as ω^2 for $\omega\tau \ll 1$ and is constant for $\omega\tau \gg 1$, behaviour obviously at variance with the linear frequency dependence under consideration. However, if we assume that the loss in amorphous semiconductors or insulators is caused by the hopping of electrons back and forth between localized defect states, which will be randomly situated in space and energy, a broad distribution of relaxation times $n(\tau)$ must ensue, instead of the fixed value necessary to produce a Debye-loss peak.

Thus, the a.c. conductivity may be written as:

$$\sigma(\omega) = \int_0^\infty a(\omega) n(\tau) \frac{\omega^2 \tau}{(1+\omega^2\tau^2)} \, d\tau \qquad [6.74]$$

where $a(\omega)$ is the polarization of a pair of sites, and we have assumed that the total conductivity may be obtained by summing over the individual contributions to the dielectric response made by *pairs* of sites, each exhibiting a Debye response.

If the distribution of relaxation times is such that $n(\tau) \propto 1/\tau$ and $a(\omega)$ is frequency independent, a linear a.c. conductivity results, since

$$\sigma(\omega) \propto \int \frac{\omega \, d(\omega\tau)}{(1+\omega^2\tau^2)} \propto \omega$$

For the condition $n(\tau) \propto 1/\tau$ to hold, the relaxation time must be an exponential function of some random variable ζ:

$$\tau = \tau_0 \exp(\zeta) \qquad [6.75]$$

where τ_0 may be regarded as being equal to an inverse phonon frequency, and ζ has a flat distribution, $n(\zeta) = $ constant (i.e. $n(\tau) = n(\zeta)(d\zeta/d\tau) \propto \tau^{-1}$). In practice, both $n(\zeta)$ and $a(\omega)$ may be weakly dependent on frequency and this results in the frequency exponent s in [6.67] being slightly sublinear.

The essential physics relating to the precise mechanism of charge transport giving rise to a near-linear a.c. conductivity is contained in [6.75], in particular in the quantity ζ. Two different mechanisms that could give rise to the form of [6.75] are the following:

(a) Classical activation of a carrier *over* a potential barrier W separating two sites, in which case:

$$\zeta = W/k_B T \qquad [6.76]$$

(b) Quantum-mechanical tunnelling (QMT) of a carrier *through* the potential barrier between the sites separated by a distance R, in which case:

$$\zeta = 2\alpha R \qquad [6.77]$$

The former mechanism in its simplest form, when W is independent of the spatial separation R, can be operative in both the simple Debye-loss case ([6.73(b)], when W is fixed and the frequency ω_{max} at which the maximum loss occurs (given by $\tau \omega_{max} = 1$) is activated, $\omega_{max} \propto \exp(-W/k_B T)$, and also in the case when there is a distribution of barrier heights in an amorphous material. Pollak and Pike (1972) suggested that the latter case holds generally in all glassy systems exhibiting low-temperature thermal anomalies, and that the loss mechanism is due to the motion of atoms (carrying at least a partial charge) between two sites forming a two-level system (see Fig. 4.19). In this case, the polarizability can be shown to depend on R and Δ via (Pollak and Geballe 1961):

$$a = \frac{e^2 R^2}{12 k_B T \cosh^2(\Delta/2k_B T)} \qquad [6.78]$$

The total conductivity can be evaluated using [6.74], and assuming the activation energies W to be uniformly distributed as $p(R, \Delta, W) = \delta(R-R_0)/(R^2 W_0 \Delta_0)$ gives:

$$\sigma(\omega) = \frac{\pi \omega e^2 R_0^2 N k_B T \tanh(\Delta_0/2 k_B T)}{6 \Delta_0 W_0} \qquad [6.79]$$

where N is the concentration of sites (cm^{-3}). Note that this atomic hopping model predicts that the frequency dependence of $\sigma(\omega)$ is exactly linear and temperature independent, at variance with the experimental observation that in glasses s is generally sub-linear and increases with decreasing temperature. Note also that both this model and the simple Debye-loss are examples of *fixed-range* hopping mechanisms, in which carrier motion takes place between sites separated by a constant distance, but only those sites which have activation energies such that $\omega \tau = 1$, i.e. $W_\omega = k_B T \ln(1/\omega \tau_0)$, contribute significantly to the total conductivity. In the random case, this may be regarded as 'variable-energy' hopping, since measurement at a frequency ω selects only those sites having energies W_ω.

If one considers now the other extreme model for carrier motion, namely quantum-mechanical tunnelling (QMT) ([6.77]), it is generally assumed that the defect centres are randomly distributed, and the probability of finding a centre at distance R is:

$$P(R) \, dR = 4\pi N R^2 \, dR \qquad [6.80]$$

Now, however, the hopping *is* variable range in nature, since the requirement that $\omega \tau = 1$ imposes the condition that only those pairs of centres separated by the hopping distance R_ω, contribute significantly to the conductivity at a frequency ω, where:

$$R_\omega = \frac{1}{2\alpha} \ln(1/\omega \tau_0) \qquad [6.81]$$

Equations 6.78 and 6.80 taken with [6.81] lead to the expression for $\sigma(\omega)$ ([6.74]) containing an R_ω^4 term. If this is evaluated, the following equation for the a.c. conductivity resulting from quantum-mechanical tunnelling is obtained:

$$\sigma(\omega) = CN^2(E_F)k_B T e^2 \alpha^{-5} \omega \ln^4(1/\omega\tau_0) \qquad [6.82]$$

where the constant C is given by $\pi/3$ (Austin and Mott (1969) or $\pi^4/96$ (Pollak 1971), and $N(E_F)$ is the density of states at the Fermi level (cm^{-3} eV^{-1}). The frequency exponent of [6.82] can be written as:

$$s = 1 - 4\ln^{-1}(1/\omega\tau_0) \qquad [6.83]$$

whereupon it can be seen that in this model s should be temperature independent but frequency dependent, having the value $s = 0.8$ for $\omega = 10^4$ s^{-1} and $\tau_0 = 2 \times 10^{-13}$ s (hence the common expression '$\omega^{0.8}$ law'). These predictions are again contrary to experimental findings, even for those systems (a-Si, a-Ge) for which the tunnelling model is expected to apply.

These considerations have led the author (Elliott 1977, 1983) to propose an alternative model for a.c. conduction, at least in chalcogenide glasses where the defects suffer a negative Hubbard U (sect. 6.2.2). This model is a combined variable-energy and variable-range hopping model, since the height of the potential barrier is correlated with the intersite separation, i.e. 'correlated barrier hopping', CBH. The detailed mechanism is shown in Fig. 6.23(a); *two* electrons are assumed to hop together from a D$^-$ centre to a neighbouring D$^+$ centre over the potential barrier between them which, because the centres are charged is reduced by a coulomb term:

$$W = W_M - \frac{8e^2}{4\pi\varepsilon\varepsilon_0 R} \qquad [6.84]$$

and this reduction is shown schematically in Fig. 6.23(b). The value of the maximum barrier height (at infinite separation) can be shown by a consideration of defect energies to be related to the optical band gap E_0 and the electrical activation energy ΔE_σ by (Elliott 1983):

$$W_M = 2(E_0 - \Delta E_\sigma) \qquad [6.85]$$

Since for most materials, $\Delta E_\sigma \simeq \frac{1}{2}E_0$, therefore $W_M \simeq E_0$. Using [6.75], [6.76], [6.80] and [6.84], together with the equation for the polarizability ([6.78]) corrected for two-electron transport, [6.74] can be evaluated to give (Elliott 1983)

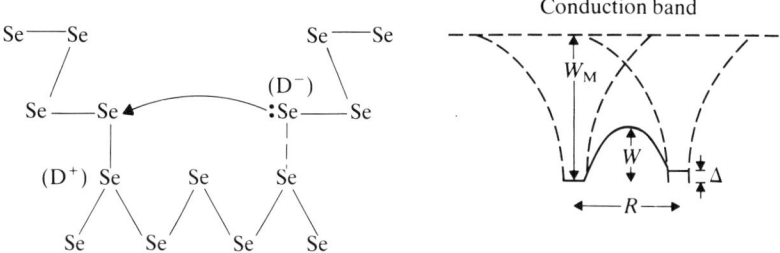

Fig. 6.23 (a) Schematic illustration of the hopping process envisaged to take place in the bipolaron correlated barrier hopping model for a.c. conduction in chalcogenide glasses (Elliott 1977). Two electrons transfer from a D$^-$ to a D$^+$ centre, which thereby exchange places (the new bond formed is shown by the dotted line). (b) Diagram of the potentials experienced by carriers distance R apart. W_M is the energy required to remove two electrons from a D$^-$ centre to form a D$^+$; W is the height of the barrier over which the carriers must hop. The disorder energy is shown as Δ.

$$\sigma(\omega) = \frac{\pi^3}{3\Delta_0} N^2 \varepsilon \varepsilon_0 k_B T \omega R_\omega^6 \tanh(\Delta_0/2k_B T) \qquad [6.86]$$

where N is the concentration of charged defects, and the hopping distance is given by:

$$R_\omega = \left(\frac{2e^2}{\pi \varepsilon \varepsilon_0}\right) \frac{1}{(W_M - W_\omega)} \qquad [6.87]$$

The term R_ω^6 can be expanded to first order as $1/W_M^6(1/\omega\tau_0)^\beta$, and so the frequency exponent is given by:

$$s = 1 - \beta \qquad [6.88]$$

where

$$\beta = \frac{6k_B T}{W_M} \qquad [6.89]$$

Two points of interest emerge from this analysis: a decreasing temperature dependence of s is predicted from [6.88] and [6.89], in accord with experiment, and a connection between the a.c. conductivity and d.c. conductivity is predicted for these materials through [6.85]. Such a correlation has been observed in experimental data (see Mott and Davis 1979), and the theoretical prediction is shown for comparison in Fig. 6.24.

Most treatments consider the hopping centres to be randomly distributed, with a probability distribution given by [6.80]. However, for melt-quenched chalcogenide glasses containing oppositely charged defects this is hardly likely to be the case because of the preferential pairing between D^+ and D^- centres due to their mutual coulombic attraction. In this case, for a quenched glass for which the atomic configurations corresponding to $T = T_g$ will be frozen-in for $T < T_g$, an approximate expression for the probability of two paired centres being at a separation R, will be:

$$P(R) = 4\pi N R^2 \exp(-4\pi N R^3/3) \exp(e^2/4\pi\varepsilon\varepsilon_0 k_B T_g R) \qquad [6.90]$$

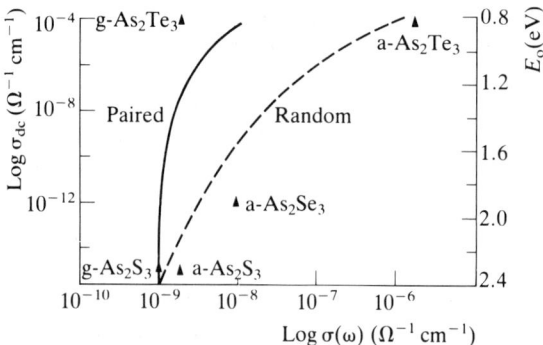

Fig. 6.24 AC conductivity (at $\omega = 10^6$ s^{-1} and $T = 300$ K) plotted as a function of optical band gap E_0 (or equivalently d.c. conductivity σ_{dc}) for random centres (dashed line) and paired centres (solid line). Experimental points are labelled 'a' (amorphous thin film) or 'g' (glass) (Elliott 1980).

where the first exponential factor is an exclusion factor and the second represents the coulombic attraction. Because of the final term, $P(R)$ is a rapidly *decreasing* function of R at small R (corresponding to the hopping distance R_ω), and this can profoundly affect the frequency dependence of $\sigma(\omega)$ (Elliott 1979b), giving $s=1$ under certain circumstances and even superlinear frequency behaviour for certain values of W_M, ε and T_g, a circumstance sometimes observed experimentally. In this case, the frequency exponent s is found to be:

$$s = 1 - \beta + \delta \qquad [6.91]$$

where

$$\delta = T/8T_g \qquad [6.92]$$

Thus if $W_M/6k_B \simeq 8T_g$, a temperature-independent frequency exponent of value unity is predicted.

Although the pair approximation is adequate at sufficiently high frequencies and/or low temperatures, it must be recognized that it fails completely to reproduce the behaviour in the d.c. limit, i.e. as $\omega \to 0$. This is because *if* the same mechanism is operative in both a.c. and d.c. regimes, as the frequency is lowered, carrier motion between an increasing number of sites takes place, until the d.c. limit or a percolation path is established. Thus in this case, any theory must be capable of accounting for the conductivity behaviour in all frequency regimes. An example of such a system would be evaporated a-Si or a-Ge thin films, which exhibit a $T^{-1/4}$ temperature dependence of the d.c. conductivity at low temperatures, indicative of variable-range tunnelling of carriers, and so the tunnelling mechanism ought to account for the high-frequency regime, but as we have already seen, it does not. The exact mechanism therefore remains a mystery, although Long (1982) has suggested that polaron hopping might be responsible. The same problem does not arise in the case of chalcogenide glasses, since the d.c. conductivity is simply activated and due to band conduction and is therefore not defect related.

6.3.5 Drift mobility

In this chapter and the last we have discussed electrical transport in general, under steady-state conditions, omitting from the discussion mention of transient phenomena. However one form of transient electrical transport has been very valuable in giving information about states in the gap of amorphous semiconductors and the mechanism of charge transport; this is drift mobility. In these experiments, carriers are injected at one position in a sample and in the presence of an applied field E their transit time t_t to another point at a distance L is determined (see Spear 1969 for a review). The drift mobility is then defined as:

$$\mu_d = L/Et_t \qquad [6.93]$$

If the circuit time constant is much less than the experimental time-scale (i.e. $RC \ll t_t$) then the field in the sample is approximately uniform during the transit and [6.93] is valid. Highly resistive samples ($\rho > 10^7$ Ω cm) need to be used, with the electrodes in a sandwich configuration, one of which may be semi-transparent if optical excitation is to be employed. Carrier injection is then effected either by the application of a voltage pulse (of duration less than t_t) of the appropriate polarity to an electrode, in which case carriers of a single type (electron or hole) are created, or else by the application of a short pulse of strongly absorbed light or of an electron beam; in either case electron–hole pairs are created close to one electrode, and according to the polarity of the applied field carriers of only one sign can transit, the others being dissipated at the electrode.

In the ideal case, a perfect current pulse whose leading profile is given by a step-function, would drift unaltered across the sample and the time at which the carrier sheet reached the back electrode would define the transit time, t_t. In practice, dispersion of the charge sheet takes place to a greater or lesser extent. In the simplest case, the charge sheet spreads out as it drifts across the sample but retains its identity as a well-defined pulse. The broadening of the pulse arises because the drift velocity of the carriers obeys gaussian statistics; the dispersion of the gaussian pulse as measured by the FWHM, σ, increases with time as $\sigma \propto t^{1/2}$, and the mean displacement of the pulse of course is proportional to the elapsed time, $l \propto t$. This process leads to a rounding of the previously sharp current transit profile, but nevertheless t_t may easily be determined: this process is illustrated in Fig. 6.25(a).

In the case shown in Fig. 6.25(a), where the transit time is clearly defined, it is pertinent to ask what is the drift mobility as deduced from [6.93]: is it equal to the

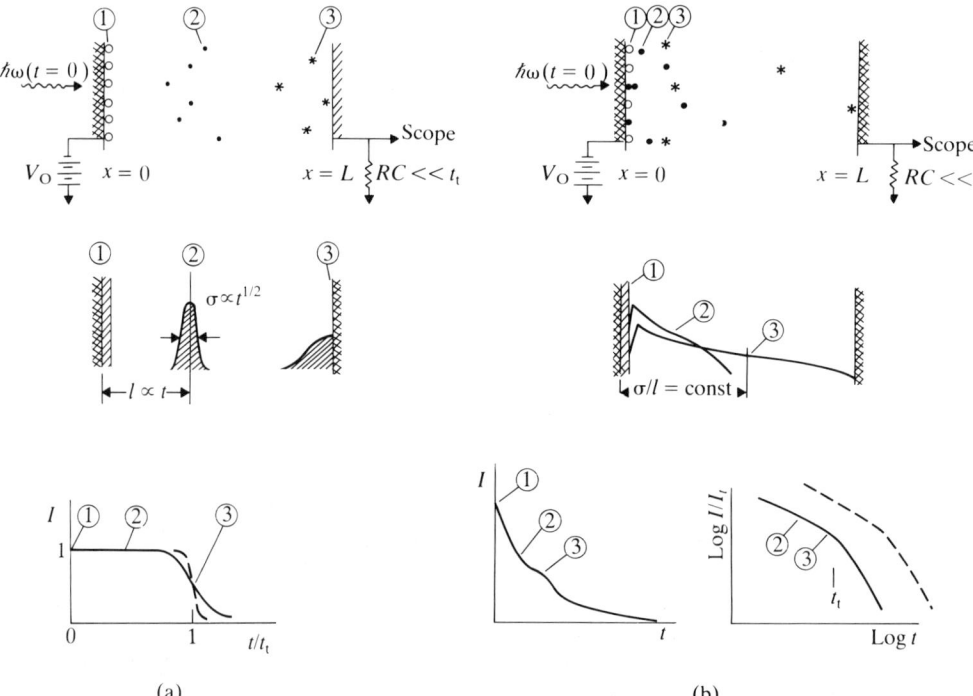

Fig. 6.25 Schematic illustration of carrier propagation under (a) gaussian and (b) non-gaussian conditions. In all cases, (1) refers to the situation at $t=0$, (2) to $t<t_t$, (3) to $t \simeq t_t$. The current pulse in the external circuit induced by the charge displacement is plotted linearly versus time, and also double-logarithmically for the non-gaussian case (where the dashed curve represents transients at a lower applied bias field, i.e. a longer transit time) (after Pfister and Scher 1978).

conductivity mobility μ_e ([5.62]–[5.64]) characteristic of conduction in extended states? In the ideal case where the carriers transport in a band in the absence of traps, μ_d and μ_e are identical, but in a real amorphous semiconductor traps are invariably present, either shallow-lying such as localized tail states below the mobility edge or deeper-lying in the form of 'defect' states in the gap. In the simplest case, when the carrier motion occurs at a mobility edge (say E_c) and is 'trap limited' by occasional trapping by a single well-defined set of traps at energy E_t below E_c, the drift mobility is given by:

$$\mu_d = \frac{N_c}{N_t} \exp(-E_t/k_B T) \qquad [6.94]$$

where N_c ($= k_B T N(E_c)$) and N_t are the density of states per unit volume at the (conduction band) mobility edge and of the trapping levels, respectively: a similar equation holds for the case of hole transport. Thus, for this case the trap level E_t may be obtained from the temperature dependence of μ_d, and the density of trap states can be deduced from a combination of drift mobility and d.c. conductivity studies: since the pre-exponential factor of the d.c. conductivity for conduction in extended states is $e\mu_e N_c$ (cf. [5.59]), this gives from [5.65]

$$N_t = \frac{\sigma}{e\mu_d} (-\gamma/k_B) \exp[(\Delta E_\sigma - E_t)/k_B T].$$

Alternatively, and more realistically, trapping is more likely to occur in a *distribution* of trap states and the most obvious examples of this are the localized tail states at the band edges. If, again, transport is at the mobility edge, and if the tail-state density of states increases as $N(E) \propto (E-E_A)^s$ above the (conduction) band edge, then it can be shown that (Mott and Davis 1979):

$$\mu_d = \mu_e \left(\frac{\Delta E}{k_B T}\right)^s \exp(-\Delta E/k_B T) \qquad [6.95]$$

where ΔE is the width of the tail-state distribution, i.e. $\Delta E = E_c - E_A$. At lower temperatures, when the transport channel moves into the tail states, transport now occurs by hopping and the drift mobility is given by [5.68]:

$$\mu_d = \mu_{hop} = \mu_0 \exp(-\Delta(E)/k_B T) \qquad [6.96]$$

where μ_0 is given by [5.69] and $\Delta(E)$ is the hopping energy, itself a function of the energy at which transport takes place because of the variable-range nature of the hopping process (see sect. 5.4.1). Thus the transition between trap-limited extended-state drift mobility and tail-state hopping drift mobility should be indicated by a marked change in activation energy, from $\Delta E = E_c - E_A$ to Δ at a certain temperature. Such a 'kink' has been observed for electron drift mobility in undoped glow discharge produced a-Si:H films by Le Comber et al. (1973), and is shown in Fig. 6.26. For each of the samples, the difference in activation energies $(E_c - E_A) - \Delta$ $\simeq 0.1$ eV; the d.c. conductivity also shows a kink in its activation energy (as in Fig. 5.26(b)) at about the same temperature (viz. $\simeq 250$ K) and of the same magnitude, which is to be expected if indeed the transport channel is moving from extended states to band-tail localized states at this temperature (cf. [5.60] and [5.71]). (This

Defects

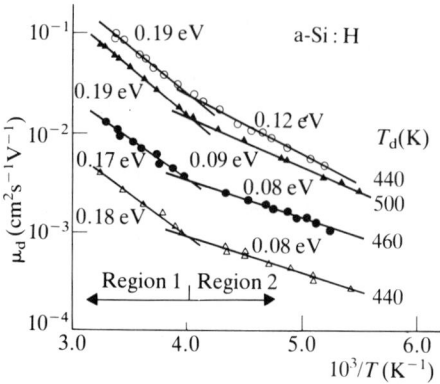

Fig. 6.26 Temperature dependence of the electron drift mobility μ_d for four samples of GD a-Si:H deposited at the substrate temperatures indicated (Le Comber et al. 1973).

kink should not be confused with the kink in σ_{dc} in the *opposite* sense observed in a-Si:H at higher temperatures ($\simeq 450$ K), which is still the subject of much debate (see sect. 5.4.2).) The value of the drift mobility for electrons in a-Si:H is $\mu_d \simeq 10^{-2}$–10^{-1} cm^2 V^{-1} s^{-1} at room temperature (Fig. 6.26) and is activated with an energy $\simeq 0.2$ eV; hole transport, on the other hand, has a much lower drift mobility, $\mu_d \simeq 10^{-4}$ cm^2 V^{-1} s^{-1}, a result of the large activation energy ($\simeq 0.4$ eV) involved (Allan 1978). The nature of the trapping levels presumed to lie $\simeq 0.4$ eV above E_v is somewhat uncertain, but their influence in diminishing the hole drift mobility sets serious limitations on the efficiencies of certain devices fabricated using a-Si:H (e.g. solar cells), where in certain configurations the hole diffusion length is so small that the photo-generated holes cannot be collected efficiently.

Another interesting example of drift mobility is that seen in a-SiO$_2$ films, where electron and hole transport exhibit startlingly different behaviour. *Electrons* have a remarkably high drift mobility at room temperature, $\mu_d^e \simeq 20$ cm^2 V^{-1} s^{-1} (Hughes 1973), which is characteristic of conduction mobilities in extended states; this observation means that the width of the localized tail states at the conduction band edge must be exceedingly small. Mott has ascribed this to the s-like character of the states (derived from both Si and O) at the bottom of the conduction band; the considerable degree of short-range order combined with the s-like nature of the electron states means that the disorder potential will be small thereby restricting the range of the localized tail states, which is the same argument invoked to explain similar behaviour in liquid rare gases (Mott and Davis 1979). The situation for the valence band edge will be completely different however; the directional character of the predominant p-states means that even small amounts of topological disorder can give rise to a considerable range of localized tail states. Indeed, the *hole* drift mobility in a-SiO$_2$ is much smaller than for electrons at room temperature $\mu_d^h \simeq 10^{-5}$ cm^2 V^{-1} s^{-1}, indicative of considerable interaction between the hole and these (or other) states. Furthermore, the transport mechanism for holes appears to be different in two time regimes: for times $\lesssim 10^{-6}$ s, the activation energy of μ_d is $\simeq 0.16$ eV, but this increases to $\simeq 0.37$ eV at longer times (Hughes 1977). The behaviour at very short times, the so-called 'prompt mobility' has been ascribed to the formation and subsequent hopping of a hole-like small polaron in the lone-pair p-states at the top of the valence band, and the polaronic nature of the carrier is substantiated by the observation of a marked decrease in activation energy of μ_d at low temperatures ($\simeq 200$ K); precisely this behaviour is expected for the polaron model, it will be recalled (sect. 5.5), where the multiphonon hopping processes, dominant at high temperature, become frozen-out with decreasing temperature leading ultimately to polaron band conduction. The drift mobility in a-SiO$_2$ at times longer than $\simeq 10^{-6}$ s cannot be described in this way, and is in fact a manifestation of the extremely dispersive transits often observed in amorphous semiconductors and which will be discussed next.

So far, we have discussed drift mobility under the conditions where some dispersion of the carrier pulse is evident, but the pulse nevertheless remains well defined and characterized by gaussian statistics. In many cases, however, the dispersion of the transit pulse is so severe that the pulse shape is no longer recognizable and defining the transit time becomes very difficult (see Fig. 6.25(b)); this situation has been termed 'anomalous transit dispersion' (Pfister and Scher 1978). This is characterized by the pulse spread, σ, and its mean displacement, l, having the *same* time dependence, i.e. $\sigma/l = $ constant (cf. $\sigma/l \propto t^{-1/2}$ for gaussian transport). In order for anomalous transit dispersion to occur, the microscopic processes which control the transport must have a wide distribution of characteristic times which extends into the experimental time domain determined by t_t (which measures the transit time of a certain fraction of the fastest injected carriers), and only then will the carrier packet grow asymmetrically as shown schematically in Fig. 6.25(b). In this anomalous regime, the drift mobility is field dependent and also is an apparent function of the film thickness.

This intriguing behaviour has been successfully accounted for by a theory developed by Scher and Montroll (1975); the reader is referred to Pfister and Scher (1978) for an extensive review. The model simulates the actual distribution of transition rates which characterize the carrier transit (due to either multiple trapping events in a distribution of trap levels or to tunnelling between randomly separated traps) by means of a single probability distribution $\psi(t)$ for a carrier moving randomly on a *regular* lattice, where $\psi(t)$ represents the probability of a carrier hopping to its next site in time t, having arrived at the original site at $t=0$; such a process is referred to as a 'continuous time random walk' (CTRW). Scher and Montroll (1975) proposed that the distribution function $\psi(t)$ should behave as

$$\psi(t) \propto t^{-(1+\alpha_d)} \qquad [6.97]$$

where the disorder parameter α_d (not to be confused with the inverse localization length) takes values between 0 and 1, and is related in a complicated way to microscopic parameters such as trap depths, radii of localized states, etc., decreasing as the amount of disorder increases. However, the power law dependence of $\psi(t)$ embodied in [6.97] does reproduce many of the features of anomalous dispersive transport as we will see, although it is not a unique distribution (and cannot be so since all the 'randomness' of the hopping system is folded into this one function); note, however, the difference between the distribution function characteristic of CTRW ([6.97]) and that which characterizes *gaussian* transport, i.e. $\psi(t) \propto \exp(-t/\tau)$, which is associated with a *single* event time τ.

Defects

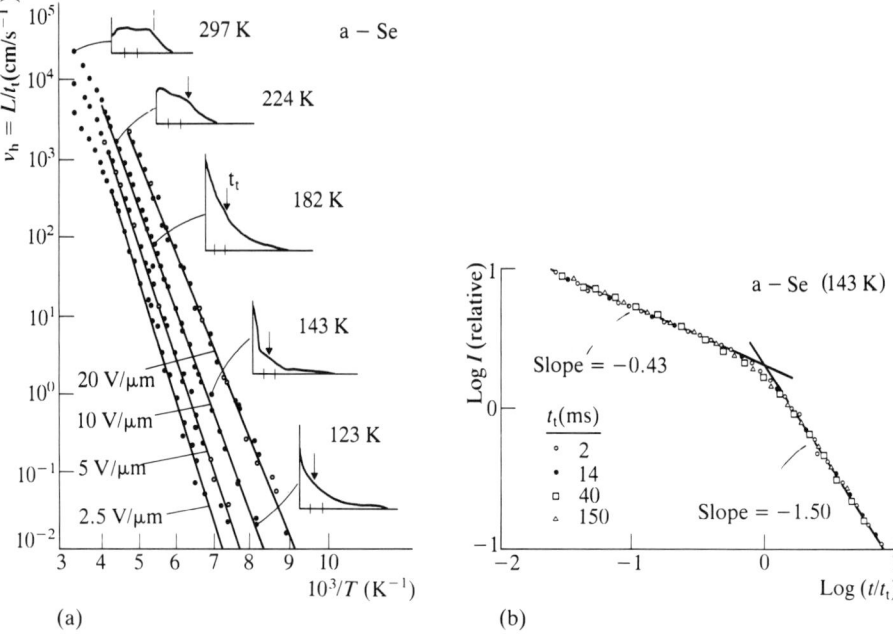

Fig. 6.27 Illustration of anomalous transit dispersion behaviour of a-Se at low temperatures. (a) Temperature dependence of hole velocity for various fields, together with representative current traces. (b) 'Universal' plot of log I v.s. log t for a-Se obtained by shifting along the axes data obtained at different bias fields (Pfister and Scher 1978). Note that the sum of the slopes is approximately -2.

The general result of the CTRW model can be summarized in the following equations for the time dependence of the current:

$$I(t) \propto t^{-(1-\alpha_d)} \quad t < t_t \quad [6.98(a)]$$

$$I(t) \propto t^{-(1+\alpha_d)} \quad t > t_t \quad [6.98(b)]$$

It should be noted that the shapes of plots of $\ln I$ v.s. $\ln t$ should sum to -2. Figure 6.27(a) shows the temperature dependence of the hole drift velocity in a-Se for different values of electric field; note particularly that as the temperature is reduced below $\simeq 200$ K the approximately rectangular current profile typical of gaussian transport broadens out, until at the lowest temperatures the transit time cannot be distinguished on a simple plot of I v.s. t. However, if the current transient is plotted double-logarithmically (Fig. 6.27(b)), a clear kink is observed at t_t and the sum of the two slopes is -1.93, i.e. close to the theoretical prediction of -2. Figure 6.27(b) clearly illustrates another aspect of non-gaussian dispersive transit, namely that since α_d is approximately field independent, changing the experimental conditions (e.g. E) at constant temperature merely results in a parallel shift of $I(t)$ along the axes; thus normalizing the curves relative to t_t should produce 'universal' curves, as is observed for a-Se, a-As_2Se_3 and several organic systems. This universality is in fact observed in a-Se only in the temperature range $140 < T < 170$ K; above these temperatures the transition to gaussian transport takes place which is not accounted for by the CTRW model, and below these temperatures the sum rule for the slopes

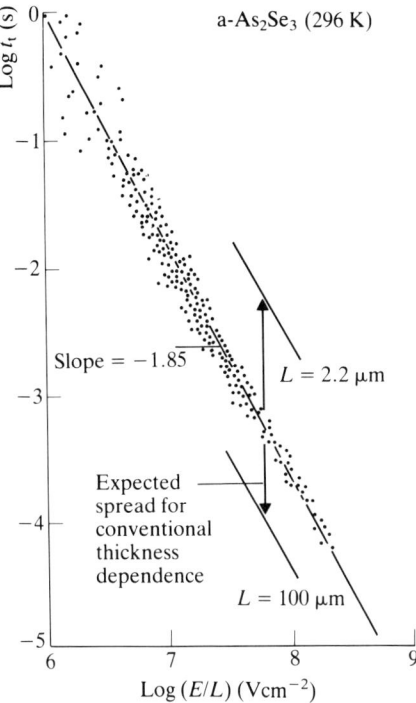

Fig. 6.28 Electric field (E) and sample thickness (L) dependence of the transit time for a-As$_2$Se$_3$. The range of thickness was 2.2–100 μm and range of fields 2–10 V μm^{-1}. Also shown is the spread in transit times expected if conventional gaussian statistics were controlling the transit (Pfister and Scher 1978).

above and below t_t fails because, it is believed, a simple algebraic expression for $\psi(t)$ no longer holds.

Finally, we consider the question of the thickness and field dependence of the transit time. To a first approximation $t_t \propto (L/E)^{1/\alpha_d}$, where L is the thickness and E the field. If this is substituted into [6.93], it is seen that the drift mobility is a superlinear function of both field and thickness. Figure 6.28 shows a plot of log t_t v.s. log (E/L) for a-As$_2$Se$_3$ which has a slope $1/\alpha_d = -1.85$, and which importantly is independent of sample thickness and applied field; the thickness dependence on such a plot expected for conventional gaussian transport is also shown for comparison and it is seen that this spread is much greater than is actually observed.

We conclude this section with a discussion of the Xerox process which, apart from the use of silica-based glasses as optical elements, must be the most widespread current application of an amorphous material (i.e. a-Se); in addition, the considerable amount of study that this process has received has spawned many of the ideas and results on drift mobility discussed previously.

The first demonstration of xerography (the name derives from the Greek for 'dry writing') was by Carlson and Kornei on 22 October 1938 in Astoria, NY. These first experiments used sulphur instead of selenium for the electro-photographic medium, but the principle is the same. Carlson described the discovery thus (Pfister 1979):

Defects

I went to the lab that day and Otto [Kornei] had a freshly-prepared sulphur coating on a zinc plate. We tried to see what we could do toward making a visible image. Otto took a glass slide and printed on it in Indian ink the notation '10-22-38 Astoria'.

We pulled down the shade to make the room as dark as possible, then he rubbed the sulphur surface vigorously with a handkerchief to apply an electrostatic charge, laid the slide on the surface and placed the combination under a bright incandescent lamp for a few seconds. The slide was then removed and lycopodium powder was sprinkled on the sulphur surface. By gently blowing on the surface, all the loose powder was removed and there was left on the surface a near-perfect duplicate in powder of the notation which had been printed in the glass slide.

Both of us repeated the experiment to convince ourselves that it was true, then we made some permanent copies by transferring the powder image to wax paper and heating the sheets to melt the wax. Then we went out to lunch and to celebrate.

The principles underlying the modern xerographic process using a-Se (or an organic photo-receptor) are precisely the same as those described by Carlson (see Fig. 6.29). The surface of the photo-conducting film is first charged positively to a surface potential $V_0 \simeq 700$ V by moving the film under a corona charging device (a). The charged photo-receptor is now exposed to light reflected from the document to be copied; light is reflected preferentially from the white areas of the page and these photons on striking the film are strongly absorbed near the surface and create electron–hole pairs (b). The photo-electrons neutralize the positive charges locally at the surface, while the holes drift across the film under the action of the field caused by the surface potential and neutralize the negative charge induced on the Al substrate (c). Thus a negative image is formed in the surface charge distribution. The latent image is developed by means of negatively, triboelectrically charged toner particles (each being a carbon particle $\simeq 10$ μm in diameter surrounded by a low-melting plastic carrier bead $\simeq 100$ μm in diameter); these are attracted to the remaining positively charged areas on the film (d). Finally, the developed image is transferred and fixed by attracting the toner particles to a sheet of paper, which has been corona charged to the opposite polarity of that of the toner, whereupon the paper is heated which melts the toner particles and fixes them in position (e).

Certain conditions must be met for a material to be a suitable photo-receptor. It must be able to be produced in defect-free large areas, and the material must be mechanically robust and chemically inert in order to withstand, respectively, both abrasion and chemical degradation in the presence of intense light and electric fields; an additional advantage is for the film to be flexible so belt operation may be used. As far as electrical properties are concerned, the material must be highly resistive ($\rho \gtrsim 10^{10}$ Ω cm) in order that the surface potential V_0 does not decrease appreciably after the initial exposure, and that image contrast is not lost by lateral charge flow on the surface. Furthermore, the carrier generation or 'quantum' efficiency, η_q (which is the number of *free* carriers generated per incident photon) must be high at the output wavelengths of suitable lamps. The carrier generation rate can be written in terms of the quantum efficiency as (Mott and Davis 1979):

$$G = \eta_q \left[\frac{I_0(1-R)\{1-\exp(-\alpha d)\}}{d} \right] \quad [6.99]$$

where the quantity in square brackets is the number of photons absorbed per second in a sample of thickness d in the direction of the incident radiation, I_0 is the incident

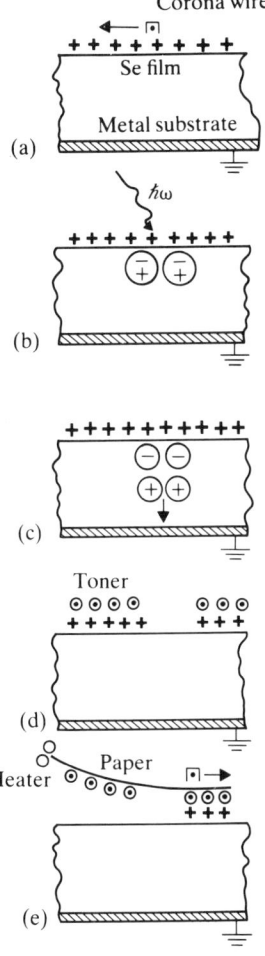

Fig. 6.29 Schematic illustration of the Xerox process, using a-Se as the photo-receptor.
(a) Positive charging of the surface by a corona discharge.
(b) Exposure of the photo-receptor by photons reflected from the document to be copied.
(c) Discharge of the surface potential locally by the photo-generated electrons, the holes drifting across the thickness of the film.
(d) Development of the latent image by means of negatively charged toner particles which are attracted to the remaining areas of positive surface charge.
(e) Transfer of the image to paper by means of a second corona discharge and subsequent fixing by heating the paper to melt the toner particles into place (after Mott and Davis 1979).

light flux density, R the reflectivity and α the absorption coefficient.

Amorphous Se (doped with small amounts of As and other additives to retard crystallization) meets all these conditions and has been found to be an excellent photo-receptor. The behaviour of its quantum efficiency is rather complicated however, being a function of electric field, temperature and of course photon energy. It is found that η_q reaches unity at high fields and at photon energies $\simeq 3$ eV and decreases sharply at smaller energies, even though the optical gap is nearer 2.2 eV; this is in marked contrast to the case for crystals where unit quantum efficiency is achieved at the band-gap energy. The higher than band-gap energy required for n_q to become unity in a-Se has been ascribed to the mutual coulomb interaction of the photo-induced electron–hole pairs; whether or not the carriers can escape from each other, and so contribute to the quantum efficiency, depends on whether their Brownian (diffusive) motional energy is sufficient to overcome the potential barrier resulting from the perturbation of a coloumb well by an applied external field E, viz. $e^2/4\pi\varepsilon\varepsilon_0 r + eEr$. This is essentially the same process as that considered by Onsager in a discussion of dissociation in weak electrolytes, and it has been successfully applied by Pai and Enck (1975) to the behaviour of η_q for a-Se.

The importance of the field dependence of η_q for the use of a-Se as a photoreceptor lies in the fact that the decay in time of the surface potential is consequently much more gradual than if η_q were constant (for which case $V(t) = V_0 - (e\eta_q/C)I_0$, where C is the capacitance of the film); the non-linear decay of $V(t)$ when $\eta_q = \eta_q(E)$ means that the difference between two discharge curves $V(t)$ for two light fluxes, I_1 and I_2, the so-called 'contrast potential' (which is developed ultimately by the toner particles) becomes a broad, peaked function of time (unlike the linear variation if η_q = constant) and so offers a considerable latitude in exposure times to give approximately the same degree of contrast.

Problems

6.1 Show by using an analysis similar to that employed in section 6.3.3 that the temperature dependence of the d.c. hopping conductivity in the case when the thickness of the film is less than the hopping length obeys the law $\ln \sigma \propto -T^{-1/3}$ ([6.65]). Hence estimate the thickness of a film for which a transition from 2-D to 3-D behaviour is expected at 300 K (take $N(E_F) \simeq 10^{19}$ cm^{-3} eV^{-1} and $\alpha^{-1} \simeq 10$ Å).

6.2 Discuss how thermopower and conductivity measurements may be used to decide whether the conduction channel in an amorphous semiconductor is through the extended states at a mobility edge or through the localized states at the Fermi level.

6.3 Derive the expressions which relate the position of the Fermi level to the electron occupancy of a defect, suffering both positive and negative U_{eff} ([6.21] and [6.22]).

Bibliography

R. Balian, M. Kleman and J.-P. Poirier (eds) (1981), *Physics of Defects* (North-Holland).
M. H. Brodsky (ed.) (1979), *Amorphous Semiconductors* (*Topics in Applied Physics*, vol. 36) (Springer-Verlag).
N. F. Mott (1980), *J. Phys.*, **C13**, 5433.
N. F. Mott and E. A. Davis (1979), *Electronic Processes in Non-Crystalline Materials* (OUP).
J. Robertson (1982), *Phys. Chem. Glasses* **23**, 1.

7 Amorphous metals

7.1 Introduction
7.2 Structure
7.2.1 Experimental methods
7.2.2 Structural data
Transition metal–metalloid alloys
Metal–metal alloys
7.2.3 Structural models
Dense random packing of spheres
Random packing of trigonal prisms

7.3 Electronic properties
7.3.1 Electronic structure
7.3.2 Optical properties
7.3.3 Electrical transport
7.3.4 Superconductivity

7.4 Magnetic properties
7.4.1 Magnetic moments and saturation magnetization
7.4.2 Curie temperature
7.4.3 Magnetic anisotropy and magnetostriction

7.5 Mechanical properties

Problems

Bibliography

7.1 Introduction

Thus far in this book, we have tried to give a description of a variety of physical properties exhibited by solids in the non-crystalline state, and to this end, we have illustrated our account with examples which for the most part have been chosen from the insulating or semiconducting class of amorphous materials. In so doing we have omitted from the discussion any significant mention of an important class of non-crystalline solids, namely amorphous metals. By the term amorphous metal we mean those materials, principally alloys, composed primarily, but not necessarily exclusively, of metallic elements and which exhibit characteristic 'metallic' properties, for instance, in their electrical, magnetic or optical behaviour. Thus by this definition we exclude materials such as a-As or a-Sb (metallic in the crystalline state, but semiconducting when amorphous), or highly doped a-Si:H (which may have an electrical conductivity close to that characteristic of metals, but which contains no metallic elements).

Amorphous metals merit a chapter to themselves in view of the fact that many of their properties differ completely from those of their insulating counterparts, even though certain features do appear to be shared in common, for instance the low-temperature thermal anomalies caused by the presence of two level systems and structural relaxation properties where $\ln t$ kinetics are obeyed. We will therefore attempt to give a unified account of these scientifically interesting and technologically most important materials in what follows.

Amorphous metals can be prepared using a variety of the techniques discussed in Chapter 1. Vapour deposition methods, such as evaporation or sputtering, almost invariably produce amorphous thin films, although in many cases the substrate temperature needs to be held at a very low temperature in order to avoid the production of a polycrystalline film. In this way, for example, pure Bi or Ga can be rendered amorphous by deposition on to substrates held at $\simeq 4$ K. However, many amorphous films prepared in such a manner are fairly unstable and tend to crystallize at temperatures not much above the deposition temperature, although the presence of some impurities does tend to stabilize the films somewhat.

The most common method of producing amorphous metals, and certainly the most important technologically, is by rapid quenching from the melt. However, to avoid the production of polycrystalline material, extremely high rates of cooling must be employed, necessitating the use of techniques such as splat quenching, melt spinning or melt extraction (see sect. 1.3.5 and Fig. 1.10). In these ways, a variety of alloy systems can be produced in a glassy state, albeit in the form of a thin ribbon. The thickness of quenched glassy metal ribbons is determined by the rate of cooling, in turn governed by the heat transfer between melt and substrate, and the thickness and thermal conductivity of the liquid layer. We may estimate the maximum thickness obtainable (d_m) from the equation $d_m \simeq (D_t T_m/q_c)^{1/2}$, where D_t is the thermal diffusivity of the melt, T_m the melting temperature and q_c the critical quenching rate. For liquid alloys $D_t \simeq 0.2$ cm^2 s^{-1} and $T_m \simeq 1000$ K; thus for $q_c \simeq 10^6$ K s^{-1}, thicknesses of $\simeq 100$ μm result. Note however that the cooling rate varies with the thickness of the liquid layer, thereby leading to structural and other inhomogeneities within the glassy metal foil.

Several distinct metallic glass-forming systems have been established. These belong essentially to the following four binary alloy families: (a) the transition metal–semimetal (or 'metalloid') system; (b) the inter-transition metal system; (c) alkaline-earth-containing alloys; (d) the actinide–transition metal system.

The first group, often called the metal–metalloid system, is perhaps the most widely studied type of metallic glass because of their potential technological applications. Typically the glass-forming compositions consist of 80% transition metal (such as Pd, Fe, Ni) alloyed with 20% metalloid (such as Si, B, P), although more complicated alloys containing more than a single type of transition metal, or metalloid, also form glasses if the overall ratio of transition metal (T) to metalloid (M) is maintained. The second group consists of alloys between early transition metals (TE), e.g. Nb, Zr, Ti, and late transition metals (TL), e.g. Ni, Cu, Rh, typically

Table 7.1 Classification of binary glass-forming metallic systems

Group	Class	Typical composition of glasses	Most stable intermetallic compounds
a	T–M	$Pd_{80}Si_{20}$ $Ni_{80}P_{20}$ $Fe_{40}Ni_{40}P_{14}B_6$ (Metglas 2826)	Pd_3Si, Fe_3P (cementite) Pd_2Si, Fe_2P
b	TE–TL	$Nb_{60}Ni_{40}$ $W_{45}Fe_{55}$ $Zr_{76}Fe_{24}$	'NbNi' disordered phase Fe_7W_6 phase
	TL–RE	$Co_{33}Gd_{67}$ $Ni_{30}Gd_{70}$	$CdCO_2$ Laves phase $GdCo_5$ ⎫ ⎬ Frank–Kasper Gd_2Co_{17} ⎭ phases
c	AE–AE AE–S AE–T S–RE	$Ca_{67}Mg_{33}$ $Mg_{70}Zn_{30}$ $Ca_{65}Pd_{35}$ $Al_{30}La_{70}$	$CaMg_2$ Laves phase $MgZn_2$ Laves phase $CaPd_2$ Laves phase $LaAl_2$ Laves phase
d	AC–T	$U_{70}Cr_{30}$	

T = transition metal; M = metalloid; TE = early transition metal; TL = late transition metal; RE = rare earth; AE = alkaline earth; S = simple metal; AC = actinide.

in the ratio of 60:40%. In addition, we may include in this group alloys between transition metals (T) and rare earths (RE) such as $Co_{33}Gd_{67}$. The third group consists of alloys of group IIA alkaline earth (AE), e.g. Be, Mg or Ca, either with themselves or other simple metals (S) such as Al or Zn, or with transition metals. In this category we may also include alloys of simple metals with rare earths (e.g. $Al_{30}La_{70}$). The final group consisting of actinide (AC) and transition metal alloys contains systems such as $U_{70}Cr_{30}$ or $Np_{80}Ni_{20}$. Table 7.1 summarizes these metallic glass-forming families and lists some of their members. Also included in the table are the corresponding most stable crystalline intermetallic compounds for comparison.

Many of these glass-forming alloys have certain common features, namely a deep eutectic lying in the glass-forming composition region, and strong interactions between the constituent atoms. The influence of these interatomic interactions is indicated by the negative free energy of mixing exhibited by many alloy systems in both groups (a) and (b) (see sect. 2.5.2), in the formation of many stable intermetallic phases and in the fact that eutectics tend to occur at discrete composition ratios of either 6:1 or 5:1 (Fig. 2.16). We will return to the question of interatomic interactions and the influence they exert on local order when we discuss the structure of amorphous metals in the next section.

The importance of the presence of a deep eutectic in determining the glass-forming region is as great (if not more so) for metallic as it is for non-metallic systems. For instance, while a cooling rate of 10^5 K s^{-1} is necessary to quench, say, binary Pd–Si alloys into the glassy state (restricting perforce the thickness of samples to be of the order of tens of microns), the addition of a third component can significantly lower the eutectic temperature, thereby drastically enhancing both the glass-forming ability and stability against crystallization. Thus, ternary alloys such as

Pd–Cu–Si have been vitrified in the form of cylindrical rods having diameters of several millimetres at quenching rates as low as $\simeq 10^2$ K s^{-1}. Furthermore it becomes understandable that four or even five element compositions form the basis of the very stable T–M alloys marketed as Metglas®, where two or three of the elements are different transition metals and the remaining elements are different metalloids.

We conclude this introductory section on metallic glasses with a further discussion of those factors which determine glass formation and subsequent stability. We have discussed a number of these already in a general sense in section 2.5, and of course some of them are equally applicable to metallic glasses as to covalent glasses. For instance the observation of Wang and Merz (1976) that glass formation is prevalent in materials having a number of crystalline polymorphs (sect. 2.5.3) is valid for many metallic systems too. Criteria such as the Zachariasen hypothesis based on the continuous random network model for the structure of glasses are obviously inappropriate for metallic glasses which have very little, if any, directional bonding. Nevertheless, a criterion for glass formation and stability based on structural grounds has been proposed (Polk 1970). This is based on the assumption that a Bernal-type dense random packing (DRP) of hard spheres model is a good approximation to the structure of glassy metals (for further discussion and a critique of this, see sect. 7.2.3). It was pointed out that the large holes in such a DRP model made up of metal atoms might be able to accommodate the smaller and softer metalloid atoms in forming a T–M alloy. The presence of the smaller atoms filling the interstices would, it was argued, stabilize the random configuration. While this conceptually appealing model also predicted the ratio of transition metal to metalloid atoms found in many glassy T–M alloys (namely 4:1), it was later pointed out that in fact the Bernal holes were too small to accommodate most metalloid atoms and that therefore the number of metalloid atoms that could be incorporated in an unmodified DRP model was seriously overestimated (Cargill 1975). Furthermore, the glass-forming composition region of some T–M systems falls well outside the predicted range, and in some cases the metalloid atoms are in fact *larger* than the metal atoms, e.g. $Pt_{65}Sb_{35}$. The detailed microscopic structure of glassy metals remains therefore a contentious matter (see sect. 7.2).

The final glass-forming and stability criterion that we shall discuss involves the electronic structure as well as the atomic structure. Nagel and Tauc (1975) proposed that a metallic glass alloy is stabilized against crystallization for the composition in which the Fermi level lies in a minimum in the electronic density of states. If the assumption is made that the electron character is s-like near the Fermi level (E_F) and that the nearly-free electron approximation is valid, then there will be a decrease in the otherwise parabolically increasing free-electron density of states at $|k+Q|=|k|$, where k is the electron wavevector and Q, for the case of a *crystalline* metal, is the wavevector spanning the zone boundaries. For a crystal, of course, a gap opens up at the zone boundary and the density of states decreases to zero there. Nagel and Tauc assumed that for a liquid or amorphous metal a 'pseudo-gap', rather than a true gap, opens up at E_F (Fig. 7.1(a)). Furthermore, they associated the 'zone-boundary' wavevector with the wavevector corresponding to the position of the first peak of the structure factor $S(Q)$, i.e. Q_p. Thus, the pseudo-gas occurs for all states at $|k_F|=\frac{1}{2}Q_p$ because of the spherical symmetry inherent in non-crystalline systems, (and hence in

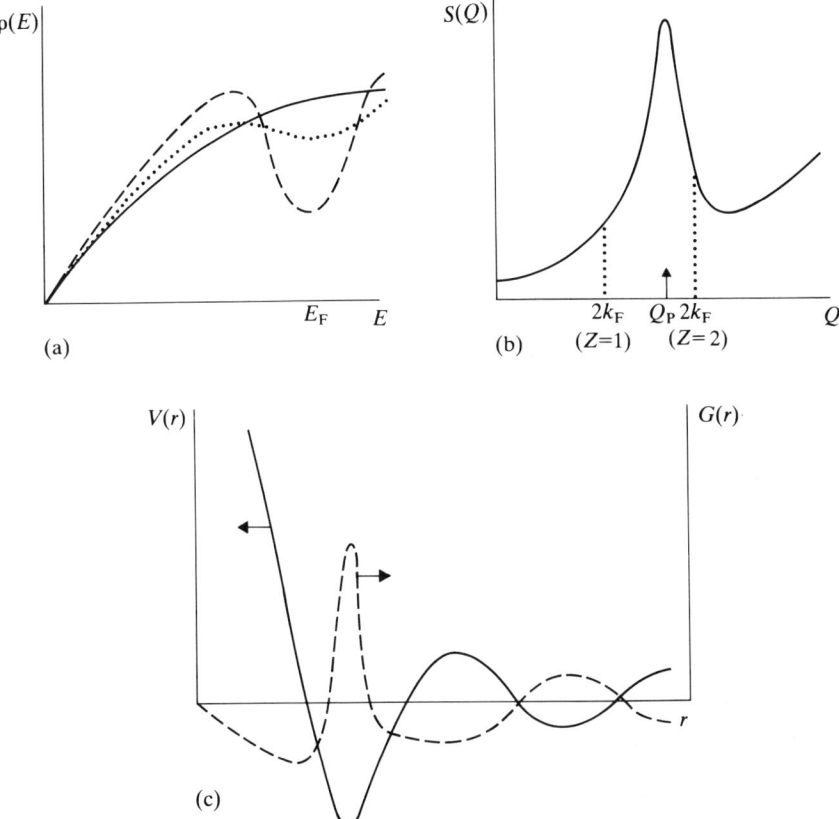

Fig. 7.1 Different models for metallic glass stability.
(a) Nagel–Tauc (1975) model in which E_F is postulated to lie at or near the minimum of a pseudo-gap in the density of states. The solid line is the free electron parabolic dependence of $\rho(E)$ with E, the dashed curve represents the density of states for an isotropic (i.e. glassy) system, and the dotted line illustrates the changes in $\rho(E)$ expected if the system becomes non-isotropic (i.e. starts to crystallize).
(b) Schematic illustration of the first peak in the structure factor of an amorphous metal at wavevector Q_P. Also shown are the positions of $2k_F$ in relation to Q_P for monovalent and divalent metals.
(c) Hafner (1981) model in which maxima in the atomic potential $V(r)$ of a metallic system coincide with minima in the (reduced) RDF, and vice versa, for optimum glass-forming ability.

$S(Q)$ and the density of states). The first peak in $S(Q)$ in liquid or amorphous metals is much narrower than the corresponding peak for liquid covalent materials (see sect. 7.2), because the random atomic packing in the metals means that the hard-sphere diameter, and multiples of it, are the dominant features in the RDF, and hence to a first approximation a single Fourier component (of wavevector Q_P) dominates $S(Q)$. Equivalently, the sharpness of the first peak is due to the near-periodic oscillations in the RDF at distances beyond about twice the nearest-neighbour separation. Thus, in the liquid metal case, Q_P takes on the significance of the reciprocal lattice vector in crystalline materials: indeed, Q_P is inversely proportional to the hard-sphere diameter, even for alloys, where the following phenomenological equation holds (Blétry 1978)

$$Q_{ij}^p = \frac{1}{\sigma_{ij}}\left[7.64 - 4.32\left(\frac{\bar{\sigma}}{\sigma_{ij}} - 1\right)\right] \qquad [7.1]$$

It is well established (see, e.g., Mott and Davis 1979) that for monovalent metals $2k_F$ lies slightly below, and for divalent metals $2k_F$ lies slightly beyond, Q_p (Fig. 7.1(b)). Nagel and Tauc argued that on alloying of elements of higher valence to a monatomic metal (and in this regard transition metals are treated as being monatomic), $2k_F$ would move through the first peak in $S(Q)$, and at the composition corresponding to $2k_F = Q_P$, the Fermi level would lie in the centre of the pseudo-gap. Stability against crystallization is supposed to arise through an electronic effect: if a fluctuation towards crystalline order should occur, this would destroy the isotropy of $S(Q)$, thereby filling in the pseudo-gap, raising the electronic energy and destabilizing the structure.

Attractive though this theory is, experimental tests have shown it to be lacking. The s-like character of states near E_F in say $Pd_{0.8}Si_{0.2}$ has not been verified by photoemission experiments, nor have psuedo-gaps in the density of states been found, and the relationship $2k_F = Q_P$ for the binary alloy systems Ca–(Mg, Zn, Al or Li) is only satisfied for Ca–Li, the only non-glass-former of the four alloys. Nevertheless, the electronic basis for metal glass stability has been retained (see Hafner 1981) in theories which ascribe relative stability to those materials having minima in the spatial dependence of the pair potential coinciding with peaks in the RDF, and vice versa (Fig. 7.1(c)). The pair potentials can be calculated accurately from pseudopotential theory, using realistic values of atomic radii and taking account of the variation in size on alloying (e.g. Ca effectively contracts when alloyed with Al, but expands when alloyed with Li); in this way *quantitative* assessments of glass stability can be made, at least for the case of simple metals.

7.2 Structure

7.2.1 Experimental methods

The structure of amorphous metals can be investigated experimentally by many of the techniques discussed in section 3.2. Diffraction methods have obviously been widely applied, but since to all practical purposes amorphous metals are invariably alloys, determination of the individual partial pair distribution functions is essential for a complete description of the structure, necessitating the use of techniques such as anomalous dispersion (sect. 3.2.2) or EXAFS (sect. 3.2.3). The various options available in the conventional diffraction techniques, such as isotopic or isomorphous substitution, are also obviously suitable for the determination of partial pair distributions of amorphous metals.

The technique of isotopic substitution in neutron diffraction is particularly useful for the study of amorphous metals because, for example, both Ni and Ti have isotopes with *negative* scattering lengths (see Table 7.2), thereby offering the possibility of forming 'zero alloys', i.e. ones for which there is no coherent scattering from atom pairs containing either Ni or Ti, and the scattering is therefore from those pairs containing only other atoms (see Problem 3.1). For the case of a binary alloy

Table 7.2 Neutron scattering lengths for two potential 'zero-alloy' components

	$b\ (10^{-12}\ \text{cm})$		$b\ (10^{-12}\ \text{cm})$
Ni (natural)	1.03	Ti (natural)	−0.335
^{58}Ni	1.44	^{46}Ti	0.48
^{60}Ni	0.282	^{47}Ti	0.33
^{61}Ni	0.76	^{48}Ti	−0.58
^{62}Ni	−0.87	^{49}Ti	0.08
^{64}Ni	−0.037	^{50}Ti	0.55

such as Ni–B, the B–B partial RDF is therefore obtained directly by Fourier transformation of $S(Q)$ for the zero alloy. We illustrate the potential of this technique for the case of a-Ni$_{81}$B$_{19}$, in which natural Ni, ^{62}Ni and the 'null' element $^{\phi}$Ni, were used (Lamparter et al. 1982); the partial structure factors and the resulting pair distribution functions are also shown (Fig. 7.2).

A different option for neutron diffraction also becomes feasible for amorphous metals. This makes use of the fact that the neutron is a spin $\frac{1}{2}$ particle, being therefore polarizable (by means of a reflecting crystal) such that its spin is either up or down, and if the amorphous metal is ferromagnetic, its magnetic moment may also be oriented by an external magnetic field, with the result that the neutron scattering length is different (b^+ or b^-) depending on whether the neutron spin polarization is parallel or antiparallel, respectively, to the metal magnetic moment direction. The polarized scattering length is related to the normal scattering length b_0 by:

$$b^{\pm} = b_0 \pm P \qquad [7.2]$$

where the magnetic scattering amplitude, P, is given by:

$$P = \frac{r_0 \gamma_n M_n f(Q) \sin \delta}{2} \qquad [7.3]$$

where r_0 is the classical electron radius ($= e^2/m_e c^2$), γ_n and M_n are the neutron and atomic magnetic moments, respectively, δ is the angle between the atomic magnetic moment and the scattering vector Q, and $f(Q)$ is the so-called magnetic form factor which decreases rapidly with increasing Q (like the X-ray form factor) because the scattering is from the unpaired electrons which are spatially non-isotropic, rather than being constant in Q like b_0 (a result of scattering from the spatially isotropic charge distribution of electrons and nucleus). The resultant drop-off in intensity at large Q values for this magnetic scattering as a consequence restricts the spatial resolution achievable.

For a binary alloy, three independent experimental measurements are needed to determine the separate partial pair correlation functions, and in a study of a glassy Co$_{80}$P$_{20}$ alloy, Sadoc and Dixmier (1976) used the following three separate measurements, two involving neutron diffraction, (one polarized and one unpolarized) and the third being an X-ray measurement, i.e. the scattering intensities $(I^+ - I^-)$, I_0 and I_x respectively. The power of this technique is illustrated in Fig. 7.3 for the case of a-Co$_{80}$P$_{20}$ where the total interference functions are shown, together with the partial interference functions and distribution functions derived from them.

Amorphous metals

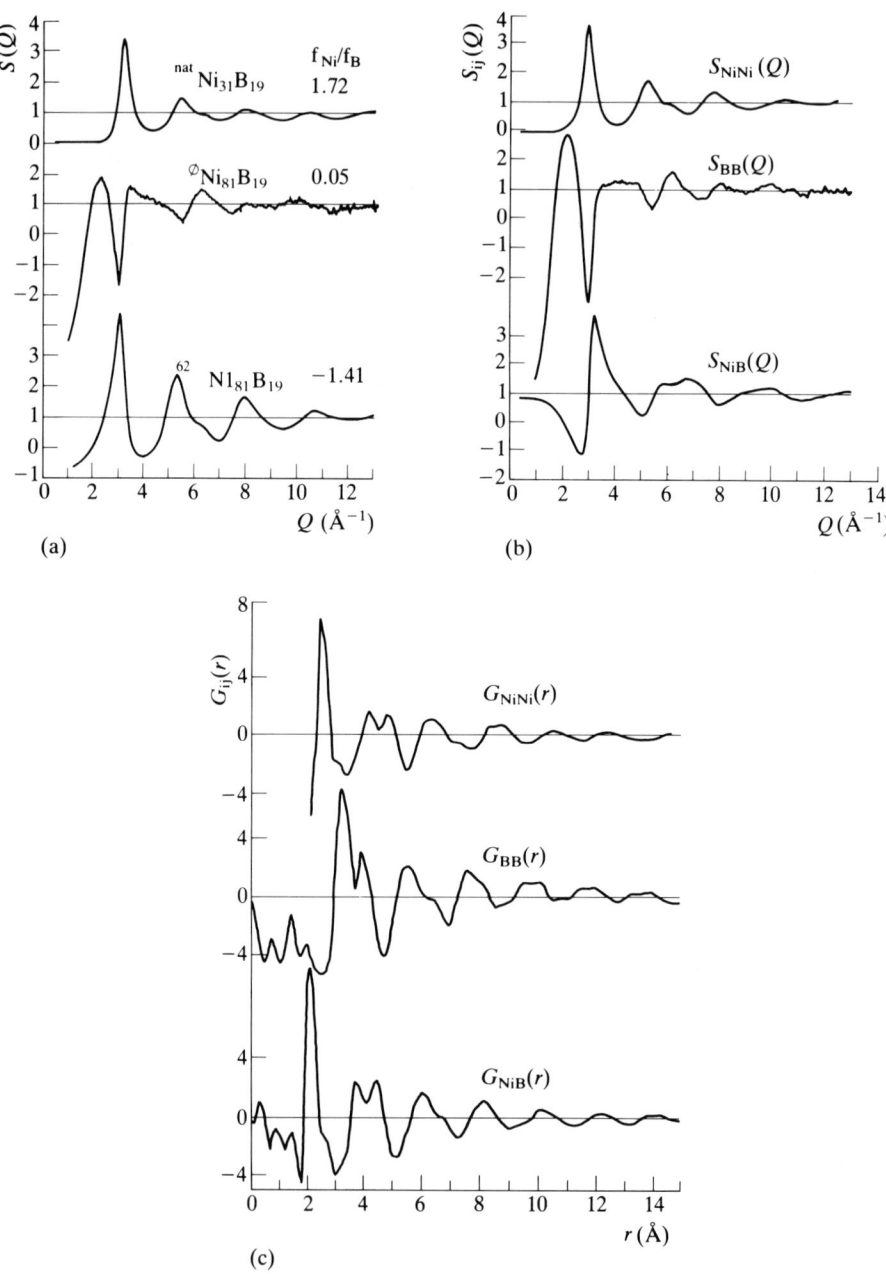

Fig. 7.2 Structural studies of glassy $Ni_{81}B_{19}$ using isotopic substitution and neutron diffraction (Lamparter *et al.* 1982).
(a) Structure factors for different isotopically substituted alloys, including the 'null element' isotopic mixture, $^{\phi}Ni$.
(b) Partial structure factors obtained from those shown in (a).
(c) Partial reduced RDFs obtained by Fourier transformation of the $S_{ij}(Q)$ shown in (b).

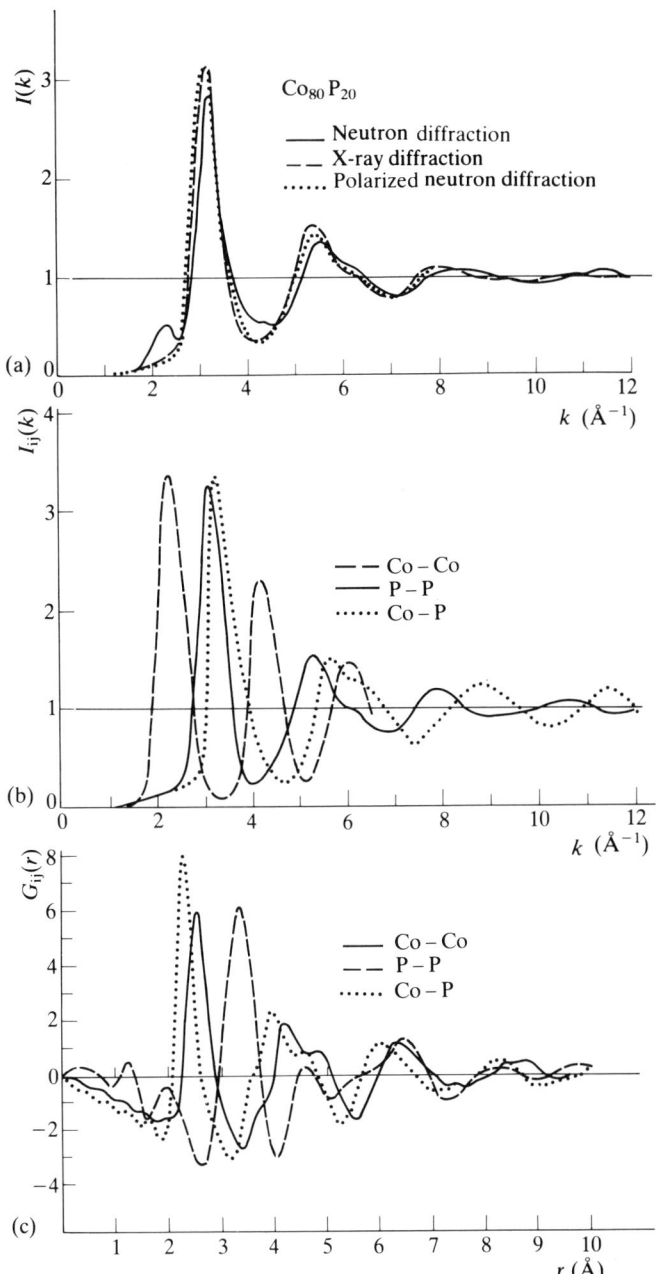

Fig. 7.3 Structural studies of glassy $Co_{80}P_{20}$ using polarized neutron scattering (Sadoc and Dixmier 1976):
(a) total scattering intensities $I(Q)$ for polarized, unpolarized neutron and X-ray diffraction;
(b) partial scattering intensities $I_{ij}(Q)$ obtained from the curves in (a);
(c) partial reduced RDFs obtained by Fourier transformation of curves in (b).

Note from Figs 7.2(c) and 7.3(c) that there is a well-defined metal–metalloid nearest-neighbour distance, which is shorter than the metal–metal distance, and in particular that there are *no* metalloid–metalloid near neighbours in these glasses. We will discuss these experimental findings, and the structural models that can account for them, in a future section.

Note that in X-ray scattering data of T–M alloys, practically all the scattering is due to metal–metal correlations, because of two factors: the metal atoms are generally much heavier than the metalloid atoms and this is reflected in the relative magnitudes of the scattering factors, and the concentration of metal atoms is usually some four times that of the metalloid content in typical glass-forming compositions, and both of these factors reinforce each other. Thus it is generally very difficult to obtain information concerning metalloid–metalloid correlations using conventional X-ray diffraction (although feasible using anomalous dispersion), and for this reason neutron diffraction is preferable, both because the scattering length of metalloid atoms is not necessarily much smaller than that of the metal atoms and also because, at least in certain systems, the possibility exists of manufacturing zero alloys which give information on the M–M correlations directly. The disadvantage, of course, is that neutron diffraction requires relatively large volumes of material, a circumstance often difficult to attain for glassy metals obtainable only as thin foils or strips.

7.2.2 Structural data

In this section we will present a selection of experimental structural results pertaining to various amorphous metals chosen from most of the major classes of glass-forming alloys listed in Table 7.1, and which will then form the basis for discussion of several proposed structural models in the next section.

Transition metal–metalloid alloys

We begin by considering the transition metal–metalloid class of glass-forming alloys, and since these have been studied perhaps the most intensively, a wealth of experimental data exists for such alloys. Figure 7.4(a) shows X-ray diffraction data for several representative binary and multicomponent amorphous T–M alloys (Cargill 1975), whereupon it can be seen that there are marked similarities between all the interference functions. In particular they are distinguished, as we remarked earlier, by a single dominant sharp peak at about $k = 2\,\text{Å}^{-1}$. The total reduced RDFs, $G(r)$, for these materials, obtained by Fourier transformation of the total scattering intensity curves, are shown in Fig. 7.4(b), and they too are very similar, exhibiting a noticeably split second peak with the feature at lower r having the greater intensity.

The origin of the sharp first peak in $I(k)$ (or $S(Q)$) is due to the near-periodic oscillation of $G(r)$ at distances beyond $\simeq 2r_1$ evident in Fig. 7.4(b). This can be seen most clearly by approximating the first peak in $I(k)$ (of, e.g., a-$Ni_{76}P_{24}$) by a gaussian, and on Fourier transformation the resulting real-space function, which is just a damped sinusoid, matches the experimental RDF remarkably well at distances beyond $\simeq 6\,\text{Å}$ (Fig. 7.5).

The ratio of the positions of the first sub-peak of the split second peak and of

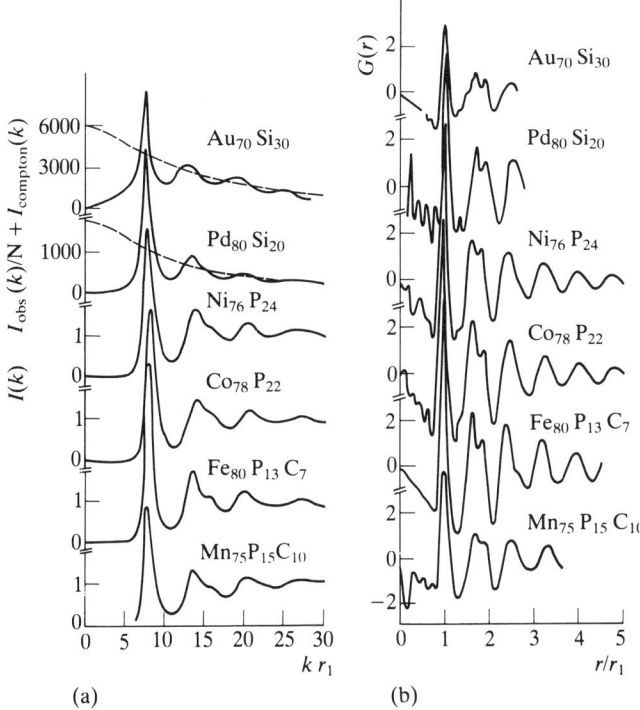

Fig. 7.4 X-ray diffraction data for a variety of T–M glassy alloys (Cargill 1975).
(a) Scattering intensity $I(k)$ (and in one case including the Compton contribution) plotted versus a reduced scattering vector kr_1, where r_1 is the position of the first peak of $G(r)$.
(b) Reduced RDF $G(r)$, obtained by Fourier transformation of data in (a), plotted versus a normalized distance r/r_1.

the first peak in the RDF for amorphous T–M alloys (as well as for pure amorphous transition metals) is $r'_2/r_1 \simeq 1.67$ and that for the second sub-peak and the first peak, $r''_2/r_1 \simeq 1.9$; the corresponding value for r'_2/r_1 is the *liquid* state is $\simeq 1.86$. The value of r'_2/r_1 for the amorphous state is approximately the average of the c/a ratio in the h.c.p. structure ($=1.63$) and the second neighbour distance for a group of four atoms in a planar parallelogram configuration (i.e. $\sqrt{3} = 1.73$) (see Fig. 7.8). Thus, it is likely that both types of clusters of (metal) atoms occur in the glassy structure.

The split second peak in the total (X-ray) RDF of T–M glasses must reflect an actual splitting in the second T–T coordination shell, i.e. in the packing of the metal atoms, because the scattering from the metalloid atoms is so relatively weak as discussed in section 7.2.1. It is of interest to note, therefore, that the partial RDFs of the T–M alloys Ni–B and Co–P shown in Figs 7.2 and 7.3 *do* exhibit just such a splitting in the T–T second shell, although there are also correlations due to T–M and M–M pairs at about the same position which will make some contribution, albeit small, to the total RDF there. Possible structural origins of this feature will be discussed in the next section.

The total nearest-neighbour coordination number, as deduced from the area under the first peak in the total RDF, lies in the range 12–13 for most amorphous T–M alloys (see, e.g., Cargill 1975), compared with the value of 12 for close-packed

Amorphous metals

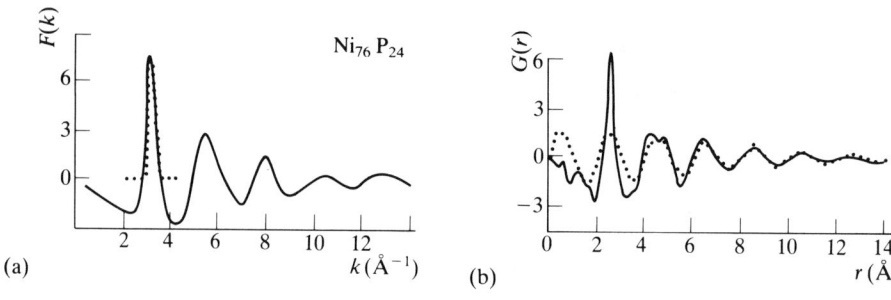

Fig. 7.5 (a) $F(k)$ for a-$Ni_{76}P_{24}$ (solid line) and gaussian fitted to first maximum (dotted line); (b) $G(r)$ (solid line) and Fourier transform of the gaussian peak shown in (a) (dotted curve) (Cargill 1975).

crystalline pure metals, although this figure must include metal as well as metalloid neighbours of any given metal atom because of the difficulty in deciding upon the upper-limit cut-off for the first peak. This ambiguity is removed of course if the *partial* RDFs are examined. For example, for the case of a-$Co_{80}P_{20}$ whose partial functions are shown in Fig. 7.3, the relevant nearest-neighbour coordination numbers are as follows: Co–Co (10.1), Co–P (2.1), i.e. a total of 12.2, and P–Co (8.9), P–P (0). Extended X-ray absorption fine structure (EXAFS) experiments on another T–M system, namely a-$Pd_{78}Ge_{22}$ alloys, by Hayes *et al.* (1978) using the Ge K-edge have shown that each Ge is surrounded only by 8.6 ± 0.5 Pd neighbours and by no Ge neighbours, in agreement with the neutron data on a-$Co_{80}P_{20}$. This avoidance of metalloid–metalloid nearest-neighbours is a reflection of a considerable amount of chemical ordering in these materials.

The final experimental aspect of the structure of glassy T–M alloys that we wish to discuss here concerns the density, or equivalently the packing fraction (i.e. the ratio of the occupied volume to the total volume for a structure viewed as a packing of rigid spheres). Calculation of the packing fraction η_p, from the measured density ρ_0 requires that the metal and metalloid atomic radii be specified, a somewhat arbitrary procedure since atoms are not in reality rigid spheres. However, if 12-coordinated Goldschmidt radii for metal atoms and tetrahedral covalent radii for metalloid atoms are used, and the relation

$$\eta_p = \frac{4\pi}{3} \langle R^3 \rangle \rho_0 \qquad [7.4]$$

is employed, where $\langle R^3 \rangle$ is the appropriate average cube of the radius, a surprisingly constant value for $\eta_p = 0.67$ is found, which varies only by about 5% between all amorphous T–M alloys (Cargill 1975). This value is to be compared with the packing fractions for crystalline arrays of identical spheres, i.e. 0.74 for h.c.p. and f.c.c., 0.68 for b.c.c. and 0.52 for s.c. lattices, respectively.

Metal–metal alloys

Since there is less experimental structural information available for metal–metal glasses than for transition metal–metalloid glasses, we therefore do not consider it worthwhile to subdivide this section according to the various classes of metal–metal alloys listed in Table 7.1. The large size differences but comparable atomic numbers

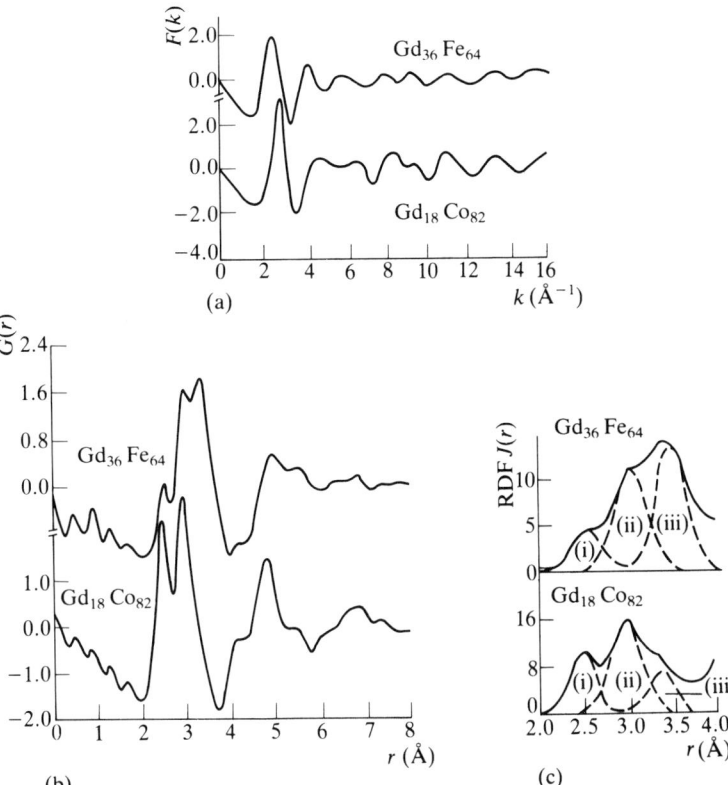

Fig. 7.6 Structural data for TE–RE and TL–RE glassy alloys (Cargill 1975):
(a) $F(k)$;
(b) $G(r)$;
(c) first peaks in the RDF, $J(r)$, showing how they may be deconvoluted into individual partial correlation functions. In each case (i), (ii), and (iii) refer to T–T, T–RE and RE–RE correlations respectively.

of the constituent metal atoms mean that the three nearest-neighbour pair functions (i.e. AA, AB and BB) can be resolved in X-ray diffraction data. As an example, we show in Fig. 7.6 X-ray data for two sputtered TL–RE amorphous alloys, a-$Fe_{64}Gd_{36}$ and a-$Co_{82}Gd_{18}$, both the total $F(k)$ and the total $G(r)$ being given. Note particularly that if the first peak in the full RDF $J(r)$ is examined (Fig. 7.6(c)), it is seen to be quite clearly split into three components, corresponding respectively to TL–TL, TL–RE and RE–RE nearest-neighbours in ascending order of distance; note also that the three deconvoluted peaks have similar widths. There is thus no evidence for strong chemical ordering as in glassy T–M alloys, and an approximately random mixture of the constituent metal atoms is sufficient to describe the structure. If the $G(r)$ for the TL–RE alloys in Fig. 7.6(b) are compared with those for T–M alloys (Fig. 7.4(b)), it can be seen that while there is some evidence for a split second peak in the TL–RE alloys as in the T–M alloys, the correlations beyond nearest-neighbours appear to be weaker for the former than for the latter. However, this is likely to be due to an artefact of the combination of the individual scattering factors (i.e. the $W_{ij} = x_i f_i f_j / \langle f \rangle^2$) to give the experimentally observed intensity, rather than because there is an increased amount of structural disorder in the TL–RE alloys. Because the W_{ij} for all pairs of atom types in the TL–RE alloys are approximately the same and hence contribute a comparable amount to the total RDF (unlike the situation for T–M glasses), partial cancellation of the individual pair distributions can occur in the summation giving the total RDF if their peaks do not coincide spatially, leading to the impression that the correlations are less at larger r.

The three partial distribution functions have been obtained for various Zr–transition metal binary alloys, and an example is given in Fig. 7.7 for the case of

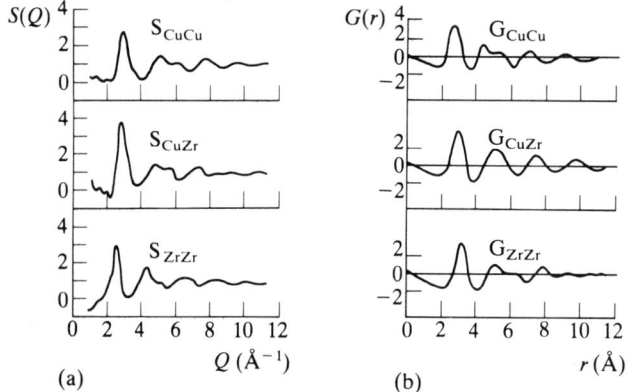

Fig. 7.7 Partial structure factors (a) and reduced RDFs (b) for a-$Cu_{57}Zr_{43}$ obtained by isotopic substitution and neutron scattering (Mizoguchi et al. 1978).

a-$Zr_{43}Cu_{57}$ obtained by isotopic substitution (Mizoguchi et al. 1978). From the partial $G(r)$ curves, it can be seen that the second peak for both Cu–Cu and Zr–Zr correlations is split, as in the case for T–M alloys. The coordination numbers about Cu are 5.4 (Cu) and 5.0 (Zr), and about Zr are 6.7 (Cu) and 5.9 (Zr); the fraction of each coordination number is $\simeq 53\%$ and $\simeq 47\%$ respectively in both cases, values which are in close agreement with the overall stoichiometry of the alloy and which again indicate that there is little chemical ordering in these amorphous metal–metal alloys.

We conclude this section by mentioning some experimental evidence that chemical ordering may in fact occur in metal–metal alloys, albeit to a limited extent. Neutron diffraction studies of a-$Cu_{66}Ti_{34}$ have revealed the existence of a 'pre-peak' in $S(Q)$ at 1.9 Å^{-1} (Sakata et al. 1979), akin to the pre-peaks observed in Ge- and As-chalcogenide glasses at $Q \simeq 1$ Å^{-1} (see sect. 3.4.1). This peak arises from correlations having a period of the order $2\pi/k \simeq 3.3$ Å over distances several times this, rather than a single correlation at a specific separation. Because the scattering in this alloy is dominated by Cu–Cu correlations due to the particular scattering lengths involved, it has been suggested that there is some chemical ordering of Cu and Ti, i.e. that Cu–Cu are preferentially second neighbours. However, the extent of this ordering is likely to be small because of the gross accentuation of the pre-peak by virtue of the negative scattering length of the Ti.

7.2.3 Structural models

A variety of structural models, both conceptual and physical, have been proposed to describe the structure of amorphous metals. As is the case with covalent amorphous solids, microcrystallites have been suggested as a possible cause of the diffuse peaks in scattering patterns of amorphous metals (see sect. 3.3.3), the broadness resulting from the assumed small size of the crystallites through the Scherrer relation ([3.63]). However, this approach has considerable difficulty in being able to fit experimental scattering data, in particular the relative sharpness of the first peak in $S(Q)$ compared with the other peaks. There is also the problem of the nature of the medium in which the microcrystallites are embedded, and many assumptions have to be made regarding the size, shape and correlation length of the microcrystallites in order to fit even approximately the experimental scattering data. For this reason, therefore, we will not consider the microcrystalline model further, but instead discuss two competing models which have been quite successful in accounting for the structure of amorphous metals, namely dense random packings of spheres and random packings of trigonal prisms.

Dense random packing of spheres

Since many amorphous metals are quenched from the liquid state and have predominantly non-directional bonding, it is a natural idea to consider the structure as being that of a frozen liquid. In this regard, the dense random packing (DRP) models of single-sized hard spheres pioneered by Bernal (1959, 1964) as descriptions of liquid structures have consequently been adopted as prototype structures for amorphous metal systems. We will describe the simple model first however, before discussing the various modifications that have been made to the model for the case of glassy metals.

The first DRP models were physical models constructed from aggregates of equal-sized ball-bearings held in flexible bags; when a non-crystalline packing had been achieved and the density maximized by squeezing the bag, the balls were glued into position and the coordinates and connectivity measured. The most ambitious project of this kind was that of Finney (1970) which involved the construction and characterization of a model containing 7934 spheres. More recently computer algorithms have been developed (e.g. Bennett 1972) which generate DRP models numerically starting from a given seed. Spheres are added sequentially such that they are in hard contact with three pre-existing spheres, choosing the particular site according to one of two criteria. Either the site which is closest to the centre of the cluster, or else that which satisfies a given 'tetrahedron perfection' condition (Ichikawa 1975), is chosen. Tetrahedron perfection is defined according to how nearly a group of four spheres approximates perfect tetrahedral symmetry, and may be quantified by considering the 'pocket' formed by three contacting surface atoms of the cluster (1, 2 and 3) and evaluating the expression (Cargill 1981)

$$\ell_{123} = \max_{\{12,13,23\}} \frac{r_{ij}}{(R_i + R_j)} \qquad [7.5]$$

where r_{ij} is the separation between centres of spheres i and j and R_i and R_j are their radii. Perfect tetrahedral arrangement is achieved if $\ell_{max} = 1$. In this manner, models

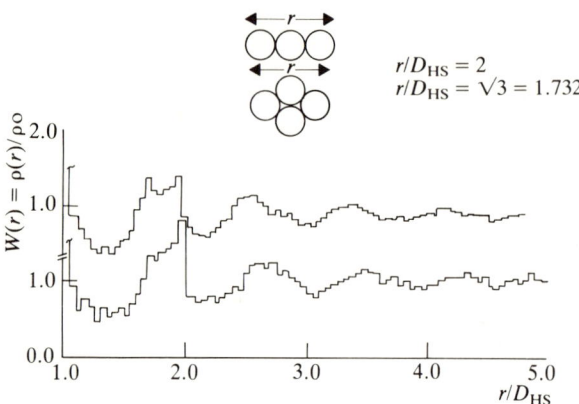

Fig. 7.8 Distribution functions for dense random packed models plotted versus hard-sphere diameter D_{HS}:
(a) DRP of single-sized balls (Finney 1970);
(b) computer generated single-sized sphere DRP (Bennett 1972). The inset shows the atomic configurations believed to be responsible for the split second peak.

containing four or five thousand atom sites have been constructed.

Characterization of both types of DRP model involves calculating the RDF and the packing fraction. The packing fractions, η_p, obtained for the physical models have been determined to be 0.6366 (cf. 0.74 for f.c.c. and h.c.p. structures); values for the computer-generated models tend to be slightly lower and Bennett found that η_p decreased with increasing radius, extrapolating to 0.61 for infinite radius. Shown in Fig. 7.8 are real-space distribution functions plotted as $\rho(r)/\rho_0$ (see sect. 3.2.1) for both the Finney and Bennett models. Points to note are the essential similarities of the two curves, a well-defined second peak following the singularity at the hard-sphere diameter, with progressively broader features at larger distances. Of particular interest is the clear splitting of the second peak in both cases, although the splitting appears to be more in physical ball-bearing than in computer-generated models and even more evident (with a reversal in the relative magnitudes of the sub-peaks of the second peak) in the much smaller models (200–1000 atoms) which have been reported. The reason for this difference is not clear, but may in some way be related to the drop-off in density that accompanies the construction of very large models. The two sub-peaks occur at 1.73 and 1.99 sphere diameters, in reasonable agreement with the values found experimentally for many T–M glasses, namely 1.67 and 1.9 respectively. The microscopic origin of this fine structure can be ascribed to certain particular recurrent local configurations as discussed earlier (sect. 7.2.2).

Although these random packing models of equi-sized spheres are dense in the sense that no internal voids exist which are large enough to accept a normal-sized sphere, this does not mean that there are no reasonably large voids present in the structure; indeed the $\simeq 10\%$ deficit in packing density between DRP models and close-packed crystal structures immediately suggests that voids of appreciable size are present. The holes found in DRP models can be classified into one of five types of polyhedron (whose vertices are defined by the sphere centres), if allowances are made for distortions of up to 20% in the edges of the polyhedra; these were termed by Bernal 'canonical holes'. These polyhedral holes are illustrated in Fig. 7.9 and of the

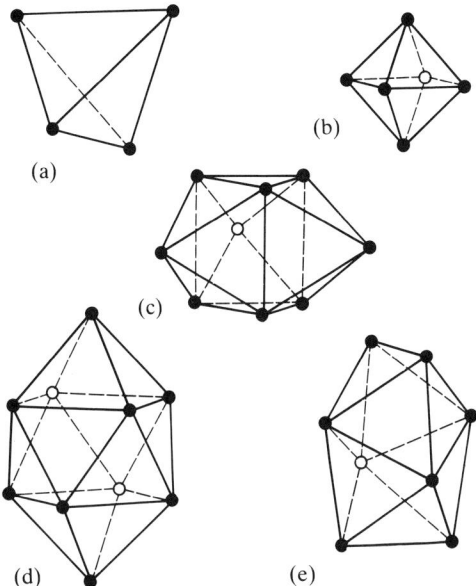

Fig. 7.9 The 'canonical holes' of Bernal (1964). They are
(a) tetrahedron;
(b) octahedron (often present as half-octahedra);
(c) trigonal prism (shown capped with three half-octahedra);
(d) Archimedian anti-prism (shown capped with two half-octahedra);
(e) tetragonal dodecahedron.

five, the smallest, the tetrahedron, occurs most often (2.9 per sphere) and this is some 25 times more frequent than the next most common polyhedra, the trigonal prism and the tetragonal dodecahedron (Cargill 1975). The topology of the DRP structures may also be described in terms of Voronoi polyhedra (or Wigner–Seitz cells – sect. 2.3.4) whose surfaces are defined by the envelope of planes which bisect perpendicularly the lines drawn from the centre of a given sphere to the centres of other nearby spheres. It is found that these have predominantly pentagonal faces, with an average number of 14.25 faces per Voronoi polyhedron (Finney 1970); for comparison, the comparable number for crystalline lattices are 12 (h.c.p.), 12 (f.c.c.) and 14 (b.c.c.).

The observation that a random packing of spheres can be described equivalently as a packing of polyhedra is not new however. An amateur botanist, Stephen Hales, then Rector of Farrington and Minister of Teddington published in 1727 a work entitled *Vegetable staticks: an account of some statistical experiments on the sap in vegetables*, in which he described his experiments on the uptake of water by plants, particularly vegetables. In one such experiment, he studied the dilation of peas upon absorption of water by observing that they could raise the heavy lid of a pot in which they were contained. In an extension of this work he carried out the following experiment (Hales 1727):

Being desirous to try whether they would raise a much greater weight by means of a lever with weights at the end of it, I compressed several fresh parcels of Pease in the same Pot, with a force equal to 1600, 800 and 400 pounds; in which Experiments, tho' the Pease dilated, yet they did not raise the lever, because what they increased in bulk was, by the great incumbent weight, pressed into the interstices of the Pease, which they adequately filled up, being thereby formed into pretty, regular dodecahedrons.

What Hales had inadvertently found was that the DRP of 'Pease' in his experiments had deformed into Wigner–Seitz cells as they expanded in a container of constant volume. A dodecahedron is distinguished by the presence of pentagonal faces and it was probably the observation of these in his random packings which prompted Hales to describe the Voronoi polyhedra as dodecahedra.

Obviously, the simple DRP models described so far are suitable as representations of the structure of pure amorphous metals (Bi, Ga, etc.), but because of the difficulties of preparation and consequent impurity content, and the problems encountered in trying to extract quantitative data from low-temperature electron diffraction measurements on such films, there is unfortunately a dearth of reliable structural diffraction data on such systems. Instead, we will consider firstly glassy T–M alloys for which no such problems exist, and which effectively act as aggregates of only transition metals as far as X-ray diffraction is concerned because of the relatively small atomic number of the metalloid atoms. In Fig. 7.10 we show a comparison of the experimental total reduced RDF for a-$Ni_{76}P_{24}$ and $G(r)$ calculated for the single atom type Finney DRP model, both scaled to the same hard-sphere diameter. Overall agreement between the two curves is good, but

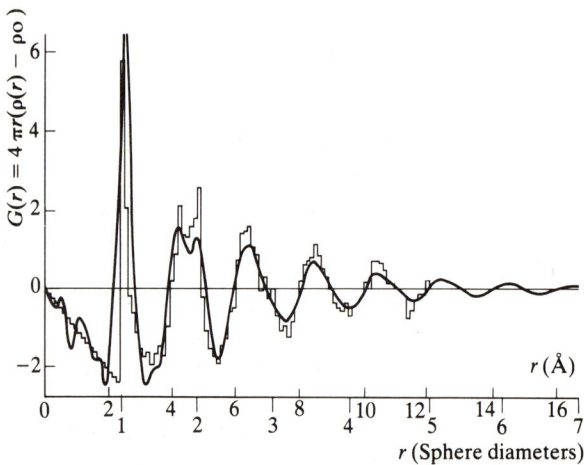

Fig. 7.10 Comparison of reduced RDF $G(r)$ for Finney's DRP (single-sized sphere) model and the experimental curve for a-$Ni_{76}P_{24}$ (Cargill 1975).

although the splitting of the second peak is reproduced by the DRP model, the relative weighting of the two sub-peaks is incorrectly predicted by the model compared with that observed generally in glassy T–M alloys (see Fig. 7.4(b)). Thus, although the simple DRP model may be a reasonable approximation to the structure if only scattering from metal atoms is considered (or equally if both T and M atoms have the same hard-sphere radius, e.g. 1.25 Å for Co and 1.28 Å for P), considerable problems arise once *partial* distribution functions are considered, since now the precise disposition of the metalloid atoms becomes important, and it is quite clear from experimental data that the metalloid atoms are *not* distributed at random, as would be the case if certain of the atomic sites in a single atom type DRP model were assigned as metalloids.

Instead, it has been suggested (Polk 1970, 1972) that the metalloid atoms occupy the larger Bernal holes in a DRP of the transition metal atoms. While this simple idea has the virtue that it predicts correctly the amount of metalloid atoms ($\simeq 20\%$) apparently needed to stabilize the structure, and also the presence of local chemical ordering, i.e. metalloid–metalloid avoidance, inherent in such alloys, subsequent work (see e.g. Cargill 1975) has shown that there is in fact insufficient void volume available to accept any but the very smallest type of metalloid atoms. However, the attractions of the interstitial model for metalloid positioning have meant that further efforts to improve the simple DRP model have been made. Among these has been the computer generation of binary DRP models having two sizes of atom (see e.g. Boudreaux and Gregor 1977), and more recently the energy relaxation of DRP models, whether comprising spheres of the same size and interstitially stuffed, or binary packings of different sizes of spheres.

Energy relaxation of DRP models is carried out in a similar manner to that already described for CRNs (sect. 3.3.2), with the exception that different interatomic potentials must of course be used. To simulate the predominantly non-directional nature of the bonding inherent in amorphous metals, Lennard–Jones potentials of the form

$$E \propto \frac{1}{2} \sum_i \sum_{j \neq i} \left[\left(\frac{R_{ij}}{r_{ij}} \right)^{12} - 2 \left(\frac{R_{ij}}{r_{ij}} \right)^{6} \right] \qquad [7.6]$$

have been used, where R_{ij} is the sum of hard-sphere radii of spheres i and j, and r_{ij} is the actual distance between their centres, and only the interactions between nearest-neighbours are included. Potentials of the form of [7.6] are obviously an approximation to the true potential, particularly so for T–M alloys where some directional character to the bonding between metal and metalloid must exist to give rise to the chemical ordering characteristic of such materials, and which is neglected in [7.6]. It should be noted also that energy relaxation using purely non-directional potentials does not necessarily preserve the topology of the model, i.e. considerable atomic rearrangement may occur during the relaxation. Thus, the relatively high degree of porosity characteristic of packings having near-ideal tetrahedral symmetry is considerably reduced upon energy relaxation. In addition, because of the 'soft' nature of potentials such as given by [7.6], energy relaxation causes the previously hard non-overlapping spheres to overlap to a certain extent; such models have been termed 'dense random packing of soft spheres', DRPSS.

As an example of this approach, we show in Fig. 7.11 a comparison of experimental scattering curves for a-$Gd_{32}Co_{68}$ with computed curves for a binary DRP model ($\ell_{max} = 1.1$) in both unrelaxed and relaxed states (Cargill 1981). Note that the overall fit to experiment is significantly improved for the relaxed model, and that the density of the model also approaches the experimental value after relaxation (as can be seen from the slope of $G(r)$ at small r). However, a similar approach to modelling the structure of a-$Nb_{67}Ge_{33}$, using again a binary DRP model rather than an interstitially stuffed one, has proved much less successful (Cargill 1981);

Amorphous metals

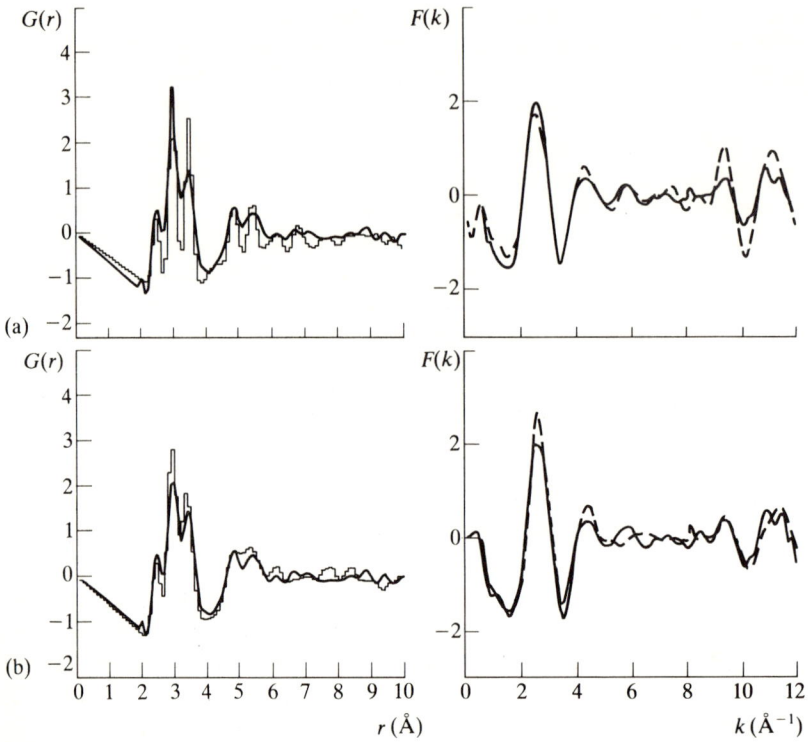

Fig. 7.11 Reduced RDF and scattering factor of a-$Gd_{32}Co_{68}$ compared with calculated curves for DRP models (Cargill 1981). In all cases, the experimental data are shown by the continuous solid lines:
(a) unrelaxed model;
(b) energy relaxed model.

insufficiencies in the form of the potential used are likely to have been a contributing factor.

Random packings of trigonal prisms

We conclude this section on the structure of amorphous metals by describing a recent model, different in spirit to the DRP hypothesis, but which addresses only the problem of the structure of amorphous T–M alloys. Gaskell (1979b) proposed that the pronounced chemical ordering exhibited by T–M alloys, compared with other amorphous metals such as TE(TL)–RE alloys, could be explained if certain structural motifs, characteristic of crystalline polymorphs, were the structural units in the amorphous phase too. There are many crystalline alloys of the general composition T_3M (i.e. close to that which readily forms glasses, i.e. T_4M), which consist of packings of MT_6 trigonal prisms. It can be seen from Fig. 7.12(a) that in these, each metalloid atom is coordinated by six metal atoms at the apices of a trigonal prism (T_I sites) and by three others in the equatorial plane at a somewhat greater distance (T_{II} sites); these trigonal prismatic polyhedra then pack by means of edge sharing in either the cementite (Fe_3C) structure (Fig. 7.12(b)) or the Fe_3P structure (Fig. 7.12(c)). Note that in this geometry, the T_I atom for the upper prism occupies a T_{II} site for the lower prism in the case of the cementite structure.

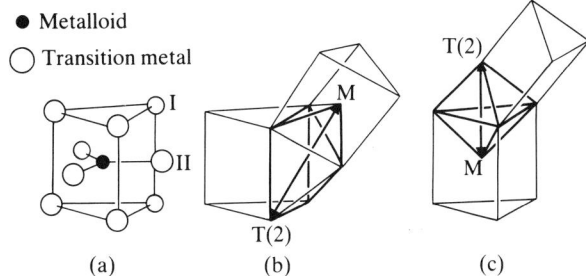

Fig. 7.12 (a) Trigonal prismatic packing of transition metals around a metalloid atom observed in many crystalline alloys.
(b) Edge-sharing arrangements in the Fe_3C (cementite) structure.
(c) Edge-sharing arrangement in the Fe_3P structure. The two arrangements may be distinguished by the metalloid–second nearest neighbour transition metal distance, M–T(2) (Gaskell 1981).

Gaskell suggested that in glassy T–M alloys, trigonal prismatic polyhedra would be the dominant structural unit, rather than the wide variety of polyhedra that make up DRP models (although the trigonal prism does appear, it accounts for only 3–4% by number in DRPs). A physical model was constructed in which 434 spheres representing T atoms formed regular trigonal prisms, connected randomly through their edges in a cementite-like fashion. This model was unrealistic in two regards: most metalloid atoms are too large to fit in the central hole in a regular trigonal prism, and the prisms must therefore be distorted; furthermore, the packing in such a model is not sufficiently dense due to the presence of large cavities, a problem which becomes more acute the larger the model. In an attempt to remedy both these failings, energy relaxation using a Lennard–Jones potential ([7.6]) was performed. The total RDFs calculated for such models agree reasonably well with experiment for glassy $Pd_{80}Si_{20}$, but quite serious discrepancies, particularly at distances beyond $\simeq 6$ Å, are revealed if instead comparison is made for the partial distribution functions (Gaskell 1981) of, for example Pd–Si or Co–P alloys (Fig. 7.13). Whether these differences can be attributed to the cementite-type packing of the trigonal prisms employed in the Gaskell model (since a-Co–P, but not a-Pd–Si, alloys crystallize in the Fe_3P structure, there is a likelihood that this configuration might be preserved in the glass too, although it would be thought that energy relaxation would destroy much of the memory of the packing type), or to a general deficiency in the trigonal prismatic model itself, is unclear at present.

7.3 Electronic properties

We turn now from a discussion of the atomic structure of amorphous metals to a discussion of their electronic structure (and concomitant electrical, optical and magnetic properties). These two topics are not unrelated however; the influence of atomic structure on the electronic structure, e.g. the density of states, is intuitively obvious, but what is more surprising perhaps, to those not versed in the theory of liquid metals, is that electrical *transport* is also closely related to atomic structure through the structure factor $S(Q)$. Furthermore, as we have seen earlier (sect. 7.1), it

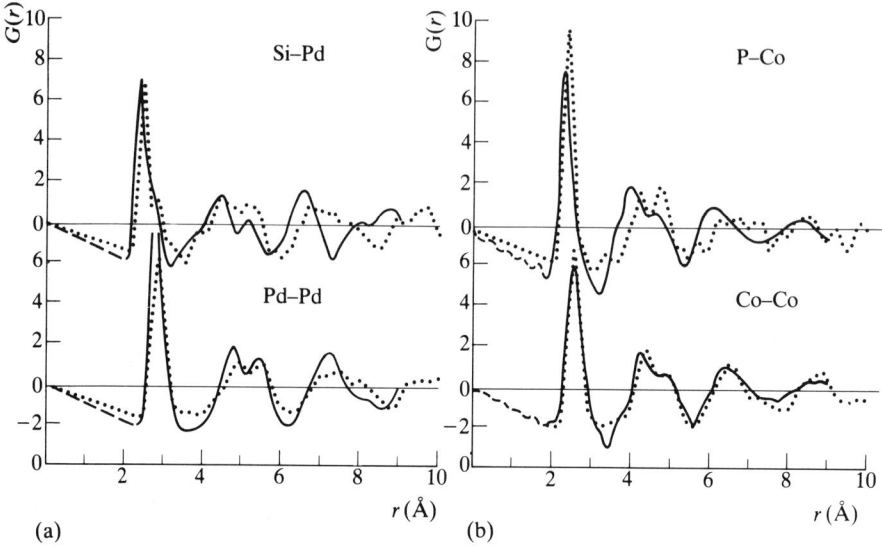

Fig. 7.13 Comparison of experimental (solid) and calculated (dotted) partial reduced RDFs for two metallic glasses Pd–Si and Co–P (Gaskell 1981). The theoretical curves are calculated for a trigonal prismatic structural model.

has been proposed by Nagel and Tauc (1975) that features in the electronic density of states should also influence glass-forming tendency and stability. Therefore, perhaps more than any other non-crystalline system, amorphous metals offer a most intriguing interrelationship between a host of disparate properties.

7.3.1 Electronic structure

Features in the electronic structure of amorphous metals can be investigated experimentally by the same techniques discussed in section 5.2.2. Measurements of the core levels by means of XPS can give information on the amount of charge transfer, and hence on the nature of the bonding, by a comparison of the shift in the positions of the core levels of the constituent atoms, either between the crystalline and amorphous forms of the same composition, or between amorphous alloys of differing compositions. Relatively few experimental photo-electron investigations of amorphous metals have as yet been reported compared with the many such studies on amorphous semiconductors. In two studies, on Nb–Ni glasses (Nagel *et al.* 1977) and on a glassy Pd–Cu–Si alloy (Nagel *et al.* 1976), the core level shifts were found to be small relative to the (crystalline) constituent species, reflecting a slight amount of charge transfer upon alloying, perhaps smaller than might be expected if there were a strong chemical interaction between metal and metalloid. Instead, it was proposed (Nagel *et al.* 1976) that the small shifts observed for the T–M alloy were a result of the filling of the Pd d-band by electrons from the Si, thereby raising the Fermi level with respect to the conduction band, and consequently causing the Pd 3d or 4p core levels to lie apparently at greater binding energies (since they are positioned relative to E_F).

Of perhaps more relevance to the present discussion are details of the valence

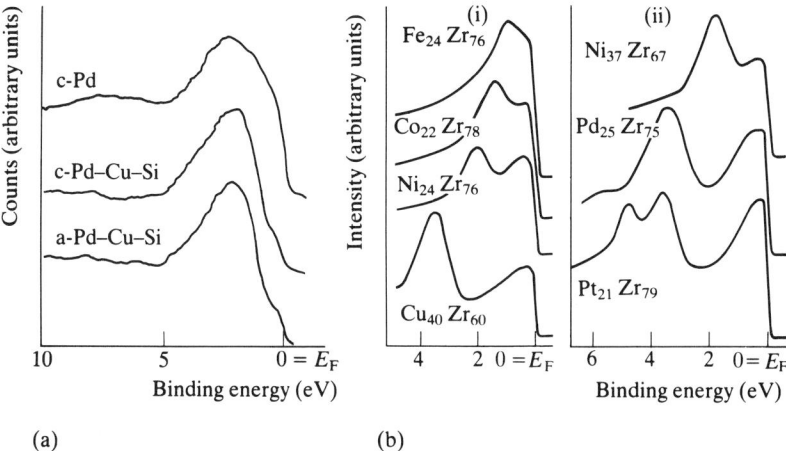

Fig. 7.14 Energy distribution curves obtained from UPS for glassy metallic alloys:
(a) data for glassy $Pd_{77.5}Cu_{6.0}Si_{16.5}$ compared with its crystalline form and polycrystalline Pd (Nagel et al. 1976);
(b) data for Zr-based glasses ($hv = 21.2$ eV): (i) Zr with a late transition metal of the same series; (ii) Zr with elements of the same group (Oelhafen et al. 1980).

band obtained by UPS and also XPS. These have been reported for the two alloy systems mentioned above, and also for a series of glassy Pd–Si alloys by Oelhafen et al. (1979). Again, the overall similarity between crystalline and glassy densities of states is striking (see Fig. 7.14(a)). There are, however, subtle, but important, differences discernible on closer examination. In particular, there is clear evidence for a shoulder in the density of states near E_F for the glass (and indeed the crystalline alloy) which is not observed in pure Pd, and in addition, the density of states is considerably higher and rises more sharply at the valence band edge for Pd than is the case for the glassy alloy. The reduction in the density of states near E_F for the glassy alloy is in accord with the assertion by Nagel and Tauc (1975) that glass formation and stability are enhanced for those alloys in which E_F lies in a local minimum of the density of states. However, there is no evidence so far from photoelectron spectroscopic experiments that E_F lies *precisely* at a minimum, and moreover some caution has to be exercised in interpreting differences in EDCs such as in Fig. 7.14 since such changes may simply reflect differences in matrix elements rather than in the actual density of states between two materials. A very interesting effect has recently been reported for glassy TL–TE alloys by Oelhafen et al. (1980) in which the valence band as measured by UPS appears to be split, the peak near E_F being due to the d-band of the TE component and the one at higher binding energy being due to the d-band of the TL component. This splitting is rather surprising since for each transition metal alone, its d-band peaks near E_F. Furthermore, for the same TE component, alloying with different TL components of increasing atomic number (i.e. Fe to Cu, Ni to Pt) causes the d-band splitting to increase (see Fig. 7.14(b)). However, the valence-band structure is almost unchanged if TL is kept constant but TE is varied within the same group. Similar effects are also seen in the crystalline analogues of these alloys. Intriguingly, Oelhafen et al. note a strong correlation between the d-band splitting and the glass-forming ability, the larger the splitting the more easy the vitrification of the alloy.

Calculations of the electronic structure of amorphous metals have only recently been attempted. Kelly and Bullett (1979) have computed the density of states for an archetypal metallic glass, $Pd_{80}Si_{20}$, using the coordinates from two different structural models, the binary DRP model of Boudreaux and Gregor (1977) and the random packing of trigonal prisms due to Gaskell (1979b), and compared these with the calculated density of states for crystalline Pd_3Si. The densities of states were calculated using the recursion method (see sect. 5.2.1) using a set of localized orbitals as a basis and assuming that only two-centre interactions are significant. Because of the higher coordination numbers and the greater number of orbitals per site for glassy metal alloys compared with amorphous semiconductors such as Si or Ge, only a relatively small number ($\simeq 250$) of atoms in a cluster can be considered in order to keep the order of the secular equation to a manageable size. The LCAO approximation is sufficiently good for the d and lower-lying s–p electron states for their features in the density of states to be a reasonable representation, but higher-lying s and p states of Pd are better described as plane waves, and therefore the conduction band is not so well described by this approach.

The overall density of states, averaged over a unit cell or formula unit, is shown in Fig. 7.15. It can be seen that there is little difference in the overall features of the density of states for the three cases; all have a broad, more-or-less featureless band at the top of the valence band comprised mainly of Pd d-states, although the states at E_F have a large amount ($\simeq 50\%$) of Si orbital character, indicative of a considerable degree of chemical bonding by Si, and finally a band of states is predicted deep in the valence band $\simeq 10$ eV below E_F due to Si s-states. Closer examination reveals subtle differences, however; in particular, the top of the valence band calculated for the DRP model (Fig. 7.15(b)) is distinctly skewed to higher energies, mainly as a result of a movement of the Pd d-levels nearer to E_F. Note too that again E_F does not appear to lie at a minimum in the density of states, as required by the Nagel–Tauc theory. However, one must conclude that such calculations for amorphous metals are capable of revealing much less information concerning the local topology of the appropriate structural models than is the case for corresponding calculations for amorphous semiconductors (see sect. 5.2.1).

7.3.2 Optical properties

Optical studies of amorphous metals have been as scanty so far as have photoelectron investigations of the electronic structure, particularly when compared with the volume of work reported on electron transport and magnetic properties. Nevertheless, a reasonably coherent picture emerges from the few studies that have been reported, namely on sputtered amorphous Au–Si films (Hauser et al. 1978, 1979) and splat-quenched glassy Pd–Si alloys (Schlegel et al. 1979).

Representative optical reflectivity spectra for these two amorphous systems are shown in Fig. 7.16, together with that for c-Pd for comparison. From the measurements of the reflectivity as a function of photon energy, $R(\hbar\omega)$, the phase may be calculated via the Kramers–Kronig relation

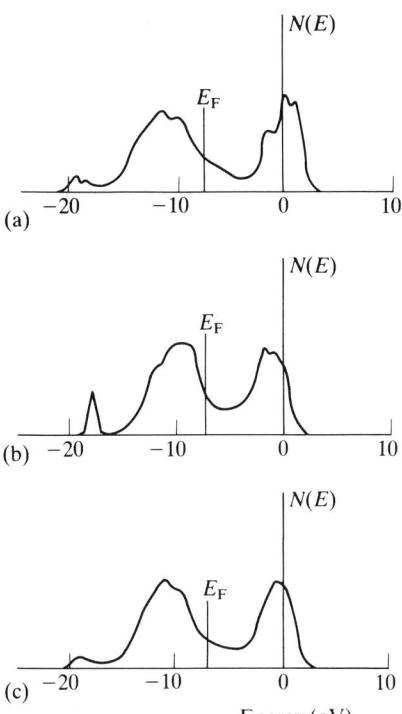

Fig. 7.15 Density of states calculations for the Pd–Si system (Kelly and Bullett 1979). Formula-unit averaged density of states for
(a) crystalline Pd_3Si, showing a unit-cell average;
(b) the Boudreaux–Gregor (1977) DRP model;
(c) the Gaskell (1979) trigonal prismatic model.

$$\theta(\omega_0) = -\frac{1}{2\pi} \int_0^\infty \frac{d \ln R(\omega)}{d\omega} \ln \left| \frac{\omega + \omega_0}{\omega - \omega_0} \right| d\omega \qquad [7.7]$$

and the real and imaginary parts of the dielectric constant

$$\varepsilon = \varepsilon_1 + i\varepsilon_2 = (n + ik)^2 \qquad [7.8]$$

can be evaluated using the equation

$$\sqrt{R} \exp(i\theta) = (n - 1 + ik)/(n + 1 + ik)$$

The optical spectra in all cases can be analysed in terms of two distinct mechanisms, depending on the photon energy. At low photon energies, *intraband* transitions take place, and the dielectric constant can then be described by means of the Drude formula:

$$\varepsilon^D(\omega) = \varepsilon_0 - \omega_p^2/\omega(\omega + i/\tau) \qquad [7.9]$$

where $\omega_p = (4\pi N e^2/m^*)^{1/2}$ is the free-electron plasma frequency, $m^* = Nm_e/N_{\text{eff}}$ is the effective mass (and N and N_{eff} are the actual and effective number of conduction electrons per formula unit, respectively), and τ is the electronic relaxation time. At

Amorphous metals

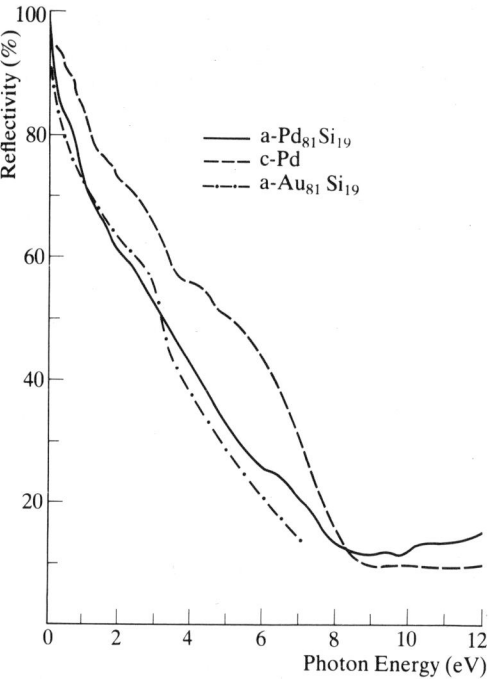

Fig. 7.16 Optical reflectivity spectra of glassy $Pd_{81}Si_{19}$, crystalline Pd and amorphous $Au_{81}Si_{19}$ (Schlegel *et al.* 1979).

higher energies, on the other hand, *interband* transitions can dominate with electronic transitions from the d-band to the Fermi level taking place; in the case of crystalline metals in particular, prominent features in the density of states of the initial d-band states as well as the final states can give rise to structure in the optical spectra.

If Drude behaviour is dominant then a plot of $1/\omega\varepsilon_2$ should vary linearly with ω^2; such a region is found for a-Au–Si alloys extending over a region of $\simeq 2$ eV, whereas for glassy Pd–Si alloys the region extends for only $\simeq 0.6$ eV (Fig. 7.16). The parameters derived from fits using [7.9] differ widely too. For Pd–Si, $\hbar/\tau = 0.75$ eV (cf. 0.16 eV for c-Pd), whereas for Au–Si, $\hbar/\tau \simeq 5$ eV (varying by a factor of 5 for a change in composition of a factor of 4), whereas it is about 0.3 eV for liquid Au and only about 10^{-3} eV for crystalline Au. Thus, the electronic relaxation times in metallic glasses are extremely short ($\simeq 10^{-15}$–10^{-16} s), because of the strong scattering; this means that equally the mean free paths ℓ are very short ($\ell = \tau v_F = \tau \sqrt{(2E_F/m_e)}$ where v_F is the Fermi velocity), becoming comparable to the interatomic spacing, and under such circumstances it becomes questionable whether the Drude theory is still valid (see Mott and Davis 1979).

The onset of interband transitions has been associated with the deviation from linearity of Drude plots; the fact that in both amorphous Au–Si and Pd–Si alloys, this onset is composition dependent and is greater than the respective values for the (crystalline) pure metals, has been ascribed to a shift towards higher energies of E_F, by virtue of the electrons donated to the conduction band by the metalloid atoms.